Advanced Manufacturing and Assembly Technologies for Aerospace Production Systems

Advanced Manufacturing and Assembly Technologies for Aerospace Production Systems

Guest Editors

Feiyan Guo

Zhengping Chang

MDPI

Basel • Beijing • Wuhan • Barcelona • Belgrade • Novi Sad • Cluj • Manchester

Guest Editors
Feiyan Guo
School of Mechanical Engineering
University of Science and Technology Beijing
Beijing
China

Zhengping Chang
School of Mechanical Engineering
Northwestern Polytechnical University
Xi'an
China

Editorial Office
MDPI AG
Grosspeteranlage 5
4052 Basel, Switzerland

This is a reprint of the Special Issue, published open access by the journal *Machines* (ISSN 2075-1702), freely accessible at: https://www.mdpi.com/journal/machines/special_issues/60O4T7V0CM.

For citation purposes, cite each article independently as indicated on the article page online and as indicated below:

Lastname, A.A.; Lastname, B.B. Article Title. *Journal Name* **Year**, *Volume Number*, Page Range.

ISBN 978-3-7258-6149-1 (Hbk)
ISBN 978-3-7258-6150-7 (PDF)
https://doi.org/10.3390/books978-3-7258-6150-7

Contents

About the Editors

Feiyan Guo

Feiyan Guo is an Associate Professor and PhD Supervisor at the School of Mechanical Engineering, University of Science and Technology Beijing. He mainly engages in research on advanced assembly and joining technologies for aerospace engineering. He has presided over multiple national-level projects and carried out numerous cooperative projects with enterprises in the aerospace industry. He has published more than 20 high-quality SCI papers, formulated 2 industry standards, and obtained/applied for over 20 patents. He has served as Guest Editor of the industry journal of *Aeronautical Manufacturing Technology* twice.

Zhengping Chang

Zhengping Chang is an Associate Professor and PhD Supervisor at the School of Mechanical Engineering, Northwestern Polytechnical University, and Deputy Director of the National Virtual Simulation Experimental Teaching Center for Mechanical Foundations and Aeronautical Manufacturing. He mainly engages in research focusing on aircraft digital assembly, intelligent manufacturing, and digital twin technologies. He has published more than 20 academic papers, including 14 SCI/EI papers, as the first/corresponding author, as well as having authored 10 invention patents, 6 software copyrights, and revised 1 aviation industry standard. He has presided over eight national and provincial-level scientific research projects, such as the National Natural Science Foundation of China (General Program, Youth Program), National Key R&D Program, Shaanxi Provincial Key R&D Program, and Aeronautical Foundation. In addition to this, he has presided over/participated in four national and provincial educational reform projects, edited one university-level 14th Five-Year Plan textbook, and won first prize in the Shaanxi Provincial Teaching Achievement Award. He has guided students to win a gold medal in the China University Student Engineering Practice and Innovation Ability Competition, and six excellent undergraduate graduation designs in the aerospace category at a university level. He is a Young Editorial Board Member of *Engineering Science and Technology*, and a Member of the Chinese Society of Aeronautics and Astronautics and the Chinese Mechanical Engineering Society.

machines

MDPI

Article

Prediction and Optimization of Interference Fit Level in Slug Riveted Structure with Deep Learning Enhanced Genetic Algorithm

Kanghe Yan [1], Lichao Wan [2], Nana Hui [3], Donghe Shan [1], Yang Zhao [1] and Zhengping Chang [1,*]

[1] School of Mechanical Engineering, Northwestern Polytechnical University, Xi'an 710072, China
[2] Wuhan Second Ship Design and Research Institute, Wuhan 430205, China
[3] AVIC Xi'an Aircraft Industry Group Company Ltd., Xi'an 710089, China
* Correspondence: chzhping@nwpu.edu.cn

Abstract

The interference fit connection with slug rivets is widely used in aircraft assembly, and an appropriate interference value is vital for aircraft structural integrity. This study proposed a prediction–optimization framework that a deep neural network (DNN) surrogate was trained on a parametric finite element dataset to regress four interference measurements (G1–G4), and the trained DNN was embedded into a genetic algorithm (GA) to search process parameters that meet prescribed target interference. An orthogonal design with range analysis was employed to rank factor importance and provide interpretable trends, while finite element model (FEM) re-runs were used for validation. Compared with support vector regression, random-forest regression, and Bayesian regression, the DNN demonstrated superior fitting accuracy and a more favorable error distribution on held-out data. GA solutions obtained using the DNN surrogate achieved the target interference with a maximum relative deviation of 9.75%, confirming the effectiveness of the proposed workflow for rapid, physics-consistent interference control. The contributions of the study were as follows: (i) an end-to-end, quick-response, reproducible FEM→DNN→GA pipeline for slug-rivet interference; (ii) quantitative factor ranking with mechanistic interpretation; and (iii) minute-scale parameter optimization suitable for engineering deployment.

Keywords: slug rivet; interference fit; deep learning; genetic algorithm; riveting parameters optimization

1. Introduction

Interference riveting technology is a commonly employed connection method in aerospace manufacturing. The interference fit at the contact interface between the rivet and the hole wall ensures a uniform load distribution, which significantly enhances the fatigue life of riveted structures and improves the sealing reliability at the connection points [1,2]. There are many factors affecting the interference fit level [3], and it is difficult to construct a multivariate nonlinear relationship between the factors and the interference value. Currently, interference values are primarily studied by conducting experiments prior to mass production and by measuring the values corresponding to known process parameter combinations. However, the above method is time-consuming and costly, while heavily relying on the experience of engineers. It is essential to conduct some research on predicting the interference values quickly and accurately, as well as optimizing the riveting parameters.

Generally, based on riveting process analysis, a mechanism model was established to predict the deformation behavior of the rivet and connected plates, thereby controlling the connection quality of the riveted joints. Scholars have conducted extensive research in this field using various methods, such as plastic mechanics, finite element analysis (FEA), and fracture mechanics. Lv et al. [4] proposed a theoretical analysis method for the nonlinear dynamic plastic behavior of riveted structures, applicable to predicting the forming quality of riveted structures under varying extrusion forces, speeds, and materials. Liang et al. [5] investigated the effect of die angle on the fatigue life of riveted joints through a combination of FEA and experimental validation, and found that variations in rivet angles resulted in different residual stress distributions, which significantly affected the fatigue performance. Tian et al. [6] analyzed the fatigue life of riveted joints with different countersunk depths using the finite element method. Derijck et al. [7] established a mapping relationship between the riveting force and the driven head size according to the plastic deformation assumption, and validated it through experiments under various materials. Huang et al. [8] proposed a novel fatigue crack growth model based on structural load to predict the fatigue life and the final crack aspect ratio of riveted joints. These studies demonstrated that riveting quality could be effectively predicted from the initial processing conditions. However, in practice, most of these methods involve complex calculations, making it challenging to rapidly predict the riveting quality under new combinations of process parameters.

Data-driven methods are capable of rapidly providing results when confronted with entirely new combinations of process parameters, significantly reducing computational resources and time costs [9]. Deep learning, as an emerging data-driven method, could build complex input–output relationships. The typical structure was composed of a multilayer perceptron with multiple hidden layers, which gradually transforms initial "low-level" feature representations into "high-level" feature representations through multiple layers of processing. Compared with traditional regression model construction methods, the unique advantage of deep learning lies in its high sensitivity to relevant details [10,11]. Belhouchet et al. [12] demonstrated that an Artificial Neural Network (ANN) exhibited higher accuracy in fitting complex relationships compared to mathematical models based on numerous assumptions, and the ANN could provide more reliable prediction results under various process parameter conditions. In recent years, many scholars have conducted related research, demonstrating that deep learning techniques could leverage large datasets to predict defects in the product manufacturing process. Wang et al. [13] proposed a method to evaluate the hydrostatic guideway straightness errors online based on a long short-term memory network model. Deshpande et al. [14] used deep learning for image segmentation in additive manufacturing to detect defects. Additionally, deep learning also serves as a high-precision, low-computational-cost surrogate model to guide the optimization of product process parameters. Sun et al. [15] proposed a computational framework combining ANNs and a genetic algorithm (GA), where ANNs were trained as surrogate models for FEA to accelerate the GA optimization process. Nejad et al. [16] utilized ANNs to fit the relationship between various process parameters and the fatigue life of riveted joints.

The GA is a well-known metaheuristic algorithm that mimics Darwin's theory of natural selection to perform selection and iteration within a population to achieve the optimal solution [17]. Using metaheuristic algorithms, such as GAs, to select the optimal combination of process parameters is a commonly used method [18]. The advantage of this approach lies in its minimal reliance on detailed structural information for the optimization problem, enabling rapid identification of optimal process parameter combinations. Shi et al. [19] proposed an intelligent matching system for countersunk hole riveting based on the fruit fly optimization algorithm, which significantly improved the control level of the riveting interference value compared to traditional tolerance zone methods. Choudhary et al. [20] designed a hybrid optimization method combining multiple metaheuristic algorithms to optimize the submerged arc welding process parameters, resulting in more favorable parameter settings. Rajendran et al. [21] introduced various metaheuristic algorithms to optimize the process parameters of wire-cut electrical discharge machining, achieving superior results compared to traditional optimization algorithms. Metaheuristic algorithms are also integrated with other techniques for process parameter design. Ni et al. [22] combined the GA and the ant colony algorithm with a finite element model (FEM) to optimize process parameters such as rivet upsetting direction and assembly sequence, thereby controlling assembly errors during the large-scale riveting process. Combined with a deep learning prediction model, Wu et al. [23] employed a GA to generate a Pareto set as an alternative for selecting machining process parameters. Oliveira et al. [24] examined the integration of metaheuristic algorithms with machine learning techniques for optimizing searches in large datasets. The mentioned hybrid approach, through learning from prior iterations, demonstrating learning from prior iterations enhances both efficiency and effectiveness compared to traditional methods.

Interference fit riveting enhances fatigue performance and sealing, while its multi-factor, nonlinear interactions are challenging to explore using finite element analysis or experiments alone. Prior studies have elucidated mechanisms, they remain limited in generalizability and incur high computational costs for new parameter combinations. The challenge was addressed using a data-driven surrogate: a DNN was trained on parametric FEM data to predict G1–G4, which was then embedded in a GA for target-seeking optimization. Factor importance was further quantified via orthogonal design and range analysis, and the optimized settings were validated by FEM re-runs. The contributions of the study were as follows: (i) a reproducible FEM→DNN→GA pipeline, (ii) accuracy gains over SVR/RFR/Bayesian baselines, (iii) interpretable factor ranking showing the leading role of plate yield strength, and (iv) minute-scale optimization to meet target interference levels.

2. Basic Prediction and Optimization Framework

The basic structure of the process parameter optimization framework proposed in this study is illustrated in Figure 1. The specific steps are as follows: (1) Develop a FEM for the riveting process and validate its effectiveness by comparing with experimental results; (2) Conduct orthogonal experiments on selected process parameters to analyze the impact of each parameter on the rivet interference fit level; (3) Design the topology of the DNN, set hyperparameters, train the predictive model, and make comparisons with other regression models to verify its accuracy; (4) Construct a fitness function using the predictive model and the targeted interference fit level, and apply a GA to optimize the process parameters. The quality of the optimized parameter is evaluated through experiments.

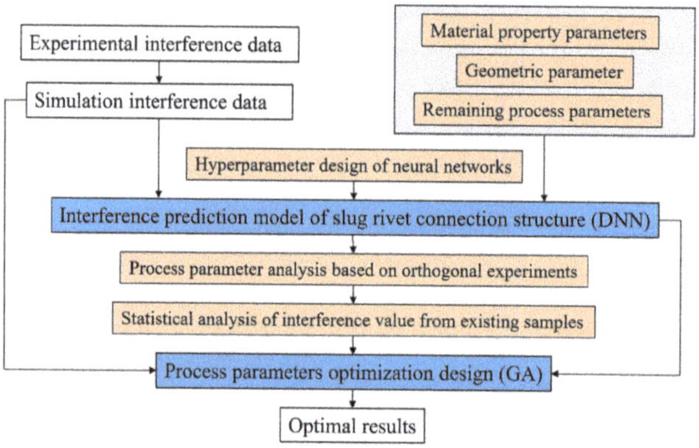

Figure 1. Flowchart of riveting process parameters optimization.

2.1. Modeling Process

A FEM for the riveting process was established, and its validity was verified through experimental testing. Potential factors included hole diameter tolerance, surface finish, friction, temperature, force profile, plate stack-up, material property parameters, and geometric protrusion. Eight factors dominating interference in automated cells, which are controllable or characterizable in FEM, were selected for analysis. The design space comprised riveting force (F_S), dwell time (t), protrusion (h), clamping force (F_C), as well as material parameters of the connected plates: elastic modulus (E), initial yield stress (A), hardening exponent (n), and hardening constant (B). The locations for interference measurement are shown in Figure 2, and the interference level is represented by four measurement values, G1, G2, G3, and G4, respectively. Corresponding experiments were conducted with the same parameters used in the FEM, and the experimental results were compared with the simulated results to validate the effectiveness of the FEM.

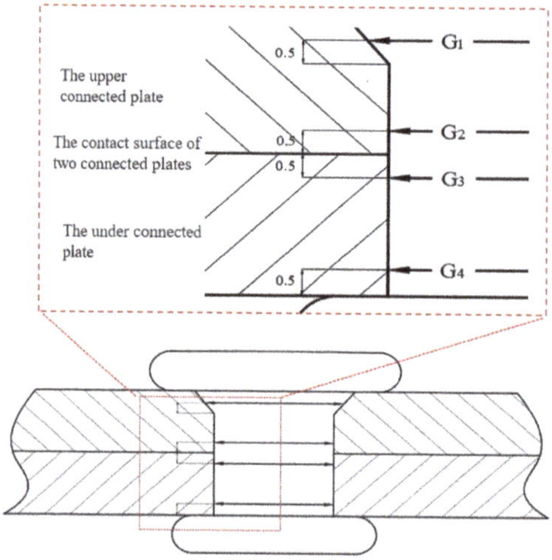

Figure 2. The specific measurement position of riveting interference quantity.

2.2. Analysis of Process Parameters

The orthogonal experiments were designed to determine the effect of the selected process parameters on the interference fit of the riveted structure. The effects of riveting parameters on the interference fit were analyzed by establishing FEMs corresponding to the orthogonal experimental parameter sets. After completing the FEM calculations, the interference values for each group parameter set were obtained. A range analysis was then performed to summarize the influence patterns of process parameters on the rivet interference fit.

2.3. Construction of Regression Model

The training of the DNN model involves minimizing the loss function value by adjusting the weights of the connections between neurons and the biases of each layer, which are continuously updated during the process. However, hyperparameters, such as the neural network topology and the type of loss function, must be pre-designed to determine the initial network structure. The design of neural networks is empirical and requires multiple iterations to avoid overfitting or underfitting. In this module, the specific design of the DNN regression model for predicting the interference fit level of riveted joints was presented. The trained DNN regression model was tested and compared with traditional regression models to verify its superiority.

2.4. Parameter Optimization Design

The DNN model could rapidly and accurately predict the interference values based on given process parameters. However, due to the end-to-end nature of the DNN model, it is difficult to understand the effect of each parameter and their coupled relationships on the final interference value. Metaheuristic algorithms such as the GA do not require a specific functional form for optimization. Therefore, the DNN model was applied to construct the fitness function for the GA. In this module, the specific process of GA in combination with a DNN to optimize the interference value of riveted joints was presented.

3. Prediction and Optimization Process of Interference Fit Level

The specific implementation methods for each module within the framework were introduced, with detailed descriptions of the algorithms used. Additionally, alternative algorithms were suggested as recommendations, allowing readers to consider them based on the actual sample size, sample labels, and feature dimensions. The framework was independent of any specific software environment and imposed no strict requirements on the software or scripts involved.

3.1. Modeling of FEM

The prediction accuracy of data-driven regression models depends on the quality and size of the sample dataset. There are two existing methods for obtaining data: physical experiments and simulation analysis. Physical experimental data are constrained by actual material conditions, making it challenging to obtain uniformly distributed samples when varying material parameters. In contrast, the simulation analysis method provides greater flexibility in the design of process parameters, resulting in more uniformly distributed samples [25].

Given the symmetry of the riveted joint structure, a 1/4 model was applied to simplify the calculations. A slug rivet with a diameter of d = 4.78 mm and a base length of 16 mm were selected. The rivet had an additional protrusion length, which served as one of the optimization parameters. Thus, the protrusion of the rivet below the connected plates increased with extended length. The rivet hole diameter D is 4.88 mm, and the thicknesses of the two connected plates were $t_1 = t_2 = 3$ mm, with a width of $L/2 = 15$ mm. A flat die was employed as a riveting tool.

Material properties were set as follows: The slug rivet was 2117-T4, while the upper and lower connected plates were made of aluminum alloy. The Johnson–Cook model

($\sigma = A + B\varepsilon^n$) was adopted for the constitutive relationship, with the effects of strain rate and temperature temporarily neglected. Since the density and Poisson's ratio of different aluminum alloys were basically consistent, the influence of the elastic modulus E, initial yield stress A, hardening constant B, and hardening exponent n of the connected plates on the riveting interference were primarily focused on in this study. The detailed material settings were described in the subsequent data preparation section.

The rivet was manufactured with aluminum alloy, and its constitutive relationship was modeled using a plastic hardening model in Equation (1).

$$\sigma(\varepsilon_p) = A + B\varepsilon_p^n \tag{1}$$

where A is the initial yield stress, B denotes the strength coefficient, and n refers to the hardening exponent.

The upper and lower pressure feet were assigned a density of 8×10^{-9} t/mm^3, a Young's modulus of 200,000 MPa, and a Poisson's ratio of 0.3. Under actual production conditions, the upper and lower rivet dies possessed extremely high stiffness, which was assumed to be non-deformable during the riveting process. Therefore, they were usually constrained as rigid bodies.

The actual riveting process included three stages of upsetting, dwelling, and unloading. The simulation model also consisted of three dynamic explicit analysis steps to replicate the above stages. The overall riveting structure was meshed using C3D8R hexahedral elements, applying reduced integration and hourglass control methods. Local mesh refinement was applied around the rivet and the hole of the connected plates, while coarser meshing was used in distant regions from the hole to reduce the number of nodes, decrease computational load, and improve efficiency. Figure 3 shows the boundary conditions of FEM. Contact relationships and mesh generation were set according to the actual riveting process.

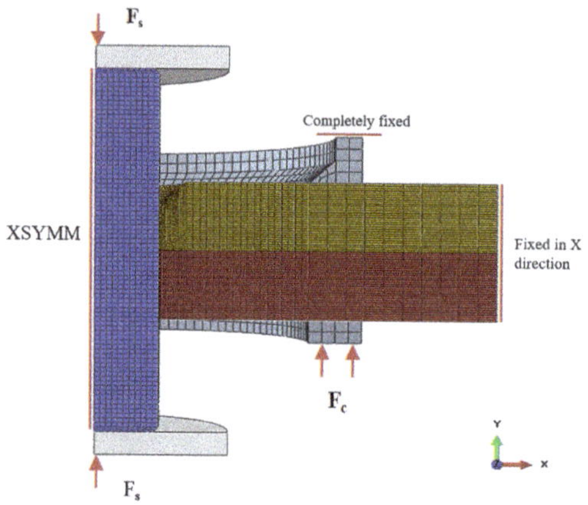

Figure 3. Boundary condition setting of FEM.

To validate the FEM, AA2117-T4 slug rivets were installed into 2024-T351 and 7055-T76511 plates using an automated drilling–riveting cell, riveting experiments, and simulated analysis. Each configuration underwent three repeated experiments. One set of experimental and corresponding simulation results is shown in Figure 4, the numbers in the figure denote the perpendicular distance from the corresponding hole-wall locations to the central axis of the hole.

During the actual riveting process, the specimens were subjected to a riveting force of 38.4 kN and a clamping force of 19.64 MPa. Owing to the use of a quarter-symmetry FE model, the total riveting force of 38.4 kN corresponded to ~9.6 kN per symmetry sector.

The rivet heads at the countersunk end were milled flat after riveting to ensure the accuracy of the aircraft profile. Therefore, the same treatment was applied in the experiments. The analysis results are shown in Table 1. Dimensional calibration was performed prior to measurement, and the reported values represented the mean of three repeated tests.

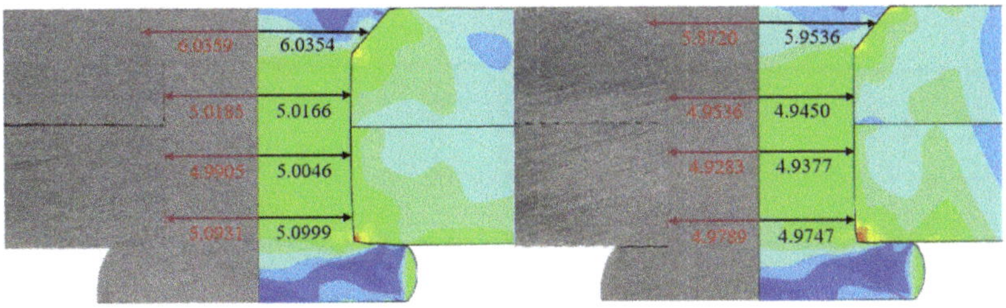

Figure 4. Comparison between the experimental results and the simulation results [26].

Table 1. Comparison of simulated results and experimental values.

Characterization of Interference Quantity	2024-T351			7055-T76511		
	Experiment (mm)	Simulation (mm)	Error (mm)	Experiment (mm)	Simulation (mm)	Error (mm)
G1	0.275	0.286	0.011	0.111	0.115	0.004
G2	0.150	0.136	0.014	0.069	0.065	0.004
G3	0.121	0.124	0.003	0.056	0.057	0.001
G4	0.217	0.219	0.002	0.103	0.094	0.009

For interference connection, numerous parameters affect the interference fit level. To balance predicted accuracy and computational complexity, a subset of parameters was selected for sample data preparation. The material property parameters included the elastic modulus, initial yield stress, hardening exponent, and hardening constant of the connected parts. And rivet protrusion length was included as the geometric parameter. The process parameters chosen comprised the riveting force, die dwell time, and clamping force of the pressure feet.

Subsequently, the selected process parameters were studied using the orthogonal experiment to analyze their effects on the four interference values. Additionally, a certain number of random process parameter combinations were generated to expand the sample datasets. Ranges for E, A, B, and n follow published aluminum-alloy data for aerospace plates, while the process windows for force, dwell, protrusion, and clamping were determined from the automatic drilling–riveting cell capability and prior shop-floor settings. An orthogonal array L50 (5^{11}) was employed to discretize each factor into five evenly spaced levels within these windows, and random samples was used to densify the interior for DNN training.

3.2. Orthogonal Experiment Design

In the practical application, the combinations of process parameters are critical for maintaining the stability of the riveting quality. The ranges of each process parameter are listed in Table 2. Each process parameter had five equidistant levels within the selected range. Therefore, an orthogonal array L50 (5^{11}) was used to design the experiments, resulting in 50 sets of data. The process parameter combinations are shown in Table 3.

Table 2. The selected range of process parameters.

Parameter	Elastic Modulus (MPa)	Initial Yield Stress (MPa)	Hardening Constant (MPa)	Hardening Exponent
Maximum value	72,000	500	540	0.4
Minimum value	60,000	300	300	0.2
Parameter	Rivet protrusion length (mm)	Riveting force (N)	Die dwell time (s)	Clamping force (MPa)
Maximum value	5.5	9800	0.0021	21.64
Minimum value	4.5	9400	0.0009	17.64

Table 3. Design of process parameter combinations for orthogonal experiments.

Riveting Force (N)	Die Dwell Time (s)	Rivet Protrusion Length (mm)	Clamping Force (MPa)	Elastic Modulus (MPa)	Initial Yield Stress (MPa)	Hardening Constant (MPa)	Hardening Exponent
9400	0.0009	4.5	17.64	60,000	300	300	0.2
9400	0.0012	4.75	18.64	63,000	350	360	0.25
9400	0.0015	5	19.64	66,000	400	420	0.3
9400	0.0018	5.25	20.64	69,000	450	480	0.35
9400	0.0021	5.5	21.64	72,000	500	540	0.4
9500	0.0009	4.75	19.64	69,000	500	300	0.25
9500	0.0012	5	20.64	72,000	300	360	0.3
9500	0.0015	5.25	21.64	60,000	350	420	0.35
9500	0.0018	5.5	17.64	63,000	400	480	0.4
9500	0.0021	4.5	18.64	66,000	450	540	0.2
9600	0.0009	5	21.64	63,000	450	480	0.2
9600	0.0012	5.25	17.64	66,000	500	540	0.25
9600	0.0015	5.5	18.64	69,000	300	300	0.3
9600	0.0018	4.5	19.64	72,000	350	360	0.35
9600	0.0021	4.75	20.64	60,000	400	420	0.4
9700	0.0009	5.25	18.64	72,000	400	540	0.3
9700	0.0012	5.5	19.64	60,000	450	300	0.35
9700	0.0015	4.5	20.64	63,000	500	360	0.4
9700	0.0018	4.75	21.64	66,000	300	420	0.2
9700	0.0021	5	17.64	69,000	350	480	0.25
9800	0.0009	5.5	20.64	66,000	350	480	0.3
9800	0.0012	4.5	21.64	69,000	400	540	0.35
9800	0.0015	4.75	17.64	72,000	450	300	0.4
9800	0.0018	5	18.64	60,000	500	360	0.2
9800	0.0021	5.25	19.64	63,000	300	420	0.25
9400	0.0009	4.5	20.64	72,000	450	420	0.25
9400	0.0012	4.75	21.64	60,000	500	480	0.3
9400	0.0015	5	17.64	63,000	300	540	0.35
9400	0.0018	5.25	18.64	66,000	350	300	0.4
9400	0.0021	5.5	19.64	69,000	400	360	0.2
9500	0.0009	4.75	17.64	66,000	400	360	0.35
9500	0.0012	5	18.64	69,000	450	420	0.4
9500	0.0015	5.25	19.64	72,000	500	480	0.2
9500	0.0018	5.5	20.64	60,000	300	540	0.25
9500	0.0021	4.5	21.64	63,000	350	300	0.3
9600	0.0009	5	19.64	60,000	350	540	0.4
9600	0.0012	5.25	20.64	63,000	400	300	0.2
9600	0.0015	5.5	21.64	66,000	450	360	0.25
9600	0.0018	4.5	17.64	69,000	500	420	0.3

Table 3. *Cont.*

Riveting Force (N)	Die Dwell Time (s)	Rivet Protrusion Length (mm)	Clamping Force (MPa)	Elastic Modulus (MPa)	Initial Yield Stress (MPa)	Hardening Constant (MPa)	Hardening Exponent
9600	0.0021	4.75	18.64	72,000	300	480	0.35
9700	0.0009	5.25	21.64	69,000	300	360	0.4
9700	0.0012	5.5	17.64	72,000	350	420	0.2
9700	0.0015	4.5	18.64	60,000	400	480	0.25
9700	0.0018	4.75	19.64	63,000	450	540	0.3
9700	0.0021	5	20.64	66,000	500	300	0.35
9800	0.0009	5.5	18.64	63,000	500	420	0.35
9800	0.0012	4.5	19.64	66,000	300	480	0.4
9800	0.0015	4.75	20.64	69,000	350	540	0.2
9800	0.0018	5	21.64	72,000	400	300	0.25
9800	0.0021	5.25	17.64	60,000	450	360	0.3

3.3. Construction of DNN Model

Within the range of process parameters determined in the previous orthogonal experiment, additional samples were generated by randomly selecting and combining process parameters.

Let z denote the input factors [forming load, dwell time, protrusion, clamping force, E, A, B, n] and Y the targets [G1, G2, G3, G4] in mm. In total, 336 datasets were collected for training the DNN regression model. A part of these samples was set aside as a test set, while the remaining samples were randomly divided into training and validation sets with a 3:1 ratio. All process parameter values of the samples were normalized using Equation (2).

$$z = \frac{x - x_{\min}}{x_{\max} - x_{\min}} \quad (2)$$

After preparation of the dataset, the problem could be abstracted into a multivariate nonlinear regression problem with eight inputs and four outputs. In this context, a DNN regression model was constructed using the PyTorch 2.1.0 [27] package in Python 3.11.5. The model consisted of an input layer, multiple hidden layers, activation functions, and an output layer, and its network structure was fully connected. To ensure the stability of the regression prediction, dropout layers were not used during the model training process. The DNN model training includes forward propagation, loss calculation, backpropagation, and parameter updates, with the parameters of each network layer being updated using an optimization method.

The ReLU was employed as activation function, and the Adam optimizer (β_1 = 0.9, β_2 = 0.999) was used with an initial learning rate 1×10^{-3}, decaying to 1×10^{-5} via cosine schedule. The mean squared error function was chosen as the loss function. Due to the existence of multiple hidden layers in the DNN model, there was a potential risk of gradient explosion and gradient vanishing during training. When the input was greater than zero, the activation function ReLU maintained a linear relationship and kept a constant gradient in this region, which helped to mitigate the problem of gradient vanishing. Additionally, the ReLU function outputs zero for negative input values, effectively truncating gradients that exceed a certain threshold to zero, thereby helping to suppress gradient explosion. The activation function ReLU is defined as follows:

$$f(x) = \begin{cases} 0 & x < 0 \\ x & x \geq 0 \end{cases} \quad (3)$$

The output of each layer of DNN was as shown in Equations (4) and (5).

$$a_i^l = relu(z_i^l) \tag{4}$$

$$z_i^l = \sum_{i=1}^{n^l} \omega_i^l a_j^{l-1} + b_i^l \tag{5}$$

The formulation of the mean squared error (MSE) function was in Equation (6).

$$MSE = \frac{1}{n}\sum_{i=1}^{n} (Y_i - \hat{Y}_i)^2 \tag{6}$$

The Adam algorithm is an optimization method that combines the advantages of both AdaGrad [28] and RMSProp [29], and the parameters of a neural network are optimized by adaptively adjusting the learning rate.

During the training process, additional hyperparameters were adjusted, and then the final DNN hyperparameters were determined. The network consisted of four hidden layers. Based on the number of input and output parameters, the number of neurons in each layer was 8, 16, 32, 16, 8, and 4, respectively. The mini-batch size was set to 20.

Upon completing the training, two quantitative performance metrics were introduced to evaluate the prediction accuracy of the model, namely, the mean absolute error (MAE, Equation (7)) and the coefficient of determination (R^2, Equation (8)). Typically, MAE was used to describe the difference between the predicted values and the actual values, assessing the model's accuracy. R^2 was used to evaluate the fitting degree between the predicted values and the actual values, with values closer to 1 indicating a better fit of the predictive model.

$$MAE = \frac{1}{N}\sum_{i=1}^{N} \left| y_{act}(i) - y_{pre}(i) \right| \tag{7}$$

$$R^2 = \frac{\sum_{i=1}^{N} (y_{pre}(i) - \overline{y})^2}{\sum_{i=1}^{N} (y_{act}(i) - \overline{y})^2} \tag{8}$$

where y_{act} represents the measured interference value, y_{pre} is the predicted interference value from the model, and N refers to the number of test samples.

3.4. GA Optimization of Riveting Process

A genetic algorithm is a computational model that simulates the biological evolution process based on Darwin's theory of natural selection and genetic mechanisms, serving as a method for searching optimal solutions by mimicking natural evolution. The specific flowchart is shown in Figure 5. By replacing the FEA calculations with a DNN model, the computational time required for GA optimization was significantly reduced.

To analyze the relationships among the four interference measurements conveniently, a stacked histogram of the interference data for partial samples was created, as shown in Figure 6, which contributed to the optimization of target interference using GA.

It was observed that the ratio between the four measured interference values remained roughly constant. The average interference values for all samples are calculated and shown in Table 4.

Based on the interference ratio and relevant production standards of the slug rivet, the absolute range of interference calculated for this rivet size was from 0.0488 mm to 0.1952 mm. The targeted values of interference are shown in Table 5.

In practice, material property parameters are often difficult to modify, whereas process parameters such as riveting force are easily adjusted. For optimization validation, the plate material was fixed to 2024-T351 (E = 67.5 GPa, A = 330 MPa, B = 596 MPa, n = 0.52).

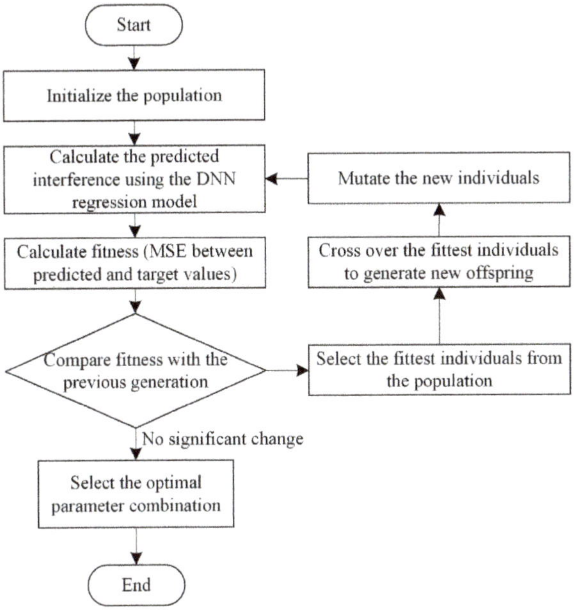

Figure 5. Flowchart of process parameter optimization.

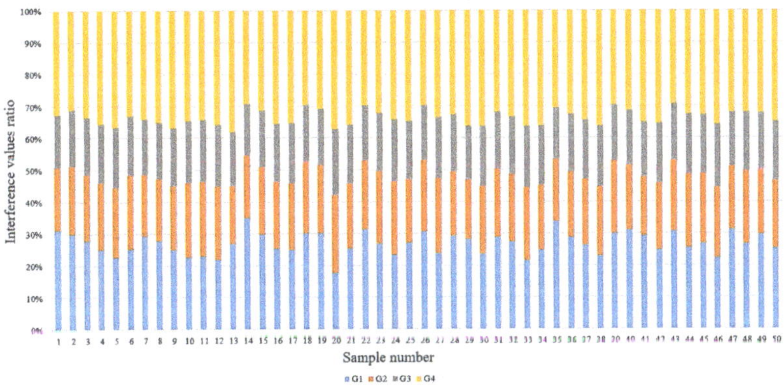

Figure 6. Stacked histogram of samples.

Table 4. Mean of interference values.

G1 (mm)	G2 (mm)	G3 (mm)	G4 (mm)
0.278743737	0.209038047	0.182087466	0.330130749

Table 5. Target values of interference.

Group	G1 (mm)	G2 (mm)	G3 (mm)	G4 (mm)
1	0.084	0.063	0.054	0.099
2	0.112	0.084	0.072	0.132
3	0.10	0.08	0.07	0.12

To obtain the desired interference value, GA was employed to optimize the process parameter combined with the pre-trained DNN prediction model. A fitness function of the optimization algorithm was constructed with the difference between the predicted results and the target interference value. The specific steps include the following four steps.

Step 1. Population Initialization: a set of individuals was initialized, and each of them contained process parameters such as riveting force, die dwell time, protrusion, and clamping force. The initial population size was set to 100.

Step 2. Fitness Calculation: combined with the material parameters of 2024-T351 aluminum alloy, the aforementioned four process parameter values of each individual were the input of the prediction model to calculate four interference values. The mean squared error between the predicted values and the targeted interference value served as the fitness function, determining the fitness score of each individual.

Step 3. Selection, Crossover, and Mutation: the fittest individuals from the population were selected, and portions of their parameter values were recombined to generate new individuals. Additionally, certain parameters of new individuals were randomly modified to generate the next generation of the population. The crossover probability and mutation probability were set as 0.5 and 0.2, respectively.

Step 4. Optimized Result Selection: The optimization procedure would terminate when there was no significant improvement in the best fitness values across the two most recent generations. Thus, the parameters of the optimal individual were selected as the final optimized result.

4. Results and Discussion

In this section, the application effect of the proposed framework on interference value prediction and process parameter optimization was discussed in detail. The influences of process parameters on the interference values were presented in Section 4.1 by orthogonal experiment results. Section 4.2 described the prediction accuracy of the constructed DNN model and its comparison with three other traditional regression models. The optimized results of the process parameter with a GA based on the DNN model were shown in Section 4.3.

4.1. Analysis of Orthogonal Test Result

Based on an orthogonal experimental design, a parametric modeling approach was developed to construct the FEM for various combinations of process parameters. Simulations were then performed to obtain the simulated values of four interference fits for each group. The simulated results are shown in Figure 7. After investigating the magnitudes of the interference values at the above four points, it was found that the interference fit levels (G1 and G4) around the rivet head were greater than the other two positions (G2 and G3). The above phenomenon could be explained as the result of the material flow of the slug rivet during the riveting process. In the initial stage, the protruding parts of the rivet shank at the top and bottom underwent plastic deformation, with most of the materials flowing into the rivet hole. Once the rivet head made contact with the connected plate, further material flow into the hole was restricted, resulting in larger interference values near the rivet head.

A range analysis was performed on the above data to evaluate the relative influence of each process parameter on G1, G2, G3, and G4 with Minitab 21 software. The results of the range analysis are presented in Table 6, while Figures 8–11 show the impact trend of each factor on the interference. A steeper slope of the trend curve indicated a greater impact of the factor on the interference. As shown in Table 6, it is observed that the range values of yield strength A are the largest for G1, G2, G3, and G4. The range values of elastic modulus E were the minimum, which indicated that the yield stress has the greatest impact on the interference fit level of the slug rivet, whereas the elastic modulus has the least impact.

The ranking of factors by their influence was as follows: initial yield stress > hardening exponent > hardening constant > elastic modulus.

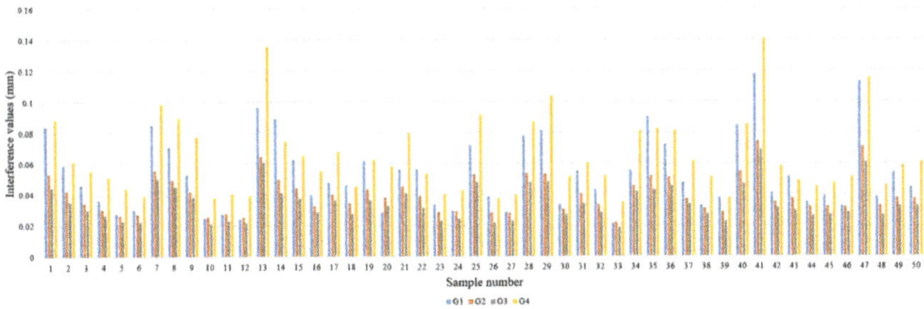

Figure 7. Partial simulation results of the orthogonal experiment.

Table 6. Range analysis of an orthogonal experiment.

Parameter	*R* (G1)/mm	*R* (G2)/mm	*R* (G3)/mm	*R* (G4)/mm	*R* (Average)/mm
Riveting force	0.00656	0.00338	0.00321	0.00519	0.004585
Die dwell time	0.00462	0.00246	0.00216	0.00485	0.0035225
Rivet protrusion length	0.01554	0.00449	0.00612	0.01836	0.0111275
Clamping force	0.00872	0.00290	0.00272	0.00665	0.0052475
Elastic modulus (E)	0.00324	0.00451	0.00440	0.00789	0.00501
Initial yield stress (A)	0.05321	0.02852	0.02701	0.05644	0.041295
Hardening constant (B)	0.01523	0.00628	0.00571	0.01545	0.0106675
Hardening exponent (n)	0.02413	0.01216	0.01138	0.02409	0.01794

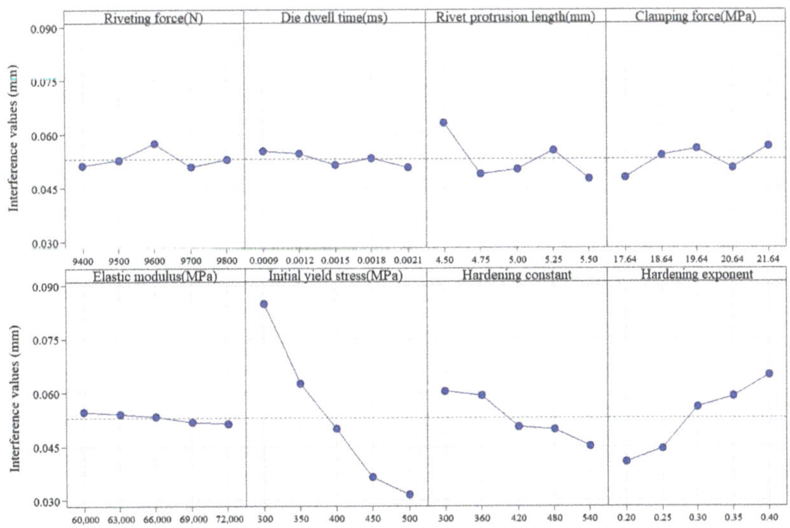

Figure 8. Influence trend on the interference fit of G1.

Initial yield stress A had the largest influence on the riveting interference, followed by the hardening exponent n, while elastic modulus E and hardening constant B had similar effects. The hardening exponent n was positively correlated with the riveting interference value, whereas the initial yield stress A was negatively correlated. According to the riveting process, the slug rivet was deformed under riveting force, then the riveting shank expanded the rivet hole, namely, the interference was primarily driven by the plastic deformation of the rivet. When the rivet is subjected to a certain riveting force, the initial yield stress of the connected plates determines the degree of deformation difficulty. The higher yield strength of connected plates results in a more difficult deformation occurring, which results in a smaller interference.

Considering the properties of the exponential function in the Johnson–Cook constitutive model [30], a larger value of n resulted in less sensitivity to increases in ε^n, and the function grows more gradually within the limits of 0 to 1. Therefore, under a given riveting force, a larger n led to greater strain in the plastic stage, which increased the interference in turn.

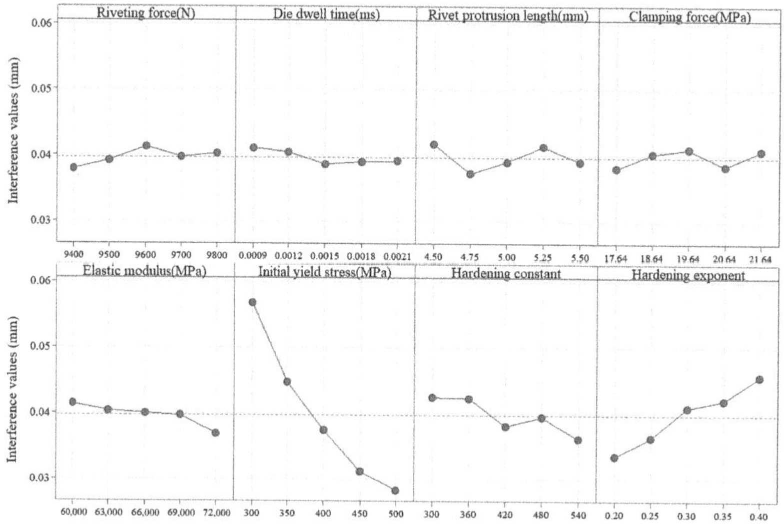

Figure 9. Influence trend on the interference fit of G2.

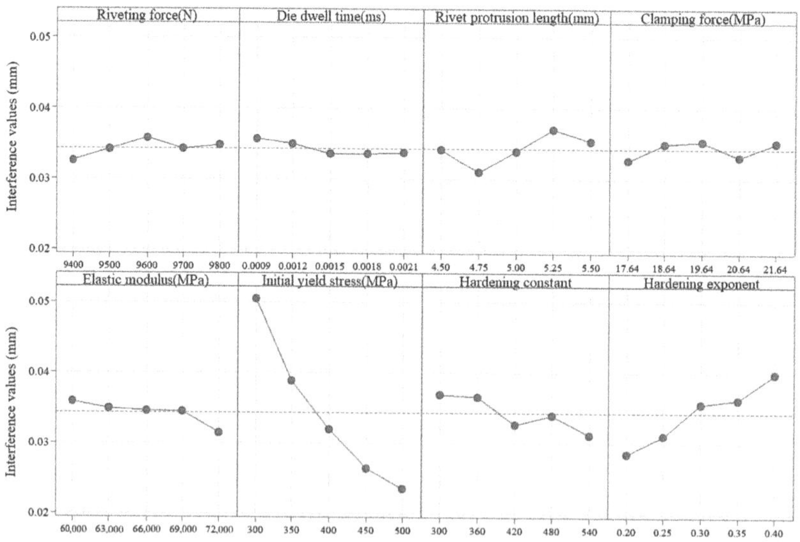

Figure 10. Influence trend on the interference fit of G3.

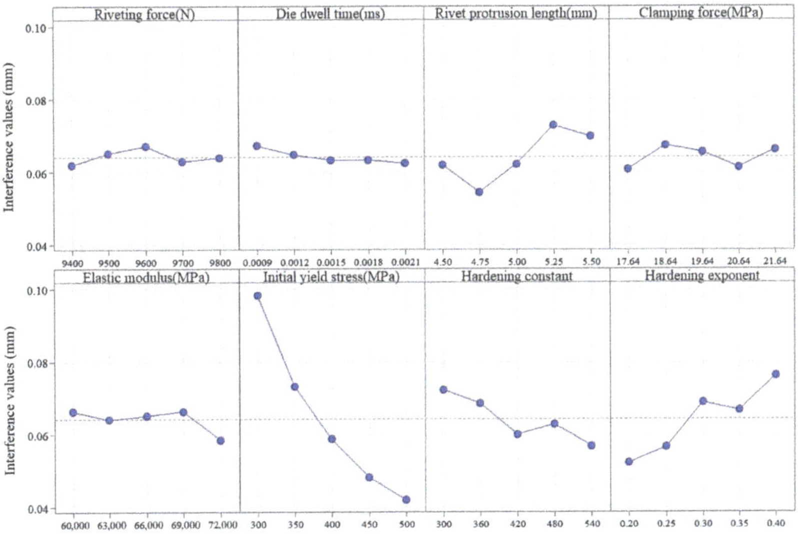

Figure 11. Influence trend on the interference fit of G4.

In the range analysis, it could be concluded that the variation in protrusion length had the greatest effect on the G1 and G4. The increase in protrusion strengthened the compression of the lower half of the hole wall during the riveting process, leading to significant changes in G4. Similarly, the increase in protrusion also had the same influence on the upper half of the hole wall through plastic flow of material, which resulted in an increased interference at other locations. Additionally, an increase in clamping force and riveting force also enhanced the material flow and deformation, thereby increasing the interference fit level. It was interesting that an extension in dwell time of riveting force led to a certain reduction in interference, possibly because of the occurrence of stress relaxation under high-stress conditions if the riveting force was applied for a prolonged period [31,32]. The stress relaxation reduced the generated residual stress during the deformation process, ultimately leading to a decrease in the interference to some extent. From the above range analysis, it was evident that G1 and G4, which were less constrained compared to the middle regions G2 and G3 of the rivet hole wall, presented a greater change in interference value when process parameters were varied.

The larger response of G1 and G4 to protrusion reflects directional material flow at early upsetting, where shank expansion was constrained by head–plate contact. The observed decrease with prolonged dwell was consistent with stress-relaxation under high contact stress, effectively reducing retained interference. These trends supported the factor ranking and explained why edge-adjacent gauges (G1, G4) are more sensitive than mid-wall gauges (G2, G3).

In summary, it was obvious that each selected process parameter has a certain degree of influence on the interference fit level of the slug rivet. The rationality and effectiveness of each process parameter as an input of the DNN regression prediction model have also been demonstrated.

4.2. Verification of DNN Model Prediction Performance

To verify the superiority and accuracy of the DNN regression model, three other data-driven regression models were established with the same dataset, including Support Vector Regression (SVR), Random Forest Regression (RFR), and Bayesian Regression (BR).

To assess the prediction performance quantitatively, the same test dataset was used to calculate the Mean Absolute Error (MAE) and the coefficient of determination (R^2) for each regression model. The results are illustrated in Figures 12 and 13.

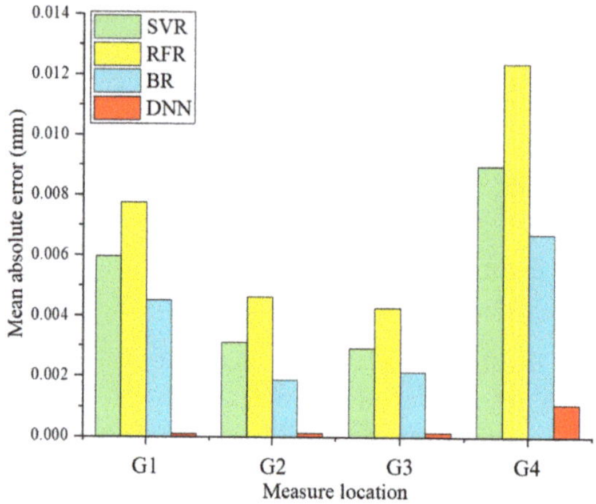

Figure 12. MAE of each regression model (mm).

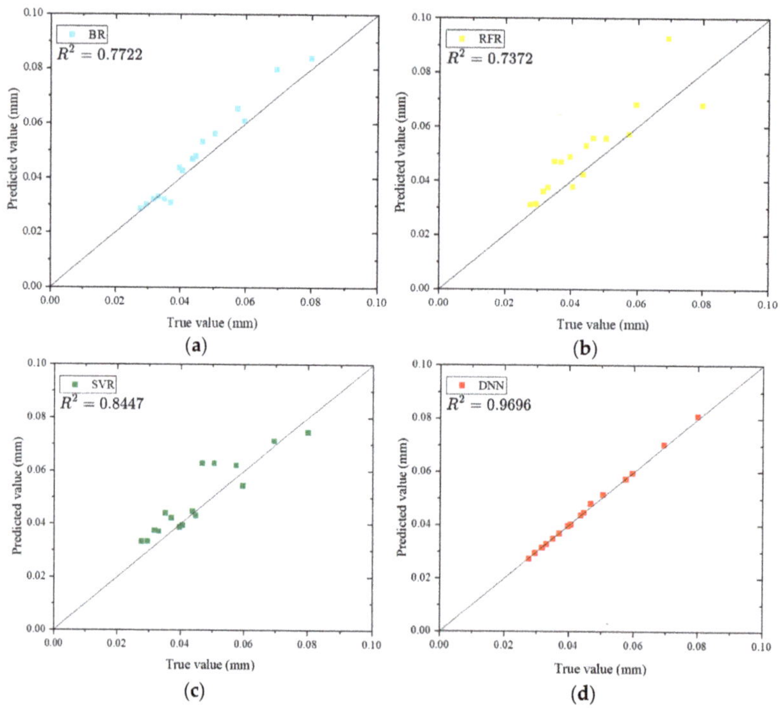

Figure 13. Prediction fitting performance of each model. (**a**) Bayesian regression, (**b**) Random forest regression, (**c**) Support vector machine regression, (**d**) DNN model.

As illustrated in Figure 13, the DNN model exhibits the smallest relative error in the test set and achieves the best fitting performance. The MAE for all test set samples was 0.00035 mm, with a mean relative error of 0.72%. Compared to three other traditional regression models, the DNN model demonstrated a clear advantage and effectiveness in serving as a substitute for the finite element simulation model.

Interference in slug-rivet joints emerges from coupled, non-separable interactions among load, dwell, protrusion, clamping, and plasticity parameters. A DNN with ReLU layers provides a high-capacity, adaptive basis that captures high-order cross-factor interactions and smooth, monotone trends within the design window. In contrast, SVR with a single stationary kernel imposes a few global length scales, which underfits anisotropic factor sensitivities and either over- or under-smooths regimes with different curvature; Random-Forest regression yields piecewise-constant predictors that bias continuous interference responses and degrade near extrema; and Bayesian regressors lack expressive nonlinearity and scale poorly with sample size while remaining constrained by stationarity assumptions. The DNN also supports multi-output learning via a shared trunk and four output heads, allowing G1–G4 to share latent structure and improving data efficiency.

4.3. Optimized Results of Process Parameters

Embed the DNN into the GA as a surrogate for the FEM, yielding a DNN-enhanced GA optimization model. After iterative optimization with DNN-enhanced GA, the optimized process parameters for each group are shown in Table 7.

Table 7. Combination of process parameters after GA optimization.

Group	Riveting Force (N)	Die Dwell Time (s)	Rivet Protrusion Length (mm)	Clamping Force of the Pressure Feet (MPa)
1	9869.440	0.00098496	5.1256	21.1568
2	10,002.356	0.00035664	4.508316	18.3692
3	9964.080	0.00060456	4.7575	17.63762

Combined with previous material parameters, the optimized process parameters were used to establish an FEA model. Then, the interference values were calculated and compared with the targeted interference values. The comparative results are listed in Table 8.

Table 8. Comparison between the optimal value and the targeted value.

	Interferometric Measurement Position	Optimal Values (mm)	Targeted Values (mm)	Relative Deviation
1	G1	0.090707899	0.084	7.99%
	G2	0.066179493	0.063	5.05%
	G3	0.0591217	0.054	9.48%
	G4	0.105208	0.099	6.27%
2	G1	0.121151878	0.112	8.17%
	G2	0.083218121	0.084	0.93%
	G3	0.0693104	0.072	3.74%
	G4	0.124837	0.132	5.43%
3	G1	0.104535649	0.10	4.54%
	G2	0.074297139	0.08	7.13%
	G3	0.0631781	0.07	9.75%
	G4	0.111384	0.12	7.18%

As shown in Table 8, the results reveals that the interference values obtained using the optimized process parameters closely matches the targeted interference values, with a maximum relative error of 9.75%, which demonstrates the effectiveness of the proposed optimization method.

A single FEM run for data generation required over 60 min on a workstation, whereas DNN inference for a single candidate took < 1 s. Given that the GA evaluated 10^2–10^3 candidates per generation, replacing FEM with DNN reduced the computational cost per generation by three to four orders of magnitude. The novelty lies in a validated FEM→DNN→GA loop with factor-ranking and target-seeking interference control, delivering minute-scale optimization while retaining physical interpretability via range analysis and partial dependence.

5. Conclusions

This paper presented an interference value prediction and optimization framework for slug rivets, employing a method based on DNN-enhanced GA to optimize the riveting process parameters. By combining the finite element simulation with an intelligent algorithm, the optimized process parameters for the targeted interference fit level were acquired. The following conclusions could be drawn:

(1) A parametric finite element model for the riveting process of the slug rivet was established and subsequently validated through experiments. The verification demonstrated that the constructed FE model could predict riveting interference value accurately.
(2) Orthogonal experiments were designed to investigate the effect of riveting parameters on the interference value. It was found that the initial yield stress of the connected plates had the most significant influence.
(3) A regression prediction model based on DNN was proposed, and its predicted results were compared with three other traditional regression models (support vector machine regression, random forest regression, and Bayesian regression). The results showed that the proposed DNN regression model revealed a superior capability for fitting the multivariate nonlinear relationship between riveting process parameters and the interference value, which achieved a determination coefficient R-squared of 0.9696 and a mean absolute error of 0.00035 mm.
(4) The riveting process parameters were optimized with a DNN-enhanced GA algorithm. Compared with simulated results, the target interference value showed a maximum relative error of 9.75%.

Author Contributions: K.Y.: Conceptualization, Methodology, Data curation, Writing—original draft. L.W.: Investigation, Project administration, Data curation, Writing—original draft. N.H.: Investigation, Data curation, Visualization. D.S.: Formal analysis, Software, Writing—original draft. Y.Z.: Formal analysis, Visualization. Z.C.: Conceptualization, Methodology, Writing—review and editing. All authors have read and agreed to the published version of the manuscript.

Funding: This research was supported by the National Natural Science Foundation of China (grant no. 52475539, 52175450), the Natural Science Foundation Project of Shaanxi Province (2024JC-YBMS-318),the Aeronautical Science Foundation of China (2024Z048053001).

Institutional Review Board Statement: Not applicable.

Data Availability Statement: All data obtained or analyzed in this work are included in this manuscript.

Conflicts of Interest: Author Nana Hui was employed by the company AVIC Xi'an Aircraft Industry Group Company Ltd. The remaining authors declare that the research was conducted in the absence of any commercial or financial relationships that could be construed as a potential conflict of interest.

References

1. Wang, Z.; Chang, Z.; Guo, F.; Kang, Y.; Li, H.; Luo, Q.; Liu, B. Interference-Fit slug Rivet Joining Technology for Large-Aircraft Skin Panels. *Aeronaut. Manuf. Technol.* **2015**, 34–39. [CrossRef]
2. Liu, L.; Li, X.; Wang, Z.; Li, W. Experimental Investigation of an Automated Drilling-Riveting Process for slug Rivets. *J. Northwestern Polytech. Univ.* **2013**, *31*, 77–82.
3. Lei, C.; Bi, Y.; Li, J.; Ke, Y. Experiment and Numerical Simulations of a Slug Rivet Installation Process Based on Different Modeling Methods. *Int. J. Adv. Manuf. Technol.* **2018**, *97*, 1481–1496. [CrossRef]
4. Lv, G.; Li, C.; Zhao, C.; Jia, D.; Zhang, X.; Zhang, H.; Huang, W. Research on Theoretical Prediction Method of Rivet-Forming Quality Considering Different Riveted Structure Parameters. *Int. J. Adv. Manuf. Technol.* **2024**, *132*, 2333–2345. [CrossRef]
5. Liang, Q.; Zhang, T.; Zhu, C.; Bi, Y. Effect of Riveting Angle and Direction on Fatigue Performance of Riveted Lap Joints. *Coatings* **2021**, *11*, 236. [CrossRef]
6. Tian, S.; Li, C.; Dai, W.; Jia, D.; Zhang, H.; Xu, M.; Yuan, Y.; Liu, Y.; Zhang, Y. Effect of the Countersunk Hole Depth on Tensile-Tensile Fatigue Behavior of Riveted Specimens of AA2024-T3 Alloy. *Eng. Fail. Anal.* **2020**, *115*, 104639. [CrossRef]
7. Derijck, J.; Homan, J.; Schijve, J.; Benedictus, R. The Driven Rivet Head Dimensions as an Indication of the Fatigue Performance of Aircraft Lap Joints. *Int. J. Fatigue* **2007**, *29*, 2208–2218. [CrossRef]
8. Huang, L.; Shi, Y.; Guo, H.; Su, X. Fatigue Behavior and Life Prediction of Self-Piercing Riveted Joint. *Int. J. Fatigue* **2016**, *88*, 96–110. [CrossRef]
9. Casolaro, A.; Capone, V.; Iannuzzo, G.; Camastra, F. Deep Learning for Time Series Forecasting: Advances and Open Problems. *Information* **2023**, *14*, 598. [CrossRef]
10. Yang, G.R.; Wang, X.-J. Artificial Neural Networks for Neuroscientists: A Primer. *Neuron* **2020**, *107*, 1048–1070. [CrossRef]
11. LeCun, Y.; Bengio, Y.; Hinton, G. Deep Learning. *Nature* **2015**, *521*, 436–444. [CrossRef]
12. Belhouchet, K.; Bayadi, A.; Bendib, M.E. Artificial Neural Networks (ANN) and Genetic Algorithm Modeling and Identification of Arc Parameter in Insulators Flashover Voltage and Leakage Current. In Proceedings of the 2015 4th International Conference on Electrical Engineering (ICEE), Boumerdes, Algeria, 13–15 December 2015; pp. 1–6.
13. Wang, Z.; Jia, Y.; Fu, P.; Li, H.; Song, L.; Yang, B.; Zhang, L.; Yuan, L.; Shi, K. Research on the Online Evaluation of the Straightness Error of Hydrostatic Guideways Based on Deep Learning. *Int. J. Adv. Manuf. Technol.* **2024**, *134*, 2023–2034. [CrossRef]
14. Deshpande, S.; Venugopal, V.; Kumar, M.; Anand, S. Deep Learning-Based Image Segmentation for Defect Detection in Additive Manufacturing: An Overview. *Int. J. Adv. Manuf. Technol.* **2024**, *134*, 2081–2105. [CrossRef]
15. Sun, H.; Cui, J.; Li, G.; Jiang, H. Genetic Sampling for Surrogate-Assisted Topology Optimization in Electromagnetic Riveting Device Design. *Mater. Des.* **2024**, *237*, 112527. [CrossRef]
16. Masoudi Nejad, R.; Sina, N.; Ma, W.; Song, W.; Zhu, S.P.; Branco, R.; Macek, W.; Gholami, A. Artificial Neural Network Based Fatigue Life Assessment of Riveted Joints in AA2024 Aluminum Alloy Plates and Optimization of Riveted Joints Parameters. *Int. J. Fatigue* **2024**, *178*, 107997. [CrossRef]
17. Michalewicz, Z.; Schoenauer, M. Evolutionary Algorithms for Constrained Parameter Optimization Problems. *Evol. Comput.* **1996**, *4*, 1–32. [CrossRef]
18. Tomar, V.; Bansal, M.; Singh, P. Metaheuristic Algorithms for Optimization: A Brief Review. In Proceedings of the RAiSE-2023, Dubai, United Arab Emirates, 13 March 2024; p. 238.
19. Shi, S.; Cheng, W.; Wang, Z. Intelligent Rivet Selection Technology for Interference-Fit Magnitude Control. *Aeronaut. Manuf. Technol.* **2021**, *64*, 74–79+85. [CrossRef]
20. Choudhary, A.; Kumar, M.; Gupta, M.K.; Unune, D.K.; Mia, M. Mathematical Modeling and Intelligent Optimization of Submerged Arc Welding Process Parameters Using Hybrid PSO-GA Evolutionary Algorithms. *Neural Comput. Appl.* **2020**, *32*, 5761–5774. [CrossRef]
21. Rajendran, S.; Ganesh, N.; Čep, R.; Narayanan, R.C.; Pal, S.; Kalita, K. A Conceptual Comparison of Six Nature-Inspired Metaheuristic Algorithms in Process Optimization. *Processes* **2022**, *10*, 197. [CrossRef]
22. Ni, J.; Tang, W.C.; Xing, Y. Assembly Process Optimization for Reducing the Dimensional Error of Antenna Assembly with Abundant Rivets. *J. Intell. Manuf.* **2018**, *29*, 245–258. [CrossRef]
23. Wu, P.; He, Y.; Li, Y.; He, J.; Liu, X.; Wang, Y. Multi-Objective Optimisation of Machining Process Parameters Using Deep Learning-Based Data-Driven Genetic Algorithm and TOPSIS. *J. Manuf. Syst.* **2022**, *64*, 40–52. [CrossRef]
24. Da Costa Oliveira, A.L.; Britto, A.; Gusmão, R. Machine Learning Enhancing Metaheuristics: A Systematic Review. *Soft Comput.* **2023**, *27*, 15971–15998. [CrossRef]
25. Demmel, F.; Hennet, L.; Jakse, N. The Intimate Relationship between Structural Relaxation and the Energy Landscape of Monatomic Liquid Metals. *Sci. Rep.* **2021**, *11*, 11815. [CrossRef]

26. Li, X.; Chang, Z.; Gao, Y.; Huo, Y.; Song, J.; Wang, Z. Prediction of slug Rivet Interference-Fit Magnitude Using a Multivariate Nonlinear Regression Model. *J. Xi'an Jiaotong Univ.* **2024**, *58*, 157–166.
27. Novac, O.-C.; Chirodea, M.C.; Novac, C.M.; Bizon, N.; Oproescu, M.; Stan, O.P.; Gordan, C.E. Analysis of the Application Efficiency of TensorFlow and PyTorch in Convolutional Neural Network. *Sensors* **2022**, *22*, 8872. [CrossRef]
28. Tian, Y.; Zhang, Y.; Zhang, H. Recent Advances in Stochastic Gradient Descent in Deep Learning. *Mathematics* **2023**, *11*, 682. [CrossRef]
29. Hong, D.; Chen, D.; Zhang, Y.; Zhou, H.; Xie, L. Attacking Robot Vision Models Efficiently Based on Improved Fast Gradient Sign Method. *Appl. Sci.* **2024**, *14*, 1257. [CrossRef]
30. Kumar, H.; Manna, R.; Khan, D. Evaluation of Johnson–Cook Material Model Parameters for Fe–30Mn–9Al–0.8C Low-Density Steel in Metal Forming Applications. *J. Mater. Sci.* **2023**, *58*, 8118–8129. [CrossRef]
31. Li, M.; Liu, Z.; Huang, L.; Chen, Q.; Tong, C.; Fang, Y.; Han, W.; Zhu, P. Automatic Identification Framework of the Geometric Parameters on Self-Piercing Riveting Cross-Section Using Deep Learning. *J. Manuf. Process.* **2022**, *83*, 427–437. [CrossRef]
32. Zhu, J.; Yuan, W.; Peng, F.; Fu, Q. Interaction of Stress Relaxation Aging Behavior and Microstructural Evolution in Inconel 718 Alloy with Different Initial Stress Status. *J. Mater. Sci.* **2021**, *56*, 13814–13826. [CrossRef]

machines

MDPI

Article

Curing Deformation Prediction and Compensation Methods for Large-Sized CFRP Components

Tiantengzi Cao [1], Chao Li [1], Lichao Wan [2], Zhongqi Wang [1] and Yang Zhao [1,*]

[1] School of Mechanical Engineering, Northwestern Polytechnical University, Xi'an 710072, China
[2] Wuhan Second Ship Design and Research Institute, Wuhan 430205, China
* Correspondence: zhyang@nwpu.edu.cn

Abstract

The residual stresses induced by the curing process for carbon fiber-reinforced polymers (CFRPs) lead to inevitable deformation, which seriously affects manufacturing accuracy, especially for large-sized CFRP components. Furthermore, deformation may cause damage or failure of components during subsequent assembly. More attention needs to be paid to improving the curing accuracy for large-sized CFRP components. In this study, a normal direction compensation algorithm based on node deformation is proposed based on the mold profile. Firstly, a finite element model was constructed to simulate the curing process of CFRPs and validated with a small-sized test piece. The results showed the effectiveness of the simulated and compensation methods. Secondly, to meet the precision requirements of large-sized CFRP components, a neural network was used to establish a mapping between curing process parameters and curing deformation, and the parameters were then optimized with a genetic algorithm for subsequent analysis. Finaly, a 15 m CFRP component was used to explore the effect of compensation methods in reducing curing deformation of large-sized CFRP components. The verified results showed that the maximum deformation after compensation was 0.99 mm in the normal direction, which was superior to the 1.258 mm compensation in the connecting direction.

Keywords: CFRP; curing deformation; deformation prediction; profile compensation

1. Introduction

Carbon fiber-reinforced polymer (CFRP) composites are extensively used in the aerospace industry due to their high specific strength and stiffness, corrosion resistance, and lightweight characteristics [1–3]. The large-sized, integrated design of main load-bearing CFRP structures has become an important development trend, aimed at meeting the requirements of high load capacity, long range, and high performance for next-generation aerospace vehicles. However, large-sized composite components exhibit uneven deformation after demolding, mainly due to thermal expansion differences between the fiber–resin interface and the CFRP, as well as between the CFRP and the mold [4,5]. These differences, combined with complex temperature variations on the large-sized mold surface, uneven internal stress distribution, and the release of residual stresses, contribute to deformation [6–10]. In aerospace composites with varying curvatures and thicknesses, non-uniform structural deformation during assembly leads to internal damage or even component failure [11–13]. Therefore, mold profile compensation is essential for improving the accuracy of large-sized CFRP composite components during the curing process.

Extensive research has focused on optimizing residual stress, process parameters, mold structure, and mold profile compensation. Rennick and Radford [14], Radford and Diefendorf [15], and Nelson and Cairns [16] reported that the curing deformation of CFRP laminates, both in-plane and through-thickness, results from mismatched thermal expansion and chemical shrinkage of the constituent materials. Svanberg and Holmberg [17] proposed a linear viscoelastic model accounting for thermal expansion, chemical shrinkage, and exothermic reactions during different stages of resin curing. The results indicated that the novel model improved computational efficiency and accuracy compared with conventional viscoelastic models. Prasatya et al. [18] used thermo-viscoelastic modeling to calculate the curing deformation of composites and concluded that resin cure shrinkage contributed over 30% to residual stresses. Zhu et al. [19] predicted residual stresses generated during curing and examined their effects on composite manufacturing accuracy. White and Hahn [20,21] developed a residual stress model to predict the curing process for composite laminates. Viscoelastic analysis revealed that chemical strains contributed less than 4% to residual stress in a typical cure cycle. Msallem et al. [22] calculated residual stresses developed during curing using finite element simulation. The results underscore the importance of characterization accuracy and model selection. Feng et al. [23] developed a multi-scale, non-orthogonal viscoelastic model to capture the constitutive behavior of preformed braided CFRP components during the curing process. Wang et al. [24] proposed a physics-based numerical model to predict and compensate for deformation induced during machining of three-dimensional woven composites.

Process parameters of CFRP composites also have a significant impact on curing deformation. Sorrentino and Tersigni [25] reported that for thick composite components, the degree of cure differs between the core and the surface, causing structural and geometric deviations. These issues mainly result from improper curing process design, underscoring the need for curing cycle design and optimization. Hui et al. [26] used a neural network to replace the conventional curing kinetics model, identify kinetics parameters, and derive the curing process of CFRP composites. The neural network model substantially improved prediction accuracy for curing behavior. Zhang et al. [27] derived the optimal compensation profile by establishing a coordinate model linking curing process parameters with mold profile compensation. Comparative results indicated that composite panel distortion was reduced by 86.2%.

In addition, curing deformation is effectively reduced by mold profile compensation. Li et al. [28] established a simple mathematical model to predict the spring-in and warpage of T-shaped structures. The results revealed that cure shrinkage and thermal expansion were the primary drivers of deformation. Fernlund et al. [29] investigated the curing process of C-shaped and L-shaped CFRP components and found that C-shaped components exhibited greater spring-in than L-shaped ones, which is attributed to the "geometric locking" effect of the part on the tooling. Furthermore, the differences in the thermal expansion coefficient between the mold and the CFRP critically influence curing-induced deformation. Wang et al. [30] integrated a finite-volume numerical model with a GA to optimize mold support structures, achieving a 17.21% improvement in curing uniformity. Dong et al. [31] proposed a compensation algorithm multiplying nodal deformation on the mold surface by a compensation factor, and the method effectively reduced curing deformation of CFRP components. Zein et al. [32] introduced a method for adjusting the mold profile with a reduced spectral basis to eliminate curing deformation. Liu et al. [33] developed a discrete curvature-based displacement adjustment method for mold profile compensation, which achieved dimensional accuracy for a composite skin in a single iteration, with a maximum surface deviation of 0.3 mm.

Most existing studies have addressed deformation prediction and compensation for small-sized CFRP components. However, methods such as profile compensation, mold structure optimization, and forming parameter optimization are inefficient and require numerous iterations for large-sized components. Traditional solutions rely on empirically determined curing parameters, followed by iterative adjustments and mold surface machining to control deformation. However, such iterative correction increases time and costs, hinders accurate mold surface design, restricts modifications to formed molds, and depends heavily on operator experience. In contrast, mold profile compensation based on node deformation remains one of the most effective and feasible approaches for controlling composite curing deformation. The purpose of this paper is to propose a curing deformation compensation method for large composite components in aerospace applications. The structure of this paper is as follows: Section 2 presents the mold profile compensation method based on node deformation. Section 3 presents the numerical analysis of CFRP curing deformation. Section 4 discusses the optimization of curing process parameters. Section 5 provides a compensation case for large-sized CFRP skin curing deformation. Finally, Section 6 concludes the paper.

2. Research on Mold Profile Compensation Based on Node Deformation

2.1. Compensation Principle

The processes of the mold profile compensation method based on node deformation are as follows: first, coordinate data for all surface nodes in both the undeformed and deformed states are obtained; then, displacement variations between the two states are calculated; finally, compensation is executed using the mold profile compensation algorithm based on deformation magnitude and direction. The resulting profile is extended to the mold boundary to account for deformation caused by the CFRP curing process, ensuring that the deformed shape aligns with the ideal profile. The compensation principle is shown in Figure 1, and the process is illustrated in Figure 2.

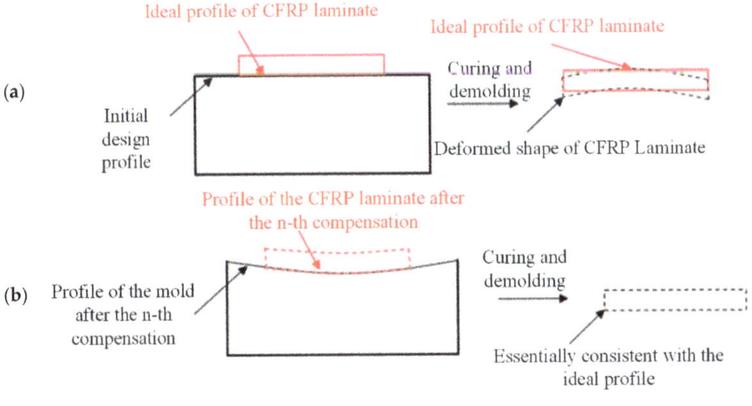

Figure 1. Schematic of CFRP component profile compensation principle: (**a**) shape of CFRP component after molding without compensation; (**b**) shape of CFRP component after molding after the n-th compensation.

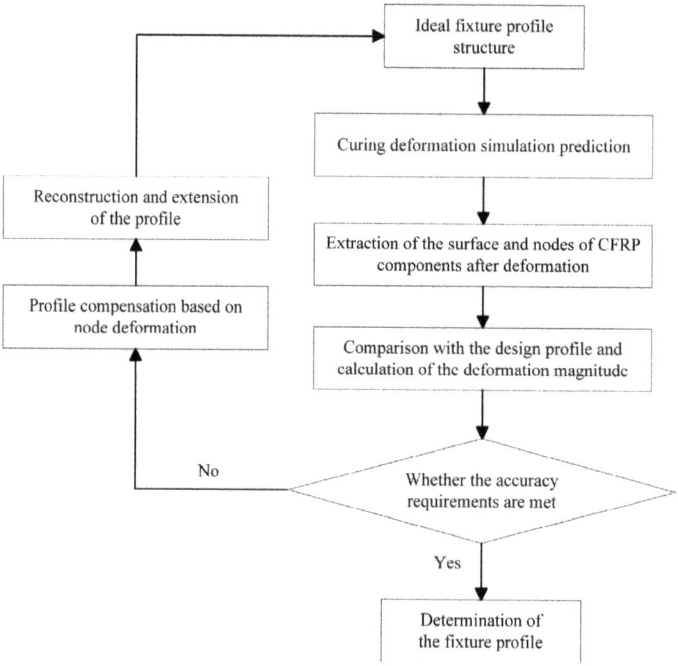

Figure 2. Compensation process for the mold profile of the CFRP component.

2.2. Mold Profile Compensation Algorithm Based on Node Deformation

The inverse deformation compensation principle was applied using curing deformation prediction data to redesign the mold profile. The method involved post-processing numerical predictions for CFRP curing deformation. The coordinates of all nodes on the lower surface, along with their deformed displacements in the x, y, and z directions, were extracted for both the pre-deformation and post-deformation states. The mold was then compensated in both the connecting and normal directions to achieve the desired deformation control for the CFRP component.

(1) Compensation process in the connecting direction

As shown in Figure 3, S_0 represents the initial profile of the component and the mold, which is the ideal profile of the component. And the node coordinates on the ideal surface were denoted as $A_i(x_i, y_i, z_i)$. S_1 represents the profile of the component after curing deformation, with the node coordinates on the surface denoted as $B_i(x_i', y_i', z_i')$. S_2 represents the final profile of the compensated mold, and the node coordinates on the surface were denoted as $C_i(x_i'', y_i'', z_i'')$.

$$\left.\begin{aligned} \Delta x &= x_i' - x_i \\ \Delta y &= y_i' - y_i \\ \Delta z &= z_i' - z_i \end{aligned}\right\} \quad (1)$$

$$C_i(x'', y'', z'') = A_i(x, y, z) - k(\Delta x + \Delta y + \Delta z) \quad (2)$$

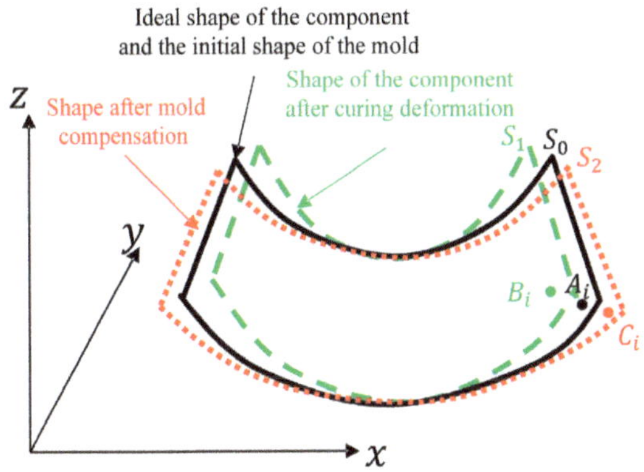

Figure 3. Schematic diagram of inverse deformation compensation.

All nodes on the lower surface of the CFRP component in the finite element model were processed according to the compensation equation $k\overrightarrow{A_iB_i} = \overrightarrow{C_iA_i}$ (where k is the compensation factor, taken as 1 for the convenience of this study). The set of all compensated nodes $V_c = \{C_i(x_i'', y_i'', z_i'') | i = 1, 2, \ldots, n\}$ (where n is the total number of nodes) was fitted to a new surface, which was used as the new mold profile. The carbon fiber prepreg was laid on the mold with the new profile, and the cured CFRP component was checked to determine whether the condition $\left|\overrightarrow{B_iA}\right| \leq \xi$ (where ξ is the dimensional accuracy of the part) was satisfied. The profile S_2 was considered as the final mold configuration when this condition was satisfied. In contrast, the optimization process of profile compensation for the mold was repeated until the accuracy of the cured CFRP component met the accuracy requirements.

(2) Compensation process in the normal direction

In normal direction compensation, the four mesh elements shared by each node were first identified, and the normal vectors of these elements were calculated. Next, the normal vector of each quadrilateral element was computed from three of its nodes, owing to the planar nature of the hexahedral mesh. Finally, these element vectors were aggregated to determine the normal direction for each node. The normal vectors of adjacent elements were summed and normalized to yield the nodal normal vector, as depicted in Figure 4, where, v_1, v_2, v_3, and v_4 denote the four edges that form a node; v_{12} represents the normal vector of the plane formed by v_1 and v_2; v_{23} represents the normal vector of the plane formed by v_2 and v_3; v_{34} denotes the normal vector of the plane formed by v_3 and v_4; v_{41} represents the normal vector of the plane formed by v_4 and v_1; and v is the vector sum of v_{12}, v_{23}, v_{34}, and v_{41}, the normal vector of the node.

The normal vector of each node is $v_i\ (a, b, c)$, and the compensated coordinates $C_i(x_i'', y_i'', z_i'')$ of each node are computed as follows:

$$\left.\begin{aligned} x'' &= x - \frac{k\Delta x \cdot a}{\sqrt{a^2+b^2+c^2}} \\ y'' &= y - \frac{k\Delta y \cdot b}{\sqrt{a^2+b^2+c^2}} \\ z'' &= x - \frac{k\Delta z \cdot c}{\sqrt{a^2+b^2+c^2}} \end{aligned}\right\} \quad (3)$$

Based on the above algorithm, a MATLAB™ (2022) script was developed to implement the compensation process. For the process of connecting direction compensation, only the deformation and compensation vector along this direction were calculated for each node, and the deformation was applied along the connection direction of the corresponding node. In contrast, for normal direction compensation, the normal vector of each node was required, and the deformation was subsequently applied along the normal direction to complete the compensation.

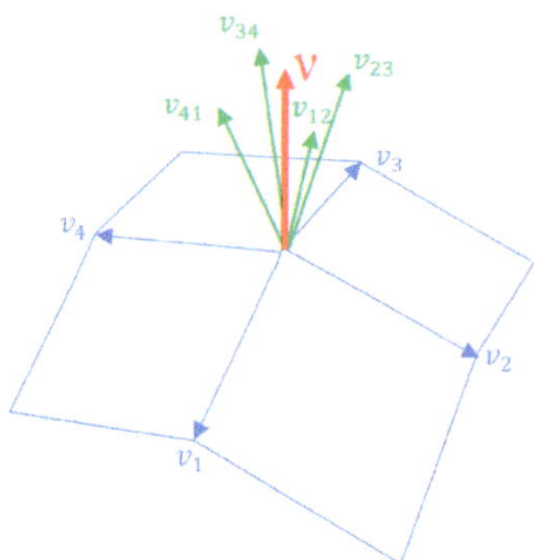

Figure 4. Determination of the nodal normal direction within hexahedral mesh elements.

3. Numerical Analysis of CFRP Curing Deformation

Temperature variations and exothermic heat from the curing reaction influenced the temperature distribution within the CFRP component. The complex temperature distribution resulted in varying degrees of curing across different locations of the component. Therefore, uneven stress distribution during cooling and stress release during demolding resulted in component deformation, a phenomenon particularly pronounced in large, curved composite structures. A numerical model of the curing deformation for CFRP components was established, and the predicted results were then validated through experiments.

3.1. Curing Deformation Prediction for CFRP Components

3.1.1. Construction of a Finite Element Model

The curing deformation model for CFRP components was established using Abaqus™ (2022), with the following as primary considerations: ① The exothermic heat of the curing reaction. During the curing process, the resin underwent an exothermic curing reaction, which caused temperature variations within the mold. ② Non-uniform thermal expansion coefficients. The resin curing reaction kinetics model was incorporated into the thermal field simulation to simulate the thermo-chemical–mechanical coupled heat transfer process for the CFRP component. ③ CFRP curing shrinkage. When the curing degree failed to reach the gel point, both the curing shrinkage coefficient and curing shrinkage strain remained unchanged. In contrast, curing shrinkage and curing strain then commenced. The detailed information on the constitutive materials, curing kinetics equations, and resin cure parameters could be found in Sections S1–S4 of the Supplementary Materials.

The T700/5051 CFRP laminates were fabricated with a stacking sequence of $[45/0/-45/0/45/90/-45/0/45/-45]_s$. The basic material parameters [34] were $v_m = 0.35$, $CHT_1 = 0.6 \times 10^{-6}$, $CHT_2 = 28 \times 10^{-6}$, $E_{f1} = 139000$ MPa, $E_{f2} = 8150$ MPa, $G_{f12} = 4430$ MPa, $G_{f23} = 3550$ MPa, $v_{f12} = 0.2$, and $v_{f13} = 0.2$, and for 6061 aluminum $E = 68900$ MPa and $v = 0.33$.

The geometric model of the mold and CFRP component are shown in Figure 5. The translational and rotational displacements at the center of the lower surface were constrained to prevent rigid motion of the CFRP component. Since this location failed to coincide with a mesh node, direct constraint application was not feasible. Given that the CFRP component was centrally positioned during modeling, the constraint was applied by restraining the four surrounding mesh nodes at the center. As shown in Figure 5, the yellow circle marks the center of the component, and the four red points denote the nodes closest to its center. The employed element was a 3D stress element. The mold was meshed with C3D4 four-node linear tetrahedral elements, while the CFRP laminates used C3D8 eight-node linear hexahedral elements.

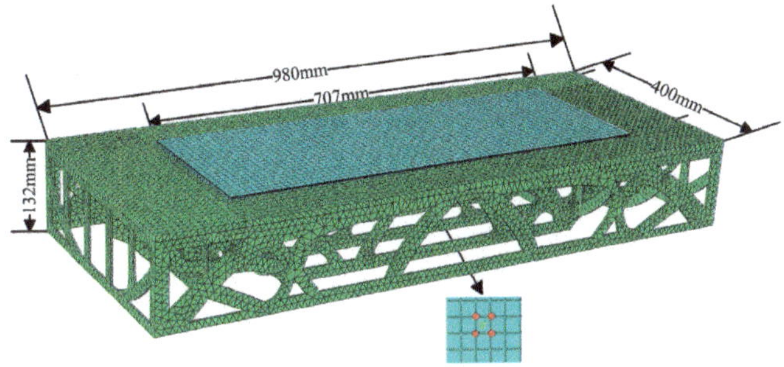

Figure 5. Geometric model of the mold and CFRP component.

3.1.2. Simulation Results for the CFRP Component

The simulation model was developed to simulate the curing and demolding processes for CFRP components. The prediction outcome is shown in Figure 6. Deformation at the center of the CFRP component was minimal after demolding, while maximum deformation occurred at the four top corners, with a maximum of 3.289 mm.

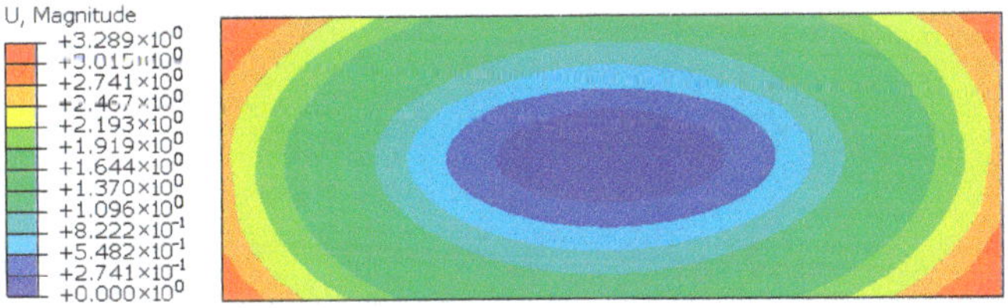

Figure 6. Demolding results for CFRP component considering curing deformation.

3.2. Experimental Verification

The prepreg was first cut to specified dimensions using a cutting machine, then hand-laid with strict alignment to the mold datum. Upon completing the layup process, the component underwent vacuum bagging to eliminate air pockets and surplus resin, followed by curing under programmed thermal cycles.

After the cured CFRP was removed and demolded, the component was measured to obtain deformation data. The point cloud data acquisition system consisted of a Handyscan700 3D scanner (resolution: 0.05 mm, accuracy: 0.025 mm), a computer processing system, and point cloud software for profile scanning. The point cloud scanning process for CFRP components after demolding is shown in Figure 7.

The imported point cloud model was aligned with the ideal plane to achieve the best fit. The profile deviations between CFRP components and the ideal profile are presented in Figure 8. The comparison of finite element deformation (Figure 6) and the scanning point cloud showed that the deformation trends of the experimental and simulated results were consistent. Specifically, the deformation at the center of the component was minimal, while the largest deformation occurred at the four corners.

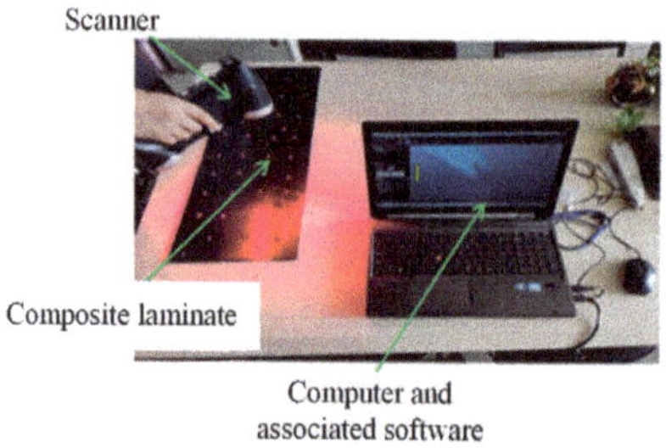

Figure 7. Point cloud scanning process for CFRP components after demolding.

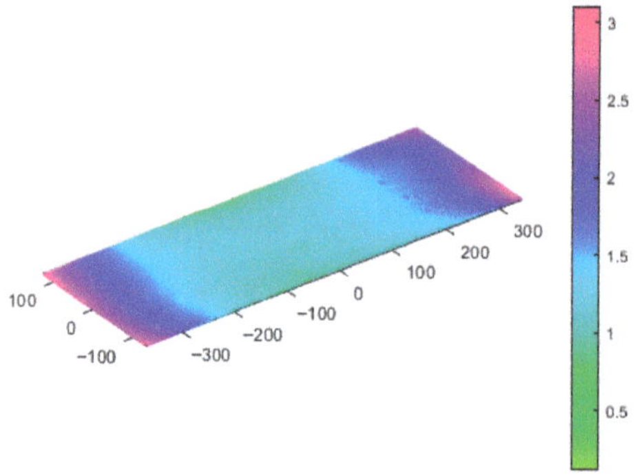

Figure 8. Profile deviations between CFRP components and the ideal profile (unit: mm).

The deformation of the CFRP component after curing manifested as warping at the four top corners, and the maximum deformation occurred at one of the top corners. Table 1 presents the comparison results for the deformation at the four corners of the CFRP component after curing. The results indicated a maximum deviation of 0.288 mm between the experimental measurements and simulated predictions, demonstrating the accuracy of the proposed CFRP curing deformation prediction method. Given the negligible deformation observed, curing process parameter optimization was deemed unnecessary, thus allowing for the direct study of profile compensation for the mold.

Table 1. Comparison of simulation and experimental results for deformation at top corner.

Vertex	1	2	3	4
Simulation value	3.007 mm	3.289 mm	3.024 mm	3.252 mm
Experimental value	2.926 mm	3.046 mm	2.820 mm	2.964 mm
Error	0.081 mm	0.243 mm	0.204 mm	0.288 mm
Relative error	2.7%	7.3%	6.7%	8.8%

3.3. *Analysis of Inverse Deformation Compensation*

3.3.1. Deformation Compensation for the CFRP Component

The normal inverse deformation compensation method was applied to validate the effectiveness of the normal compensation method. The coordinates of all lower surface nodes for the CFRP component were extracted from the inp file generated via Abaqus^{TM}

post-processing. The node coordinates of the compensated surface were derived by applying the normal compensation algorithm implemented in MATLAB™, then the node data were imported into CATIA for fitting and extension to generate the new forming mold profile. The updated mold was used for curing simulations and experiments on the CFRP component, yielding both predicted and experimental values of curing deformation. The compensation process for the CFRP component is shown in Figure 9.

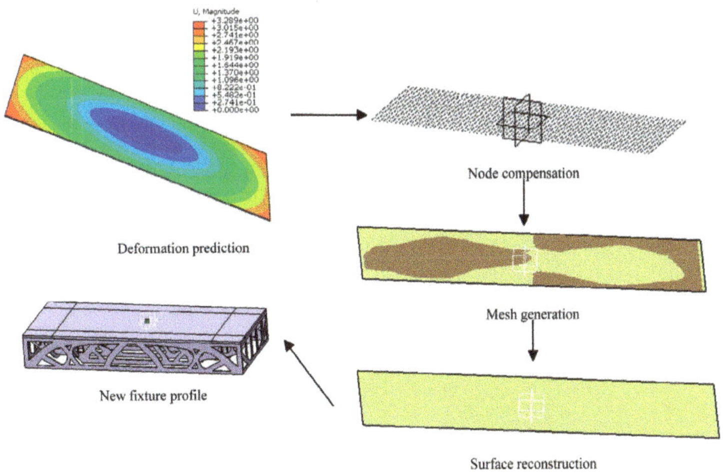

Figure 9. CFRP component deformation compensation process.

3.3.2. Results After Compensation

The compensated mold was used to simulate the curing and deformation processes for the CFRP component. The simulation results for curing deformation for the CFRP component after mold profile compensation are presented in Figure 10.

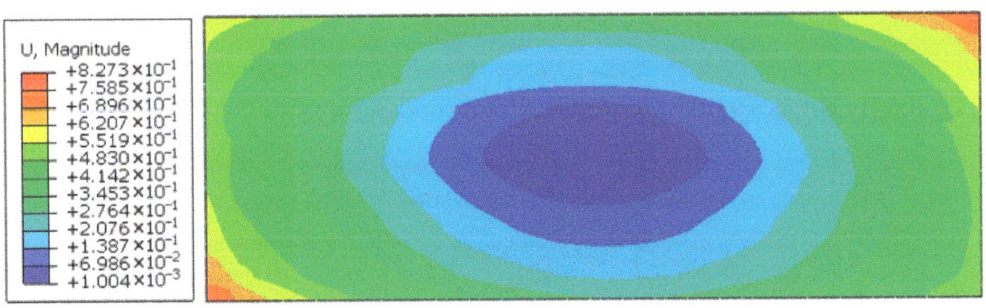

Figure 10. Simulation results for curing deformation for CFRP component after mold profile compensation.

The same CFRP, layup method, and curing process were adopted for the curing and molding experiments after compensation. Figure 11 illustrates the point cloud scanning process of the compensated CFRP components. The profile deviations between CFRP components and the ideal profile are shown in Figure 12. The comparison of the simulated and experimental results after compensation showed that the maximum deformation in the simulation was 0.8273 mm, differing by 0.0953 mm from the experimental value of 0.732 mm. The compensation method reduced the curing deformation from 3.252 mm to 0.8273 mm, representing a 74.56% reduction.

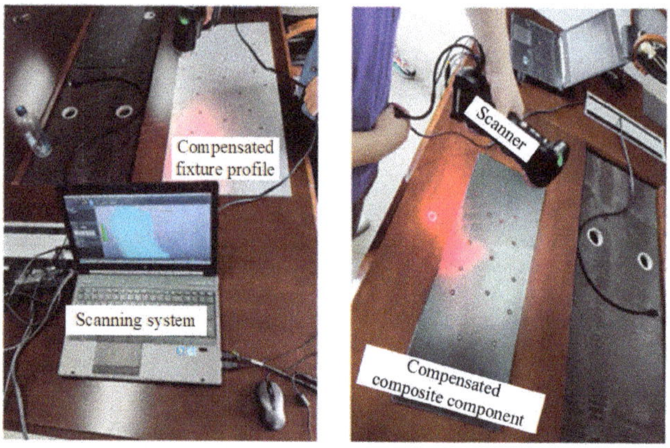

Figure 11. Point cloud scanning process for the compensated CFRP component.

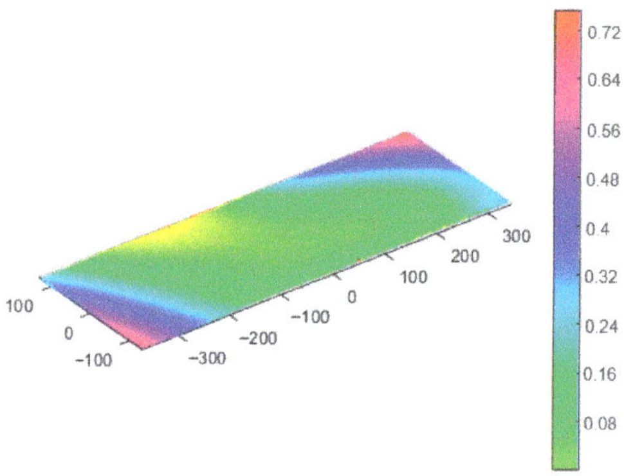

Figure 12. Profile deviations between the compensated CFRP component and the ideal profile (unit: mm).

4. Optimization of Curing Process Parameters Based on BP Neural Network and GA

Given the unclear relationship between process parameters and curing deformation, extensive simulations were conducted to generate sufficient data for establishing a mapping between curing parameters and the deformation. This mapping was subsequently applied to optimize the curing process parameters.

4.1. Construction of Curing Deformation Prediction Model Based on BP Neural Network

4.1.1. Sample Acquisition and Normalization

To avoid inconsistencies in the orders of magnitude for input variables and to reduce the occurrence of erroneous sample data, the samples were normalized to a [0, 1] range prior to neural network training. Afterward, the output vectors were denormalized to restore them to their actual values as follows:

$$z = \frac{x - x_{min}}{x_{max} - x_{min}} \tag{4}$$

where x_{max} and x_{min} represent the maximum and minimum values of x_i, respectively, and z represents the normalized value of variable x_i.

The input parameters are shown in Figure 13, where $a_1/(°C{\cdot}min())$ is the rate during the first heating process, $a_2/(°C{\cdot}min())$ is the rate during the second heating process, t_1/min is the time of the first holding platform, t_2/min is the time of the second holding platform, $T_1/°C$ is the temperature of the first holding platform, and $T_2/°C$ is the temperature of the second holding platform. The output parameter *deform* represents curing deformation, and T_{cure} denotes total curing time. The sample data are detailed in Appendix A.

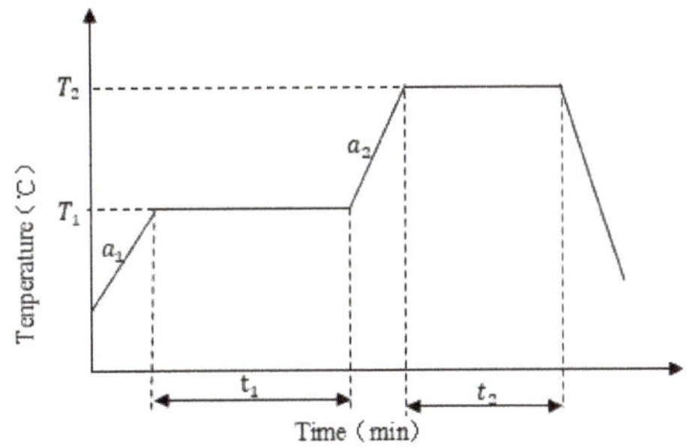

Figure 13. Schematic diagram of input parameters.

4.1.2. Optimization Results

The input test set was processed with the established neural network, and the results were then inversely normalized. These inverted normalized results were compared with the real output values of the test set for error analysis. The prediction results are shown in Figure 14. The predicted and simulated values of curing deformation are shown in Table 2, and the predicted and simulated values of curing time are shown in Table 3.

(a)

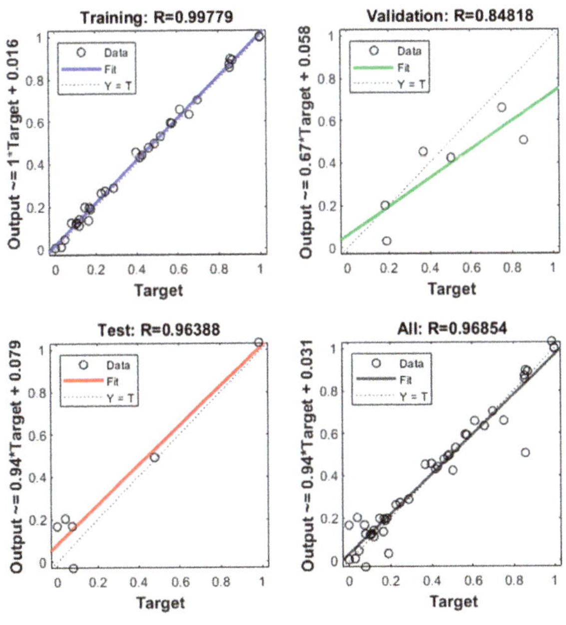

Figure 14. *Cont.*

(b)

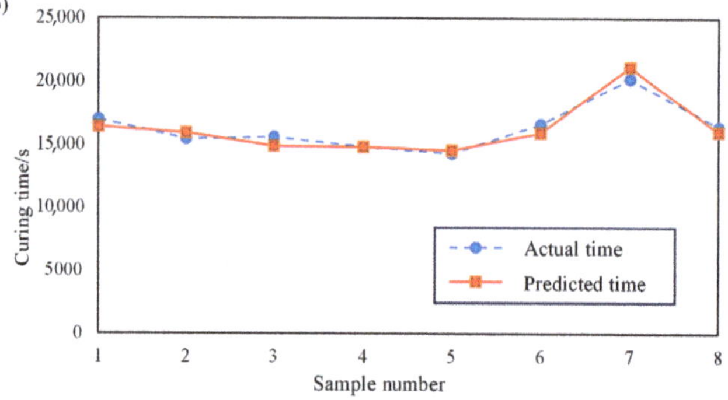

(c)

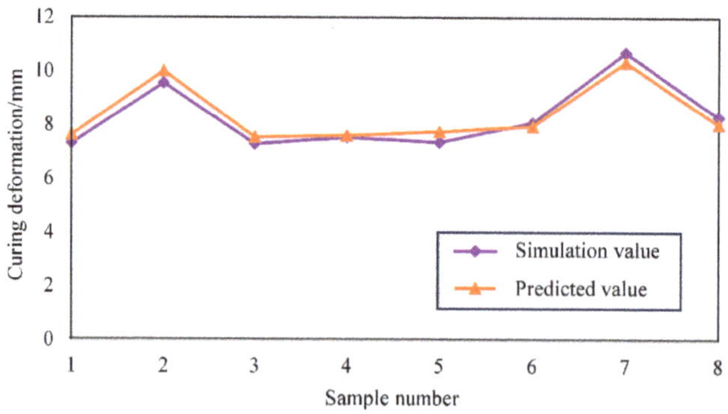

Figure 14. Comparison of training process and error: (**a**) network training process; (**b**) curing time; (**c**) curing deformation.

Table 2. Comparison of predicted and simulated values of curing deformation.

No.	Curing Deformation Test Value/mm	Curing Deformation Predicted Value/mm	Relative Error
1	7.303	7.617	4.3%
2	9.548	9.996	4.7%
3	7.279	7.529	3.4%
4	7.531	7.594	0.8%
5	7.347	7.740	5.3%
6	8.082	7.943	1.7%
7	10.693	10.332	3.4%
8	8.293	8.018	3.3%

Table 3. Comparison of predicted and simulated values of curing time.

No.	Curing Time Test Value/s	Curing Time Predicted Value/s	Relative Error
1	16,962	16,351	3.6%
2	15,415	15,889	3.1%
3	15,565	14,846	4.6%
4	14,806	14,783	0.2%
5	14,295	14,510	1.5%
6	16,578	15,889	4.2%
7	20,192	21,136	4.7%
8	16,377	15,974	2.5%

The graphical error curves and data comparison results showed that the maximum error for both curing deformation and curing time between the predicted data and the test set was less than 5%, which confirmed the effectiveness of the neural network in predicting curing deformation and curing time.

4.2. Optimization of Curing Process Parameters Based on GA

4.2.1. Algorithm Parameter Setting

To facilitate subsequent genetic operations, the solution set of the actual problem was encoded into a representation compatible with the solution space of the genetic algorithm. In this study, binary encoding was employed to convert real-valued solutions within the solution set range into binary strings consisting of 0 and 1, which facilitated the implementation of subsequent crossover and mutation operations. To avoid the loss of valuable genes due to improper selection by the fitness function, F_{min} (the minimum value of fitness) was introduced to enhance the classical roulette selection method as follows:

$$P_i' = \frac{F_i - F_{min}}{\sum_{i=1}^{N-1\Sigma} F_i - F_{min}} \tag{5}$$

A mutation operator was introduced to prevent the algorithm from converging to a local optimum during the iteration process, enabling feasible solutions selected via the selection operation to undergo mutation with a specified probability. An excessively high mutation probability causes the genetic algorithm to degenerate into a random search, while an excessively low mutation probability inhibits the generation of new individuals. Therefore, the algorithm was improved by using an adaptive crossover mutation probability. The mathematical formulation of this relationship is expressed as follows:

$$P_c = \begin{Bmatrix} \frac{k_1(f_{\max}-f)}{f_{\max}-f_{avg}}, & f \geq f_{avg} \\ k_2, & f < f_{avg} \end{Bmatrix} \tag{6}$$

$$P_m = \begin{Bmatrix} \frac{k_3(f_{\max}-f')}{f_{\max}-f_{avg}}, & f' \geq f_{avg} \\ k_4, & f' < f_{avg} \end{Bmatrix} \tag{7}$$

where P_c and P_m are the crossover probability and mutation probability, respectively; f_{max} and f_{avg} refer to the maximum fitness and average fitness of the population; f is the greater fitness value of the two parent individuals participating in the crossover operation; f' is the fitness value of the individuals undergoing the mutation operation; and k_1, k_2, k_3, and k_4 are constants that needs to satisfy the condition $k_1 < k_2$ and $k_3 < k_4$. The parameter settings for the genetic algorithm are shown in Table 4.

Table 4. Parameter settings for the genetic algorithm.

Population Size	Maximum Number of Generations	Generation Gap	Crossover Probability	Mutation Probability	Binary Bit Length
40	200	0.95	0.7	0.05	10

4.2.2. Optimization Results

As shown in Figure 15, the optimization iterations of the genetic algorithm are illustrated. These iterations demonstrated that the population progressively converged to the optimum by the 100th generation. The optimized results are presented in Table 5. The results indicated that the curing time was reduced to 8453 s, which was 31.28% lower than the curing time of 12,300 s for the original model.

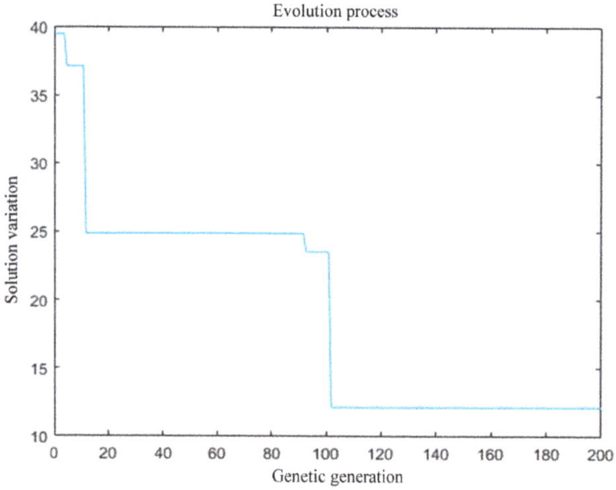

Figure 15. Convergence of optimization computation iterations.

Table 5. Optimization results.

Parameter Variable	Before Optimization	After Optimization	Magnitude of Change
$a_1/(°C·min())$	2	2.11	5.5%
$a_2/(°C·min())$	2	1.99	0.5%
t_1	1200	917	−23.58%
t_2	5400	1588	−70.59%
$T_1/°C$	80	68	−15%
$T_2/°C$	120	125	41.67%
Deformation/mm	8.834	5.253	−40.54%
T_{cure}/s	12,300	8453	−31.28%

5. Case Study

5.1. Curing Deformation of Large-Sized CFRP Skin

The forming mold consisted of 20 partitions along the length direction, each measuring 425 × 425 mm, and 6 partitions along the width direction, each measuring 500 × 500 mm. A curved skin with a length of 14,600 mm, a width of 4600 mm, a curvature of 5.18×10^{-5}, and a thickness of 10 mm was established. The geometric model of the forming mold and the geometric model of the curved CFRP component are shown in Figure 16.

The CFRP composite employed in this study was T700/5051, and the mold was fabricated from Invar steel. The performance parameters of the invar steel material are provided in Table 6.

Table 6. The performance parameters of invar steel.

Performance	Value
Thermal conductivity coefficient	11 W/(m·°C)
Density	$\rho = 8100$ kg·m^{-3}
Specific heat capacity	515 J/(kg·°C)
Modulus	E = 142,000 MPa
Poisson's ratio	$\nu = 0.25$

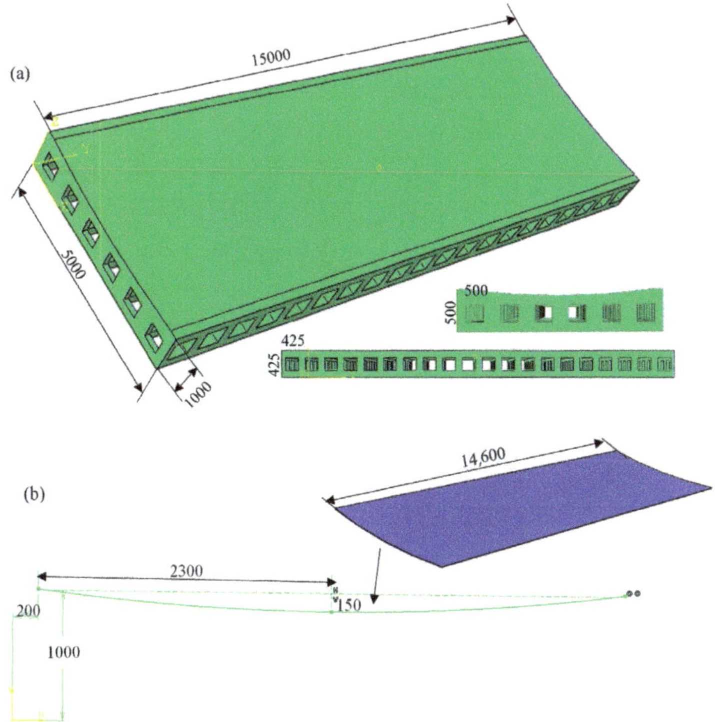

Figure 16. Geometric model of the forming mold: (**a**) geometric model of forming mold; (**b**) geometric model of CFRP component.

As shown in Figure 17, the lower surface of the CFRP component and the upper surface of the mold were defined for contact heat transfer, while the outer surface of the composite and the outer surface of the mold were defined for convective heat transfer with the surroundings. The mold was fully constrained during curing analysis, while the central node of the composite skin was restrained during demolding analysis.

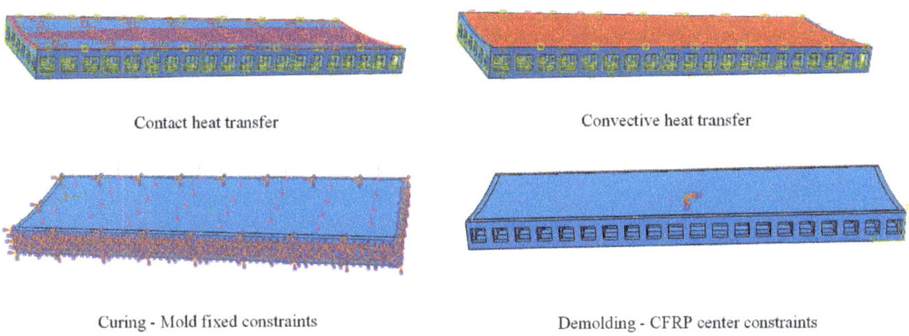

Figure 17. Boundary conditions and load settings of finite element model.

5.2. Results Analysis

The simulation results for curing deformation for the large-sized CFRP component are presented in Figure 18. It was observed that the maximum deformation of the original model was 8.834 mm, while the molding accuracy requirement specified a maximum deformation of no more than 2 mm. The difference between the simulation results and the required molding accuracy was substantial. Direct compensation would have required a large correction to the mold profile, potentially causing internal defects in the CFRP. Therefore, it was necessary to optimize the curing process parameters for further reduction in the deformation.

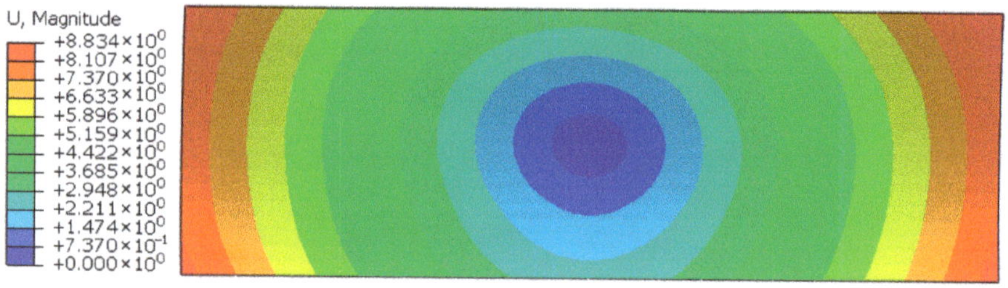

Figure 18. Simulated result for curing deformation for large-sized CFRP component.

The simulated curing deformation obtained using the optimized process parameters is presented in Figure 19. In terms of addressing curing deformation, the first holding time (t_1) and the second holding time (t_2) exert a negative influence on deformation. Specifically, extending t_1 and t_2 contributes to reduced deformation, and a moderate increase in their durations further enhances this deformation-reducing effect. In contrast, the heating rate (a_1) exerts a positive influence on deformation: a lower a_1 helps mitigate deformation, while an increase in a_1 leads to a corresponding rise in deformation [35–37]. The results indicated that the optimized curing deformation was 5.153 mm, representing a 41.67% reduction compared to the original model's curing deformation of 8.834 mm. These results demonstrated that the optimization method effectively reduced curing deformation while considering curing time. Although the optimized curing process parameters reduced the deformation of the composite skin, the resulting accuracy remained below the required threshold. As a result, additional compensation of the mold profile was required.

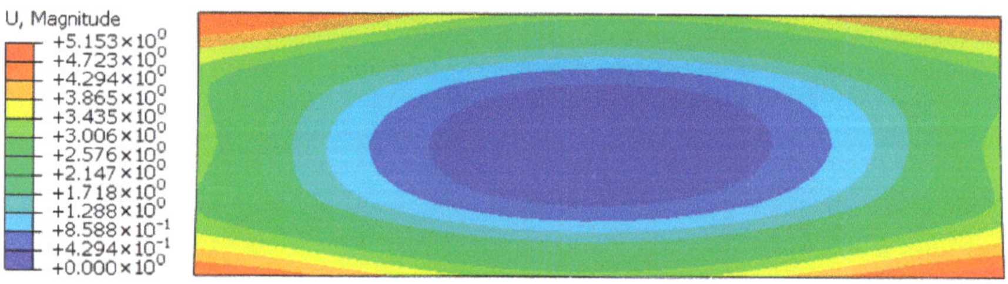

Figure 19. Optimized results for curing deformation for CFRP component.

The coordinates of the lower surface nodes of the CFRP skin were obtained using the method described in Section 3.3. These coordinates were then compensated using the connecting direction compensation algorithm and the normal direction compensation algorithm outlined in Section 2.2. The compensated results are shown in Figure 20 and Figure 21, respectively.

The compensated results for the normal direction and connecting direction are shown in Figure 22 and Figure 23, respectively. The compensation data for large-sized components are shown in Table 7. They indicated that the application of the normal direction compensation algorithm reduced the maximum curing deformation to 0.9934 mm, representing an

80.7% decrease from the original 5.153 mm and meeting the required molding accuracy for the CFRP component. However, the maximum curing deformation after applying the connecting direction compensation algorithm remained at 1.258 mm, reflecting a 75.5% reduction from the deformation before compensation. Therefore, the mold compensation in the normal direction is more effective for large-sized CFRP components.

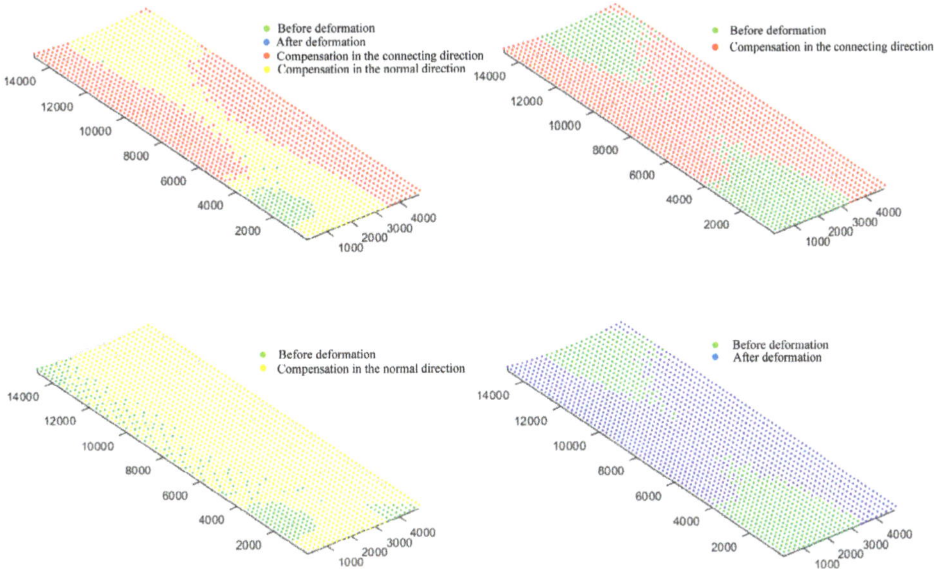

Figure 20. Point cloud maps after compensation.

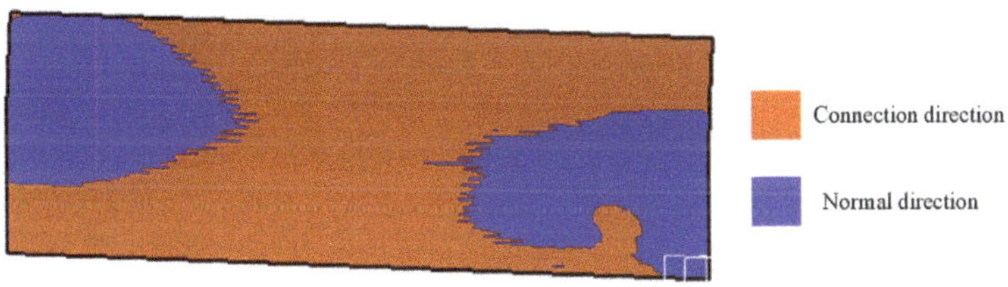

Figure 21. Profile shape of mold after compensation.

Figure 22. Simulated results for large-sized CFRP component after normal direction compensation of the mold.

Table 7. The compensation data.

Optimization Project	Before Optimization (mm)	After Optimization (mm)	Magnitude of Change
Curing process parameters	8.834	5.153	41.67%
Normal direction	5.153	0.9934	80.7%
Connecting direction	5.153	1.258	75.5%

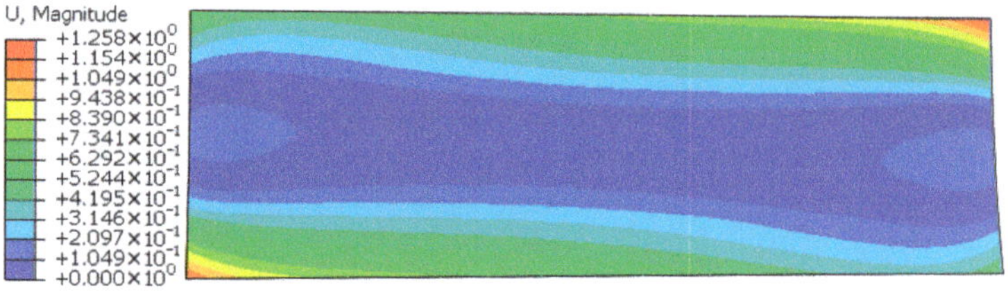

Figure 23. Simulated results for large-sized CFRP component after connecting direction compensation of the mold.

6. Conclusions

In this study, a normal direction compensation algorithm based on node deformation was proposed to establish an effective model for predicting and compensating for the deformation of large-sized CFRP components. Initially, the compensation process and the mold profile compensation algorithm based on node deformation were introduced. Subsequently, a finite element model was established to analyze the curing deformation of small-sized CFRP components and validated with a small-sized test piece. A maximum deformation of 3.252 mm was predicted by the numerical simulation, compared to an experimental measurement of 2.964 mm, yielding a deviation of 0.288 mm. The results validate the accuracy of the deformation prediction model. Next, a neural network model capturing the relationship between process parameters and curing deformation was established. A multi-objective optimization of the curing process parameters was conducted using the neural network model in combination with a genetic algorithm, with curing deformation and curing time both taken into consideration. Finally, a simulation model was developed to predict the curing deformation of 15 m CFRP components. The maximum deformation after optimization was reduced to 5.153 mm, representing a 41.67% reduction relative to the value prior to optimization. Further compensation showed that the maximum deformation in the normal direction after compensation was 0.9934 mm, demonstrating a significantly better compensation effect compared to the connection direction. Although the curing process was effectively simulated and deformation was controlled, certain limitations persisted, including limited simulation datasets and insufficient experimental validation for large-sized components. Most existing studies focus solely on curing deformation while neglecting deformation induced by demolding. Further research is needed to integrate demolding forces and methods into curing deformation analysis, thereby improving model accuracy for large CFRP components and enabling more precise deformation prediction.

Supplementary Materials: The following supporting information can be downloaded at: https://www.mdpi.com/article/10.3390/machines13100890/s1, Section S1: The CHILE (Cure-hardening instantaneously linear elastic material) model; Section S2: Curing kinetics model; Section S3: Thermo-chemical model; Section S4: Resin curing shrinkage model.

Author Contributions: Methodology, C.L.; software, C.L.; validation, L.W.; investigation, T.C.; resources, Y.Z.; data curation, L.W.; writing–original draft, T.C.; writing–review & editing, Y.Z.;

supervision, Z.W.; funding acquisition, Z.W. All authors have read and agreed to the published version of the manuscript.

Funding: This research was supported by the National Key R&D Program of China (grant no. 2021YFB3401700), the National Natural Science Foundation of China (grant no. 52475539), and the Natural Science Foundation of Shaanxi Province (2024JC-YB MS-318).

Data Availability Statement: The original contributions presented in this study are included in the article/Supplementary Material. Further inquiries can be directed to the corresponding author.

Conflicts of Interest: The authors declare no conflict of interest.

Appendix A

Table A1. Sample data.

Number	a_1	a_2	t_1	t_2	T_1	T_2	Deform	T_{cure}
1	1.42	1.06	4704	5411	97	129	7.990	18,048
2	2.06	0.93	5460	2404	97	120	7.347	14,295
3	2.45	1.07	2593	4373	93	134	8.715	14,259
4	0.70	1.59	4454	1668	67	127	6.892	15,019
5	2.82	1.28	8690	765	90	158	9.217	17,995
6	2.44	2.81	5193	4618	83	156	10.153	16,729
7	1.72	1.58	4978	3300	104	113	7.192	13,997
8	1.59	0.96	2545	3192	89	147	9.548	15,415
9	1.62	2.76	4707	5486	96	123	8.293	16,377
10	1.27	2.95	5842	3893	80	131	8.082	16,578
11	1.77	1.60	6305	3935	79	137	8.749	17,633
12	1.78	0.78	3922	5241	97	157	10.693	20,192
13	2.54	1.15	3686	4950	64	131	8.521	16,244
14	2.49	1.52	8899	3714	66	159	10.135	21,300
15	2.23	0.95	3024	2589	87	128	8.258	10,635
16	1.45	1.16	8035	1896	78	145	8.860	19,217
17	2.53	2.01	8272	5387	97	143	8.483	20,299
18	1.83	2.28	7288	755	96	137	7.531	14,806
19	1.38	1.05	1429	3245	63	145	9.391	14,592
20	2.85	0.79	2800	1507	78	143	8.645	13,914
21	2.69	1.24	3417	5885	84	119	7.585	15,129
22	1.88	1.30	6310	4449	79	116	7.303	16,962
23	2.06	1.56	1747	3303	90	160	10.647	13,689
24	1.97	1.77	6658	3144	88	119	7.279	15,565
25	1.02	0.71	1497	922	73	112	5.993	11,087
26	1.25	1.16	6092	4283	79	138	8.810	19,415
27	1.68	2.50	4751	829	61	154	7.832	12,969
28	1.08	0.57	7144	986	104	143	8.296	20,206
29	2.61	2.82	6606	3417	68	120	7.071	14,942
30	0.99	2.33	8191	1122	65	128	6.618	16,478

$a_1/(°C{\cdot}min())$ is the rate during the first heating process; $a_2/(°C{\cdot}\mathrm{min}())$ is the rate during the second heating process; t_1/min is the time of the first holding platform; t_2/min is the time of the second holding platform; T_1/°C is the temperature of the first holding platform; T_2/°C is the temperature of the second holding platform. The output parameter *deform* represents curing deformation, and T_{cure} denotes total curing time.

References

1. Zhao, Y.; Zhao, M.; Wang, A.; Chang, Z.; Wang, Z.; Zhang, K. Experimental study on the mode I interlaminar properties of self-healable vitrimeric CFRP with various interfaces. *Compos. Part B-Eng.* **2023**, *261*, 16. [CrossRef]
2. Netzel, C.; Mordasini, A.; Schubert, J.; Allen, T.; Battley, M.; Hickey, C.; Hubert, P.; Bickerton, S. An experimental study of defect evolution in corners by autoclave processing of prepreg material. *Compos. Part A-Appl. Sci. Manuf.* **2021**, *144*, 13. [CrossRef]
3. Zhao, Y.; Zhao, M.; Wang, S.; Wang, A.; Chang, Z.; Wang, Z. Experimental investigation of stacking interface on mode II interlaminar behaviour of self-healable vitrimeric CFRP. *Compos. Part A-Appl. Sci. Manuf.* **2024**, *185*, 13. [CrossRef]
4. Cheng, X.; Han, Q.; Huang, Y.; Shi, M.; Shi, H.; Ji, M.; Yang, C. Bioinspired ultra-fine hybrid nanocoating for improving strength and damage tolerance of composite fan blades in flexible manufacturing. *Compos. Sci. Technol.* **2025**, *259*, 10. [CrossRef]
5. Wang, Z.; Li, Y. Drop-weight impact and compressive properties of repeatable self-healing carbon fiber reinforced composites. *Compos. Part B-Eng.* **2024**, *283*, 10. [CrossRef]
6. Twigg, G.; Poursartip, A.; Fernlund, G. Tool—Part interaction in composites processing. Part I: Experimental investigation and analytical model. *Compos. Part A-Appl. Sci. Manuf.* **2004**, *35*, 121–133. [CrossRef]
7. Twigg, G.; Poursartip, G.; Fernlund, G. Tool—Part interaction in composites processing. Part II: Numerical modelling. *Compos. Part A-Appl. Sci. Manuf.* **2004**, *35*, 135–141. [CrossRef]
8. Twigg, G.; Poursartip, A.; Fernlund, G. An experimental method for quantifying tool—Part shear interaction during composites processing. *Compos. Sci. Technol.* **2003**, *63*, 1985–2002. [CrossRef]
9. Bavel, B.; Shishkina, O.; Vandepitte, D.; Moens, D. Reliability-based composite pressure vessel design optimization with cure-induced stresses and spatial material variability. *Comput. Methods Appl. Mech. Eng.* **2024**, *432*, 22.
10. Nagaraj, M.; Shah, S.; Sabato, A.; Maiaru, M. Validation and verification of a novel higher-order FE Framework for process modeling of thermoset composites. *Compos. Part B-Eng.* **2024**, *279*, 12. [CrossRef]
11. Ding, A.; Li, S.; Sun, J.; Wang, J.; Zu, L. A thermo-viscoelastic model of process-induced residual stresses in composite structures with considering thermal dependence. *Compos. Struct.* **2016**, *136*, 34–43. [CrossRef]
12. Chang, Z.; Ren, Y.; Li, X.; Yan, K.; Wang, Z.; Zhao, Y. Assembly deformation prediction of aircraft composite panel considering manufacturing variation and actual assembly process. *Int. J. Adv. Manuf. Technol.* **2024**, *135*, 3811–3825. [CrossRef]
13. Ma, H.; Lau, K.; Hui, D.; Shi, S.-Q.; Poon, C.-K. Theoretical analysis on the pullout behavior of carbon nanotube at cryogenic environment with the consideration of thermal residual stress. *Compos. Part B-Eng.* **2017**, *128*, 67–75. [CrossRef]
14. Rennick, T.S.; Radford, D.W. Components of manufacturing distortion in carbon fiber/epoxy angle brackets. *Int. SAMPE Tech. Conf.* **1996**, *28*, 189–197.
15. Radford, D.W.; Diefendorf, R.J. Shape instabilities in composites resulting from laminate anisotropy. *J. Reinf. Plast. Compos.* **1993**, *12*, 58–75. [CrossRef]
16. Nelson, R.H.; Cairns, D.S. Prediction of dimensional changes in composite laminates during cure. *Int. SAMPE Symp. Exhib.* **1989**, *34*, 2397–2410.
17. Svanberg, J.M.; Holmberg, J.A. Prediction of shape distortions Part I. FE-implementation of a path dependent constitutive model. *Compos. Part A-Appl. Sci. Manuf.* **2004**, *35*, 711–721. [CrossRef]
18. Prasatya, P.; McKenna, G.B.; Simon, S.L. A Viscoelastic Model for Predicting Iso-tropic Residual Stresses in Thermosetting Material: Effects of Processing Parameters. *J. Compos. Mater.* **2001**, *35*, 826–848. [CrossRef]
19. Zhu, Q.; Geubelle, P.H.; Li, M.; Tucker, C.L., III. Dimensional Accuracy of Thermoset Composites: Simulation of Process-Induced Residual Stresses. *J. Compos. Mater.* **2015**, *35*, 2171–2205. [CrossRef]
20. White, S.R.; Hahn, H.T. Process modeling of composite materials: Residual stress development during cure. I—Model formulation. *J. Compos. Mater.* **1992**, *26*, 2402–2422. [CrossRef]
21. White, S.R.; Hahn, H.T. Process modeling of composite materials: Residual stress development during cure. II—Experimental validation. *J. Compos. Mater.* **1992**, *26*, 2423–2453. [CrossRef]
22. Msallem, Y.A.; Jacquemin, F.; Boyard, N.; Poitou, A.; Delaunay, D.; Chatel, S. Material characterization and residual stresses simulation during the manufacturing process of epoxy matrix composites. *Compos. Part A-Appl. Sci. Manuf.* **2010**, *41*, 108–115. [CrossRef]
23. Feng, Y.; Han, Z.; Liu, M.; Zhang, W. Multiscale modeling for viscoelasticity of woven CFRP considering preforming and curing effects via finite element and long-short term memory analysis. *Compos. Part A-Appl. Sci. Manuf.* **2024**, *186*, 10. [CrossRef]
24. Wang, Q.; Li, T.; Yang, X.; Wang, K.; Wang, B.; Ren, M. Prediction and compensation of process-induced distortions for L-shaped 3D woven composites. *Compos. Part A-Appl. Sci. Manuf.* **2021**, *141*, 10. [CrossRef]
25. Sorrentino, L.; Tersigni, L. A Method for Cure Process Design of Thick Composite Components Manufactured by Closed Die Technology. *Appl. Compos. Mater.* **2012**, *19*, 31–45. [CrossRef]

26. Hui, X.; Xu, Y.; Zhang, W.; Zhang, W. Cure process evaluation of CFRP composites via neural network: From cure kinetics to thermochemical coupling. *Compos. Struct.* **2022**, *288*, 12. [CrossRef]
27. Zhang, Y.; An, L.; Zhao, C. Enhancing the Forming Accuracy of CFRP through a Global Compensation Method by Introducing an Optimal Profile. *Polymers* **2024**, *16*, 1792. [CrossRef] [PubMed]
28. Li, J.; Yao, X.; Liu, Y.; Chen, S.; Kou, Z.; Dai, D. Curing Deformation Analysis for the Composite T-shaped Integrated Structures. *Appl. Compos. Mater.* **2008**, *15*, 207–225. [CrossRef]
29. Fernlund, G.; Rahman, N.; Courdji, R.; Bresslauer, M.; Poursartip, A.; Willden, K.; Nelson, K. Experimental and numerical study of the effect of cure cycle, tool surface, geometry, and lay-up on the dimensional fidelity of autoclave-processed composite parts. *Compos. Part A-Appl. Sci. Manuf.* **2002**, *33*, 341–351. [CrossRef]
30. Wang, Q.; Wang, L.; Zhu, W.; Xu, Q.; Ke, Y. Design optimization of molds for autoclave process of composite manufacturing. *J. Reinf. Plast. Compos.* **2017**, *36*, 1564–1576. [CrossRef]
31. Dong, C.; Zhang, C.; Liang, Z.; Wang, B. Dimension variation prediction for composites with finite element analysis and regression modeling. *Compos. Part A: Appl. Sci. Manuf.* **2004**, *35*, 735–746. [CrossRef]
32. Zein, S.; Parmentier, A.; Dumas, D. A mold compensation method using spectral shape representation. *Compos. Struct.* **2025**, *370*, 9. [CrossRef]
33. Liu, C.; Sun, L.; Wang, J. An improved approach to compensating for cure-induced deformation to manufacture composite aircraft skin accurately. *Polym. Compos.* **2025**, *46*, 2891–2901. [CrossRef]
34. Zhao, Y.; Wang, Z.; Jiang, P.; Wang, A.; Chang, Z.; Kang, Y. Numerical simulation on mechanical properties and damage behavior of CFRP with self-healing microvascular channels. *Mech. Adv. Mater. Struct.* **2024**, *31*, 297–313. [CrossRef]
35. Jung, W.; Chu, W.; Ahn, S.; Won, M.-S. Measurement and compensation of spring-back of a hybrid composite beam. *J. Compos. Mater.* **2007**, *41*, 851–864. [CrossRef]
36. Pereira, G.C.; LeBoulluec, P.; Lu, W.; Yoshida, M.; Alves, A.P.; Ávila, A. Spring-back behavior on L-shaped composite structures: A statistical analysis of angular recovery as a function of time and residual cure. *Compos. Part A-Appl. Sci. Manuf.* **2019**, *124*, 15. [CrossRef]
37. Li, J.; Yao, X.F.; Liu, Y.H.; Cen, Z.Z.; Kou, Z.J.; Hu, X.C.; Dai, D. Thermo-viscoelastic analysis of the integrated T-shaped composite structures. *Compos. Sci. Technol.* **2010**, *70*, 1497–1503.

machines

Article

An Assembly Accuracy Analysis Method for Weak Rigid Components

Dongping Zhao [1,*], Zhe Yuan [1], Xiaosong Zhao [1] and Gangfeng Wang [2]

[1] School of Aircraft Engineering, XiHang University, Xi'an 710077, China; zheyuan67@163.com (Z.Y.); xs1464203901@163.com (X.Z.)
[2] Key Laboratory of Road Construction Technology and Equipment of MOE, School of Construction Machinery, Chang'an University, Xi'an 710064, China; wanggf@chd.edu.cn
* Correspondence: zhaodongp@163.com

Abstract

Most existing assembly accuracy analysis methods focus on rigid assemblies or assume assemblies to be rigid bodies, neglecting the influence of assembly deformation in weak rigid components (WRCs) such as thin-walled structures, cantilever structures, etc. As a result, the assembly accuracy analysis becomes inaccurate, and the accuracy of key components cannot be effectively controlled. This may lead to serious issues such as forced assembly, repair, and rework. To address these problems, this study proposes a rigid–flexible coupling-based assembly accuracy analysis method for WRCs. The stiffness matrix and assembly deformation of WRCs are calculated, and by coupling assembly deformation with other assembly deviations, a rigid–flexible coupling assembly accuracy data model is established. This model incorporates multiple deviation sources, including assembly process variations, design tolerances, and assembly deformations. Assembly deviation transfer modeling and accumulation calculation methods for WRCs are investigated, enabling assembly accuracy simulation and statistical analysis. A case study on WRC assembly accuracy analysis is conducted, and the results demonstrate that the proposed method improves the accuracy of assembly analysis for WRCs, verifying its reliability.

Keywords: assembly accuracy analysis; rigid–flexible coupling; assembly deviation transmission; weak rigid components (WRCs)

1. Introduction

Weakly rigid components (WRCs) are components in a structure or system that have relatively weak or insufficient resistance to external deformation [1]. This type of component is prone to problems such as accuracy loss, assembly difficulty, and decreased assembly performance during the assembly process.

To guarantee the product assembly quality and efficiency when considering WRCs, assembly accuracy analysis technology is applied during the assembly process preparation stage. With the continuous advancement in aircraft performance requirements, assembly accuracy analysis has progressed from two-dimensional to three-dimensional space [2]. Three-dimensional (3D) accuracy analysis accounts for spatial variability in deviation propagation and accumulation, as well as dimensional deviations, geometric deviations, positioning deviations, and their spatial relationships, making it more representative of real-world assembly scenarios. Consequently, this approach has undergone rapid development.

WRCs have the characteristics of low stiffness and easy deformation. Unlike rigid components, weak rigid components have more sources of deviation, especially under assembly forces, fixture positioning, and clamping effects, and their assembly accuracy is affected by the coupling of multiple sources of deviation. Therefore, the impact of deformation caused by assembly on the accuracy of the final product cannot be ignored. However, most existing methods for predicting product assembly accuracy are based on the assumption of rigid bodies, ignoring the weak rigidity of the structure and the impact of assembly deformation on the target accuracy. This leads to inaccurate prediction results, which can necessitate forced assembly, process modifications, rework, and on-site assembly adjustments, ultimately prolonging the assembly cycle and increasing costs [3].

To enhance the accuracy of the assembly analysis and assembly quality of WRCs while minimizing the assembly rework, this paper first develops a novel deviation representation modeling. Then, the assembly accuracy analysis method of deviation transfer and accumulation calculation of rigid–flexible coupling for WRCs is proposed. Finally, an example is demonstrated to validate the reliability of the proposed method.

2. Literature Review

Assembly accuracy analysis involves the comprehensive consideration of assembly deviations and their propagation/accumulation processes prior to production, enabling the analysis and evaluation of the final product assembly accuracy. Current research on assembly accuracy analysis primarily focuses on two key aspects:

- The development of assembly deviation representation models.
- The computational methods for deviation propagation and accumulation analysis.

2.1. Research on Assembly Deviation Modeling Methods

Assembly deviation modeling includes two aspects: deviation mathematical modeling and deviation representation modeling.

(1) Deviation mathematical modeling is a mathematical description of deviation semantics. Scholars have studied deviation mathematical modeling from the perspective of changes in geometric elements and the changing angle of deviation domain boundaries.

In terms of deviation mathematical modeling based on geometric element changes, Hillyard et al. [4] proposed a vector method to express dimensional tolerances, where the size of dimensional tolerances is represented by the displacement of vectors, and the direction of dimensional tolerances is represented by the rotation angle of vectors. Multi-dimensional vectors are required to express geometric shape tolerances. Zhang [5] proposed a tolerance mathematical representation method based on the degrees of freedom, using deviations in multiple degrees of freedom directions to express the changes in assembly features within the tolerance domain. However, since the changes in assembly features are constrained by the tolerance domain, the deviation constraints in each degree of freedom direction are complex. To address this problem, Shen et al. [6] presented an integrated Jacobian–NSMS model based on NSMS and the unified Jacobian–torsor model, which combines the advantages of efficient deviation propagation analysis and high-fidelity representation of tolerance. Zhao et al. [7] proposed an integrated characterization method for accuracy analysis. By analyzing the internal coupling relationship of the different geometrical deviations in a single part, the monomer model was established. The effectiveness of the monomer model is verified through an analysis of a simulated rotor assembly analysis, and the corresponding accuracy analysis method based on the model reasonably predicts the assembly deviation of the rotor.

Cui et al. [8] proposed a tolerance specification method based on machine learning. Laifa et al. [9] studied the variation characteristics of assembly features and proposed the use of the translation and rotation of geometric features to express the variation deviation. Finally, the deviation small displacement vector model was investigated to represent the rotation deviation and translation deviation of the feature by multivariate vectors. Jin et al. [10] studied the modeling methods of various types of tolerance domains using screw parameters based on the screw theory.

In terms of mathematical modeling of deviation based on the description of the deviation domain boundary, Wu [11] proposed a new tolerance model by representing the position variation of control points of geometric element. The control points of point, line and plane geometric element are defined as point itself, two end points of line, any three vertices of bounding box of plane and the interior constituting points of line and plane, which are used to present the dimensional error, the position error, the orientation error and the form error. Guo et al. [12] constructed an accuracy control model that includes clamping process parameters and fixture position parameters. An assembly quality control technology was researched in forced clamping and compensation processes for large and integrated aeronautical composite structures. Cao et al. [13] used tolerance specifications to uniformly describe the tolerance information in the drift model. However, Anwer et al. [14] believed that although the drift model of feature boundary expression can express common tolerances in three-dimensional models, it cannot effectively distinguish the types of geometric tolerances and geometric features. In the American Society of Mechanical Engineers standard (ASME) Y14.5 [15], the direction of the tolerance domain is described by the direction of the derived element and the position of the tolerance domain is described by the position of the derived element, the range is represented by the tolerance value and the shape is described by a vector equation. However, this standard requires that the tolerance constraints must be approximated, and the parameter vectors describing the tolerance must be independent of each other. This approximate processing will cause the deviation accumulation to be too large during the assembly process.

(2) Deviation representation modeling addresses the critical challenge of associating tolerance information with the corresponding 3D solid features in assembly systems. Currently, the research on deviation representation modeling mainly focuses on two aspects: deviation representation based on a solid model and deviation representation independent of a solid model. The former methodology directly embeds tolerance specifications within the geometric model framework, while the latter establishes deviation characterization through separate mathematical or parametric models that maintain independence from the core solid modeling architecture.

First, there are three methods of deviation representation based on solid models: constructive solid geometry (CSG), boundary representation (B-Rep) and hybrid representation (B-Rep + CSG). In terms of B-Rep model establishment, Johnson [16] proposed the evaluated dimension and tolerance (EDT) method to realize the boundary representation of tolerance information and the association between the geometric information of the solid model and the dimension and tolerance information. However, the entity features involved in CSG are not unique and there is feature redundancy in the tolerance information representation, while B-Rep is not conducive to the hierarchical expression of feature information. Therefore, Roy et al. [17] combined CSG with the B-Rep method and proposed the B-Rep + CSG method, which uses CSG to organize features hierarchically and uses B-Rep to associate dimension and tolerance information with the geometric features of the solid model. It is noted that the deviation representation method based on the solid model realizes the association between deviation information and geometric model feature information, but the expression of deviation semantic information is relatively vague.

Consequently, researchers have increasingly focused on developing solid-model-independent deviation representation methods to overcome the limitations of traditional geometry-bound approaches. Arizona State University [18] proposed a deviation T-Map representation model to express the size deviation, position deviation, shape deviation, etc., of assembly geometric features on an ideal Euclidean geometric body. With the further improvement of the T-Map model, the deviations of other geometric features such as axes, bevels, radii, etc., can also be represented in the model. Desrocher et al. [19] proposed the technologically and topologically related surfaces (TTRS) model to organize the geometric information on three-dimensional entities in the form of a binary tree and construct the minimum geometric datum element to add deviation information. The model cannot express composite tolerance information since it cannot express the relationship between features. To cope with this problem, Liu et al. [20] proposed the feature topological- and technological-related surface (FTTRS) model to identify the topological structure of part features. This model can effectively organize the geometric features and tolerance information, but it still does not support the expression of assembly deviation information.

2.2. Research on Deviation Propagation and Accumulation Calculation

The deviation transfer mechanism characterizes both the accumulation effect of deviations along transmission paths and the functional relationships between individual deviation sources and final assembly requirements. Early deviation transfer research mainly studied the transfer and accumulation of dimensional deviations. Chase et al. [21,22] proposed a second-order method for assembly tolerance analysis. From the perspective of small movement adjustment, He also proposed a general 2D tolerance analysis method. Tang et al. [23] proposed an assembly accuracy analysis method based on multi-stage linearized contact. A model of part surface asperities, considering morphological errors, was established using the linear superposition of discrete cosine transform (DCT) kernel functions and the assembly interface is simplified. By employing homogeneous coordinate transformation (HCT), the prediction of the part's pose during the assembly process with the rigid body assumption is achieved. Liu et al. [24] established a probability distribution function for estimating the target functional requirements based on the convolution method and probability theory, and solved the probability density function of the target functional requirements. Marler [25] constructed a homogeneous transformation matrix to describe deviation transfer based on the dimension chain model. The homogeneous transformation matrix consists of displacement and rotation, the size of the dimension vector in the dimension chain is represented as the displacement, and the direction is expressed as the rotation. In the deviation transfer modeling, the transfer relationship of the assembly dimension deviation is presented by connecting the end to the end of the dimension vector. Tsai et al. [26] constructed a deviation transfer dimension chain for the deviation modeling problem of a non-normal distribution, superimposed the variance, mean and other parameters of the deviation probability distribution of each component ring, and finally, obtained the deviation probability distribution parameters of the functional requirements.

From the above analysis, it can be seen that dimensional deviation transmission is mainly achieved by constructing a dimension chain model. However, the deviation of the constituent rings in the dimension chain model is represented by a one-dimensional vector, which makes it difficult to express the multi-directional deviation caused by geometric tolerances and matching structures. When the deviation target feature deviates along multiple degrees of freedom, it causes the functional target to deviate from the nominal position and direction. For the transmission of such deviations, Cao [27] described the deviation change based on the state space model and used a set of six-element vectors to describe the change in the target assembly feature. The state transfer matrix is constructed by the change vector corresponding to the target assembly feature and the change vector corresponding to the assembly function requirement feature to establish the state space model of deviation transfer. Hu et al. [28] proposed the concept of deviation flow based on the research on the deviation accumulation of multi-station thin plate assembly and constructed a state space model for the deviation transmission of multi-station thin plate assembly. The state space model uses space vectors to express the change in the assembly position of parts and obtains the accumulated result of deviation through space vector operations. However, since each space vector can only represent a position of the measured element in space, and the geometric changes under tolerance and fit constraints have uncertainty in the spatial direction, there are many space vectors and the calculation is complicated. Polini et al. [29] conducted geometric tolerance analysis by constructing a Jacobian model of rigid components based on translational deviation. Guo et al. [30], considering the non-independent relationship among multi-dimensional error sources, analyzed their internal stress function with practical error items and the extreme out of tolerance caused by the deviation's coupling and focusing function (EOCDCF) phenomenon. Then, the probability method was adopted for evaluating and improving the accuracy reliability. Yang et al. [31] presented a digital-twin-driven assembly accuracy prediction method for the high performance precision assembly (HPPA) of complex products. The state space model, Jacobi–vector pair model, direct linear method (DLM) method, and digital twin method all support the expression of geometric and positional tolerances, but they do not consider the influence of deviation sources such as the assembly sequence and fit changes on the assembly accuracy, and the current tolerance standards do not support the direct definition of fit change deviations. In addition, the true sensitivity of the deviation source to the target assembly accuracy is difficult to reflect, which is not conducive to the subsequent optimization of assembly accuracy.

The above research on the mechanism of assembly deviation transmission is mainly aimed at fixed assemblies. With the improvement of product assembly accuracy requirements, scholars have begun to study the deviation transmission in weak rigidity and even flexible assemblies. In the field of deviation transmission and accumulation in weak rigidity assemblies, the finite element analysis (FEA) method has been widely used [32,33]. Ref. [34] combined statistical variation analysis (SVA) with FEA and proposed the SVA-FEA method to achieve assembly accuracy prediction. Shi et al. [35] proposed an assembly accuracy analysis method that integrates the effects of surface morphology and non-uniform contact deformation on the error propagation process and it further adopts a phase optimization method to improve the assembly accuracy of multi-stage rotors.

Shi et al. [35] proposed use of the nonlinear FEA method to deal with deformation problems in assembly accuracy prediction and studied the influence of assembly deformation on the assembly success rate. Dupac et al. [36] studied the influence of the deformation deviation of flexible assemblies under external forces on the assembly accuracy. Yan et al. [37] proposed an assembly error analysis and optimization approach based on digital twin technology integrated with a genetic algorithm. He also developed an error transfer and accumulation model (ETAM) using small displacement torsor (SDT) theory, which accounts for manufacturing errors and positional deviations. This model was further enhanced into a DT-based ETAM (DT-ETAM) by incorporating structural deformation data from FEA simulations along with actual measurement data. Imani et al. [38] established a flexible mechanism deviation transmission model based on DLM and expressed the flexible deformation deviation. The above methods take into account the influence of assembly deformation on the position of parts and improve the accuracy of assembly accuracy analysis to a certain extent. However, assembly deformation not only causes a change in the matching position of parts but also changes the geometric shape of parts, which in turn affects the geometric shape deviation.

To address this problem, Schleich et al. [39] proposed establishing a discrete frame model based on a finite element deformation grid to describe the influence of assembly deformation on the geometric shape of parts, and also proposed the use of discrete geometric models to reflect the influence of deformation on the geometric surface of parts. Franciosa et al. [40] used the deformation grid method to analyze the transmission and accumulation mechanism of mechanism assembly deviation. This method not only considers deformation deviation but also geometric shape deviation. However, these deviation propagation models rarely take into account the characteristics of WRC assembly, and their analysis of deformation problems in the assembly of a large number of flexible wall panels is insufficient. At the same time, when these models are combined with the Monte Carlo method, the amount of calculation will be too large, resulting in an algorithm crash.

A summary of assembly accuracy analysis for weakly rigid components is as follows:

(1) Rigid–flexible coupling deviation modeling for WRCs requires further research.

Existing assembly accuracy models primarily focus on tolerance modeling, structural modeling, and relational modeling. However, the assembly accuracy of WRCs is governed by the interaction of multiple factors, including assembly deformations, positioning methods, tooling constraints, and assembly sequences. Thus, establishing a rigid–flexible coupling deviation model for WRCs is essential to enable accurate assembly accuracy analysis.

(2) Deviation propagation and accumulation in WRCs demand deeper investigation.

Most current tolerance analysis and deviation propagation methods neglect assembly-induced variations, failing to account for the unique assembly behavior of WRCs. Therefore, to improve the accuracy of the deviation transfer and accumulation calculation of WRCs., it is necessary to comprehensively consider the coupling effects of part manufacturing deviation, fixture positioning deviation and assembly deformation deviation, and to study the deviation transfer and accumulation methods of WRCs.

This study presents a comprehensive framework for accurate assembly accuracy analysis of WRCs through three key contributions. First, we establish a systematic WRC assembly accuracy analysis process incorporating deformation effects. Second, we develop a novel vector representation method that explicitly integrates assembly-induced deformations into deviation modeling. Third, we construct a multi-source deviation vector model for WRCs and implement a multi-dimensional vector-loop-based approach to characterize deviation propagation and accumulation. The proposed methodology enables precise computation of deviation transfer paths, ultimately achieving high-fidelity assembly accuracy analysis for WRC systems.

3. General Outline of WRC Assembly Accuracy Analysis

The overall technical route for constructing a deviation vector representation model for WRCs based on the equivalence of multiple deviation sources and studying the assembly accuracy analysis of WRCs based on a multi-dimensional vector ring of deviation transfer is designed and illustrated in Figure 1.

Firstly, the assembly deviation sources of WRCs are analyzed, and the assembly deviation vector representation model is established based on the equivalence of multiple deviation sources. According to the characteristics of the assembly deviation direction and the diversity of transmission paths of WRCs, the assembly deviation sources are divided into dimensional deviation, geometric shape deviation, runout deviation, kinematic pair fit deviation, and assembly deformation deviation. For deviation sources that cannot be directly represented in the deviation model, such as assembly deformation deviation, a deviation equivalent model is established based on the principle of fit change to achieve expression in the deviation model, and then the WRC assembly multi-deviation source vector representation model is established.

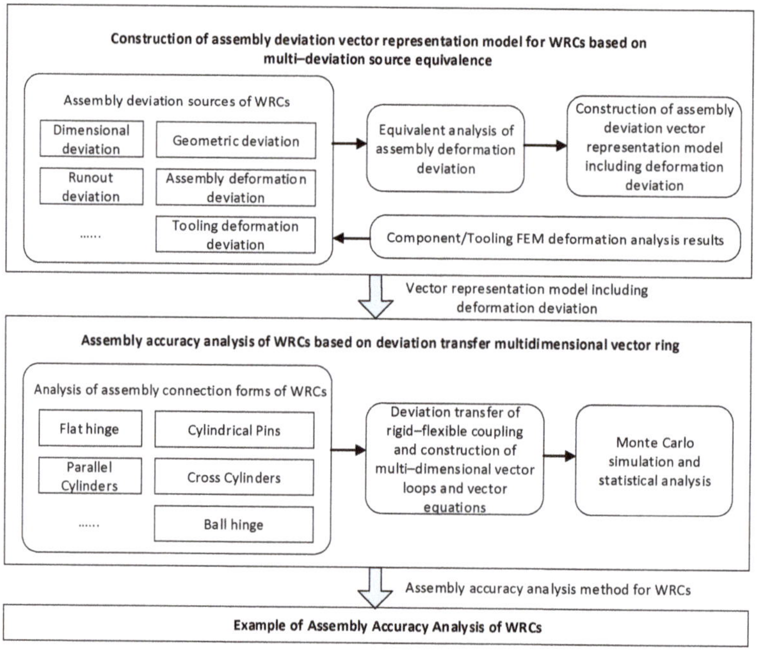

Figure 1. Overall technical route of the project research.

Secondly, the assembly deviation transmission mechanism of WRCs is studied, and the assembly accuracy analysis model is constructed based on the multi-dimensional vector ring of deviation transmission. The accurate analysis of WRC assembly accuracy is achieved through the Monte Carlo method. The assembly connection forms and degrees of freedom of WRCs are analyzed, and the concepts of the deviation transmission reference path and deviation transmission path are defined. The deviation transmission paths under different assembly connection forms are studied. In the assembly matching part, the deviation transmission model describing the deviation transmission and accumulation mechanism is established based on vector representation, so as to establish a rigid–flexible coupled deviation transmission multi-dimensional vector loop model and vector equation, laying a model foundation for the analysis of WRC assembly accuracy. After establishing the deviation transmission multi-dimensional vector loop and vector equation, the Monte Carlo method is proposed to simulate and statistically analyze the assembly accuracy of WRCs.

4. Equivalent Analysis of Assembly Deformation Deviation Sources Based on Fit Changes

Assembly deformation refers to the elastic deformation of parts caused by loads such as the positioning clamping force and self-weight during assembly. This deformation not only affects part dimensions but also changes assembly positions. The following discusses the impact of assembly deformation on the part size and assembly position, and then studies the equivalent method of assembly deformation deviation.

(1) Impact of assembly deformation on part size

The design dimensions of the part are expressed by the nominal value L and the tolerance T. The maximum and minimum allowable values of the length dimension of the part are $L + T$ and $L - T$, respectively, as shown in Figure 2a. Under the action of assembly constraints and load F, the part undergoes axial deformation, where the deformation amount is δ (positive for tension and negative for compression). At this time, the maximum length dimension of the part is $L + T + \delta$ and the minimum is $L - T + \delta$, as shown in Figure 2b.

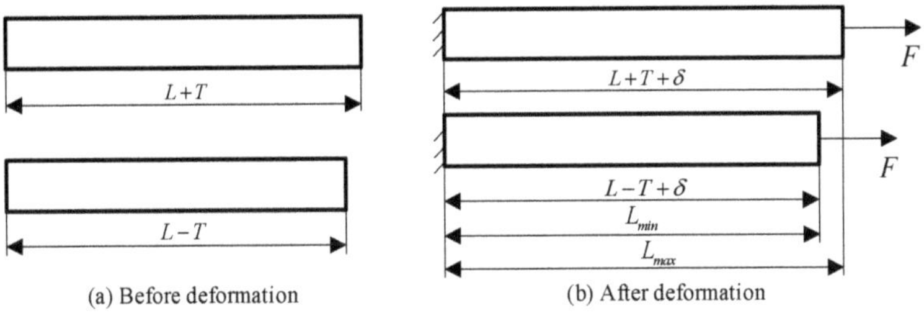

Figure 2. The influence of assembly deformation on the part size.

Under the influence of assembly deformation, the size and deviation of the part in the deformation direction change to $L' \pm T'$. Among them,

$$L' = \frac{L_{max} + L_{min}}{2} \quad (1)$$

$$T' = \frac{L_{max} - L_{min}}{2} \quad (2)$$

where $L_{\max}$ and $L_{\min}$ are the maximum and minimum size of the part in the deformation direction, respectively.

(2) The influence of assembly deformation on the assembly position of parts

Assembly deformation affects the assembly position of parts, and a hinged motion mechanism is shown in Figure 3. When a rod with a nominal size of L rotates around the hinge at an angle of θ, if the assembly deformation is not considered, the part assembly is located at the position shown by the solid line in the figure. When taking the gravity G of the part into consideration, the part assembly is located at the position shown by the dotted line in the figure. It can be seen from the finite element method (FEM) that the axial deformation of the part under gravity is δu, and the radial deformation is δv. Among them, δu has an impact on the axial size L of the part, which is determined by $L' \pm T'$. δv has an impact on the assembly position of the part, which is determined by the angle deviation of $\Delta\theta$. The angle deviation can be obtained by Equation (3) based on the geometric relationship:

$$\Delta\theta = \arctan\frac{\delta v}{L} \quad (3)$$

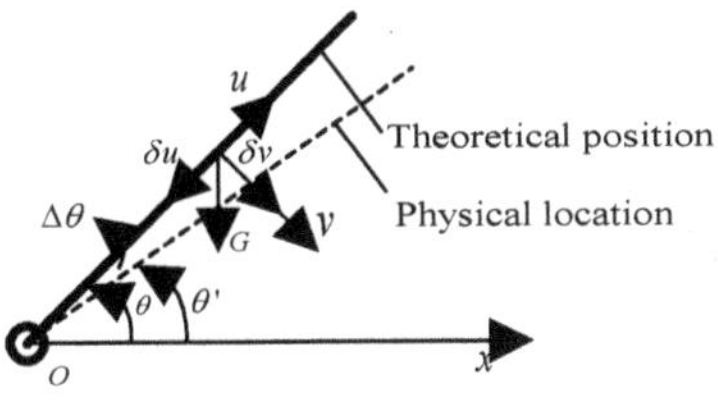

Figure 3. The influence of assembly deformation on the assembly position of parts.

Then, the angle to determine the actual assembly position of the part is:

$$\theta' = \theta - \Delta\theta \quad (4)$$

Among them, θ is the assembly position of the part without considering the assembly deformation; θ' is the actual assembly position of the part after considering the assembly deformation; and $\Delta\theta$ is the assembly position deviation of the part caused by the assembly deformation, that is, the equivalent deviation.

Using the above equivalent method for assembly deformation deviation, the deformation is converted into part-size-related deviations. This allows the deviation source to be represented by a size vector, enabling a unified expression of multiple deviation sources in the deviation model. Based on this equivalence, the WRC assembly deviation vector representation model is then constructed.

5. Construction of Rigid–Flexible Coupling Multi-Dimensional Vector Ring Model Based on Assembly Constraints

5.1. Rigid Component Assembly Deviation Transfer Model and Vector Ring Equation Construction

5.1.1. The Principles of Deviation Transfer Path Construction Between Parts

During the assembly process, the deviation is transmitted between parts along specific paths through the assembly relationship. Therefore, the assembly connection types and degrees of freedom of components significantly influence deviation propagation. In the assembly of mechanical products, positioning is achieved through the positioning datum. Thus, when constructing deviation propagation paths, the first step is to establish the positioning datums for assembly components. According to the design datum, the assembly positioning datums of the parts are defined as the datum reference frame (DRF). The path formed by connecting the nominal dimension vector from the mating point to the DRF is defined as the datum flow chain (DFC). Following these definitions, Figure 4 illustrates the established DRFs and DFCs for a sample assembly. Figure 4a shows the assembly design model consisting of two components: a thin-shell part and a block part. At location A, the block is connected to the bottom of the thin-shell part through a face-to-face fit, and the connection is marked as J_A, and the top of the thin-shell part is connected to the block part through a row of rivets, which is marked as J_B at location B. H_C is the fit clearance formed by the rivet connection, which is also an assembly functional requirement. The part DRFs are constructed in combination with the part design datum, and the datum paths DFC_{21} and DFC_{22} are established from the connection A between the two parts, and the datum paths DFC_{11} and DFC_{12} are established from the connection B, as shown in Figure 4b.

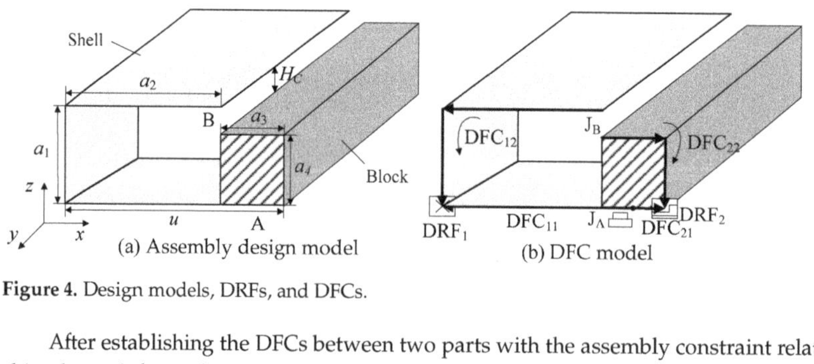

Figure 4. Design models, DRFs, and DFCs.

After establishing the DFCs between two parts with the assembly constraint relationship, the path formed by the deviation transfer along the DFC is defined as the deviation transfer path. Through the analysis of the datum reference structure, datum flow chain and deviation transfer path construction process, the following four principles for creating assembly deviation transfer paths are summarized:

(1) The deviation enters the interior of the part from the assembly connection. In the deviation transfer model shown in Figure 5, the deviation enters part i through the connection point J_1 (part $i-1$, part i) of the line–surface fit, and enters part $i+1$ through the connection point J_2 (part i, part $i+1$) of the arc–arc fit.
(2) Inside the part, the deviation reaches the DRF along the direction of the DFC. For example, inside part i as shown in Figure 5, the deviation reaches DRF_1 along DFC_{11}; inside part $i+1$, the deviation reaches DRF_2 along DFC_{21}.
(3) Inside the part, the deviation reaches another assembly connection along the direction of the second DFC. For example, inside part i as shown in Figure 5, the deviation reaches the assembly connection J_2 (part i, part $i+1$) along DFC_{12}; inside part i + 1, the deviation reaches the assembly connection J_3 (part $i+1$, part $i+2$) along DFC_{22}.
(4) The deviation exits the part from the assembly joint and enters the adjacent part.

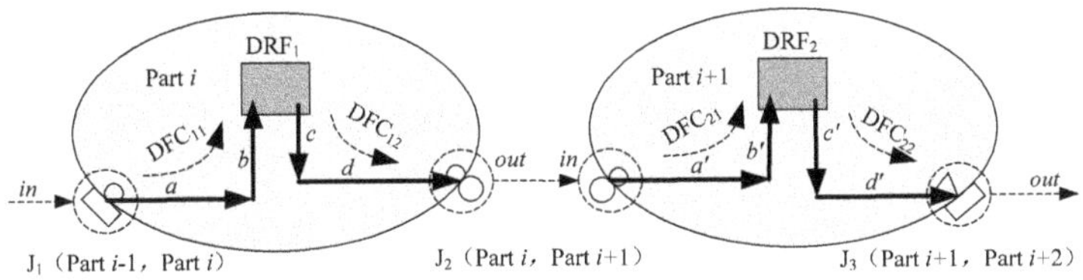

Figure 5. Deviation transformation model.

In the deviation transfer model shown in Figure 5, the two parts are connected by the arc–arc. In part i, path a and b constitute the reference path DFC_{11}, which means the path for the deviation to enter the interior of part i. Path c and d constitute the reference path DFC_{12}, which is the path for the deviation to exit the interior of part i. After the deviation exits part i through DFC_{12}, it is transferred along the reference path DFC_{21} of part $i+1$, and then exits part $i+1$ through DFC_{22}. Under the constraints of the assembly connection relationship, the deviation is transferred in sequence along the paths DFC_{11}, DFC_{12}, DFC_{21}, DFC_{22}, etc., to form a deviation transfer path. Among them, a, b, c, d, a', b', c', d', etc., are called deviation transfer vectors. Since the controlled design dimensions and adjustable assembly dimensions are the occurrence parts of the deviation source in the assembly process, the establishment of the DFC must be along the direction of the design dimension or assembly dimension vector.

5.1.2. Deviation Transfer Vector Loop Construction in Assembly Process

According to the aforementioned principles, the deviation transfer path of multiple assembly parts in the assembly process is established according to the assembly connection form given by the design and the positioning method given by the process to form a deviation transfer vector loop. Based on the assembly function requirements and the characteristics of the deviation transfer path, the deviation transfer multi-dimensional vector loop is divided into a closed loop and an open loop.

Then, based on the assembly connection relationship, the method for establishing the deviation transfer vector loop of the assembly process is studied. As shown in Figure 6, assuming that both parts are rigid parts, according to the multi-dimensional vector loop modeling principles, the deviation is transferred along the path a, b, Hc, c, d, e', e'' to form a deviation transfer vector loop, Loop 1. Since the connection J_B is achieved by multiple rivets, a series of spatial multi-dimensional vector loops can be constructed for each rivet connection point based on Loop 1 and combined with the rivet connection point spacing (f_i, g_i).

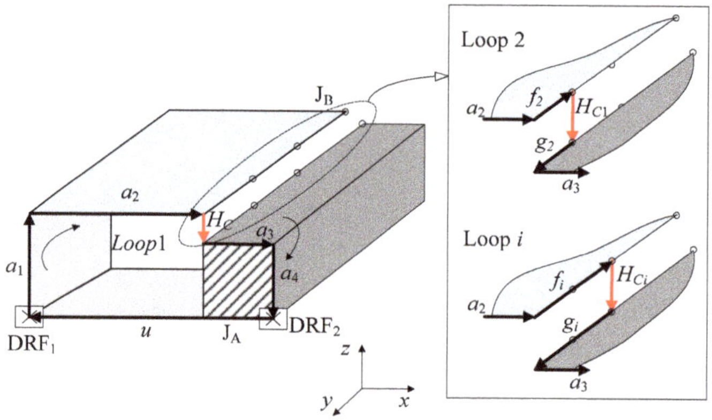

Figure 6. Deviation transfer vector loop construction.

5.1.3. Deviation Transfer Vector Equation Construction

After establishing the deviation transfer multi-dimensional vector loop model, in order to realize the mathematical expression of the deviation transfer and accumulation, the deviation transfer vector equation is further established. Let $\{\boldsymbol{a}_1,\boldsymbol{a}_2,\cdots,\boldsymbol{a}_\mathrm{n}\}$ represent the part size vector in the deviation transfer multi-dimensional vector loop, $\{\boldsymbol{b}_1,\boldsymbol{b}_2,\cdots,\boldsymbol{b}_k\}$ represent the geometric shape deviation vector in the multi-dimensional vector loop, $\{\boldsymbol{r}_1,\boldsymbol{r}_2,\cdots,\boldsymbol{r}_l\}$ represent the virtual connector size vector of the fit clearance in the multi-dimensional vector loop, and $\{\boldsymbol{u}_1,\boldsymbol{u}_2,\cdots,\boldsymbol{u}_m\}$ represent the assembly size vector in the multi-dimensional vector loop. Then, the assembly deviation transfer multi-dimensional vector equation can be expressed by a series of homogeneous matrix equations:

$$[H] = \begin{bmatrix} \boldsymbol{a}_1 & \boldsymbol{a}_2 & \cdots & \boldsymbol{a}_n \end{bmatrix}^T \cdot \begin{bmatrix} \boldsymbol{b}_1 & \boldsymbol{b}_2 & \cdots & \boldsymbol{b}_k \end{bmatrix}^T \cdot \begin{bmatrix} \boldsymbol{u}_1 & \boldsymbol{u}_2 & \cdots & \boldsymbol{u}_m \end{bmatrix}^T \quad (5)$$

where $[H]$ is the vector matrix of key feature function requirements.

5.2. Assembly Deviation Transfer Model of WRCs with Rigid–Flexible Coupling and Construction of Vector Loop Equation

The above-established assembly deviation transfer vector loop only contains the dimension vector, and it is assumed that all the parts contained in the assembly are rigid. However, in actual assembly processes, other deviation sources besides dimension deviation, such as part geometry deviation, assembly deformation and fit clearance, will affect the deviation transfer path and deviation accumulation results, consequently affecting the product's ultimate assembly accuracy. In order to establish an assembly deviation transfer vector loop model containing multiple deviation sources, the following focuses on integrating the assembly deformation deviation vector, geometry deviation vector and fit clearance virtual connector deviation vector into the existing vector loop.

As shown in Figure 7a, the thin-shell part deforms under the action of the vertical assembly force F_i. The deformation θ can be obtained through the finite element calculation, and the thin-shell part in the assembly process model is replaced by the deformed finite element model. As can be seen from Figure 7, the deformation of the thin-shell part affects the dimension a_2, forming a new dimension vector $a_2{}'$. Since the assembly deformation of the thin-shell part is within the elastic deformation range of the material and the deformation angle θ is small, $\cos\theta \approx \theta$. At this time, the new dimension vector $a_2{}'$ can be calculated by Equation (6):

$$a_2{}' = \frac{a_2}{\theta} \quad (6)$$

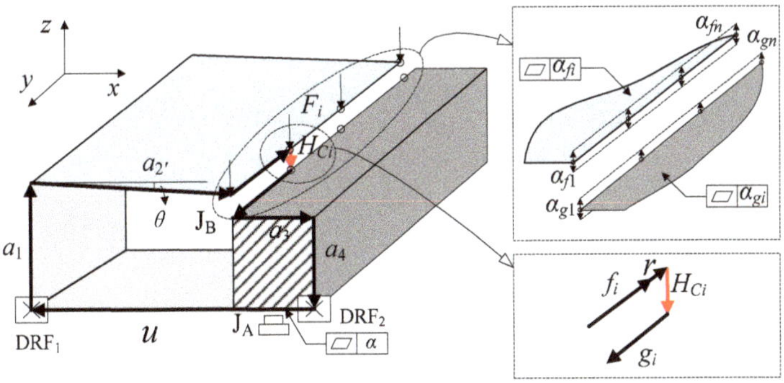

(a) Construction of deviation transmission vector loop

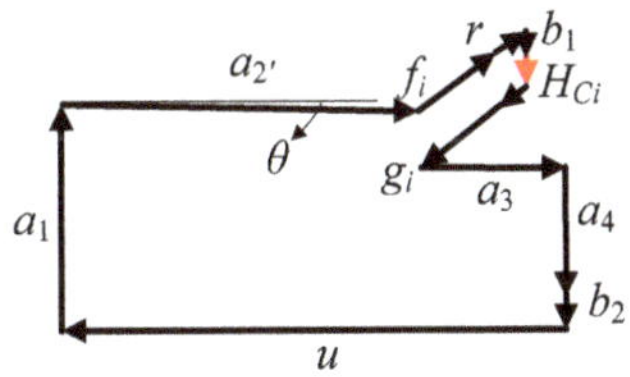

(b) Deviation transfer vector loop model

Figure 7. Construction of the deviation transfer vector loop model of rigid–flexible coupling.

Taking the construction of the deviation transmission multi-dimensional vector loop Loop i of the rivet position (f_i, g_i) as an example, the deviation transmission multi-dimensional vector loop model under the action of assembly deformation is shown on the left side of Figure 7a.

At the connection point J_B, the flatness of the surface to be connected is α_{fi}, and the flatness of the target connection surface is α_{gi}, as shown on the right side of Figure 7a. Assuming $\alpha_{fi} < \alpha_{gi}$, only the influence of α_{fi} on the deviation transfer needs to be considered, and the flatness deviation vector is half of its tolerance zone, denoted as b_1. Similarly, at the connection point J_A, the flatness deviation vector is denoted as b_2. In this example, the assembly of the thin-shell part and the block part is a clearance fit, so the fit clearance deviation is expressed by the size vector of the virtual connector, and the size vector of the fit clearance virtual connector is denoted as r. Finally, the deviation vectors b_1, b_2, and r are added to the deviation transfer multi-dimensional vector loop to establish a deviation transfer vector loop model containing multiple assembly deviation sources, as shown in Figure 7b.

After establishing the deviation transfer multi-dimensional vector loop model, in order to realize the mathematical expression of the deviation transfer and accumulation, the deviation transfer vector equation is further established. Let $\{\boldsymbol{a}_1, \boldsymbol{a}_2, \cdots, \boldsymbol{a}_n\}$ represent the part size vector in the deviation transfer multi-dimensional vector loop, $\{\boldsymbol{b}_1, \boldsymbol{b}_2, \cdots, \boldsymbol{b}_k\}$ represent the geometric shape deviation vector in the multi-dimensional vector loop, $\{\boldsymbol{r}_1, \boldsymbol{r}_2, \cdots, \boldsymbol{r}_l\}$ represent the virtual connector size vector of the fit clearance in the multi-dimensional vector loop, and $\{\boldsymbol{u}_1, \boldsymbol{u}_2, \cdots, \boldsymbol{u}_m\}$ represent the assembly size vector in the multi-dimensional vector loop. Then, the assembly deviation transfer multi-dimensional vector equation can be expressed by a series of homogeneous matrix equations:

$$\begin{aligned} [H] &= \begin{bmatrix} \boldsymbol{a}_1 & \boldsymbol{a}_2 & \cdots & \boldsymbol{a}_n \end{bmatrix}^T \cdot \begin{bmatrix} \boldsymbol{b}_1 & \boldsymbol{b}_2 & \cdots & \boldsymbol{b}_k \end{bmatrix}^T \\ &\cdot \begin{bmatrix} \boldsymbol{r}_1 & \boldsymbol{r}_2 & \cdots & \boldsymbol{r}_l \end{bmatrix}^T \cdot \begin{bmatrix} \boldsymbol{u}_1 & \boldsymbol{u}_2 & \cdots & \boldsymbol{u}_m \end{bmatrix}^T \end{aligned} \tag{7}$$

where $[H]$ is the vector matrix of key feature function requirements.

According to whether the multi-dimensional vector loop contains key characteristic requirements, the corresponding deviation transfer vector open loop and closed loop can be further established. The deviation transfer vector equation described in the above formula is decomposed in the closed-loop vector direction $\boldsymbol{\Lambda}$ and the open-loop vector direction $\boldsymbol{\Gamma}$, as shown in the following formula:

$$\begin{cases} \boldsymbol{h_c} = \Lambda(\boldsymbol{a}_1, \boldsymbol{a}_2, \cdots, \boldsymbol{a}_n, \boldsymbol{b}_1, \boldsymbol{b}_2, \cdots, \boldsymbol{b}_k, \boldsymbol{r}_1, \boldsymbol{r}_2, \cdots, \boldsymbol{r}_l, \boldsymbol{u}_1, \boldsymbol{u}_2, \cdots, \boldsymbol{u}_m) = 0 \\ \boldsymbol{h_o} = \Gamma(\boldsymbol{a}_1, \boldsymbol{a}_2, \cdots, \boldsymbol{a}_n, \boldsymbol{b}_1, \boldsymbol{b}_2, \cdots, \boldsymbol{b}_k, \boldsymbol{r}_1, \boldsymbol{r}_2, \cdots, \boldsymbol{r}_l, \boldsymbol{u}_1, \boldsymbol{u}_2, \cdots, \boldsymbol{u}_m) \end{cases} \tag{8}$$

where $\boldsymbol{h_c}$ represents the deviation vector of the key control characteristic requirement in the closed-loop S direction; $\boldsymbol{h_o}$ represents the deviation vector of the key control characteristic requirement in the open-loop $\boldsymbol{\Gamma}$ direction; $\boldsymbol{\Lambda}$ is the direction of the deviation transfer closed-loop vector; and Γ is the direction of the deviation transfer open-loop vector. So far, a multi-dimensional vector ring model of deviation transmission of rigid–flexible coupling assembly of WRCs has been established.

The established deviation transfer vector equation describes the nonlinear implicit functional relationship between the assembly key characteristic functional requirements and each deviation source. In order to obtain a linear and explicit deviation transfer function, the deviation transfer vector equation described by Equation (8) is linearized using Taylor expansion. The first-order Taylor expansion of the deviation transfer vector open-loop and closed-loop equations is:

$$\begin{aligned} \delta h &= \tfrac{\partial h}{\partial a_1}\delta a_1 + \cdots + \tfrac{\partial h}{\partial a_n}\delta a_n + \tfrac{\partial h}{\partial b_1}\delta b_1 + \cdots + \tfrac{\partial h}{\partial b_l}\delta b_k \\ &+ \tfrac{\partial h}{\partial r_1}\delta r_1 + \cdots + \tfrac{\partial h}{\partial r_l}\delta r_l + \tfrac{\partial h}{\partial u_1}\delta u_1 + \cdots + \tfrac{\partial h}{\partial u_m}\delta u_m \end{aligned} \tag{9}$$

where ∂h represents a small change in the functional requirement dimension, δa_n represents a dimension deviation, δb_k represents a small change in the geometric shape deviation, δr_l represents assembly deformation deviation, and δu_m represents a small change in the assembly dimension. At this time, the assembly deviation transfer scalar equation can be expressed as:

$$[\delta H] = [A][\delta a] + [B][\delta b] + [R][\delta r] + [U][\delta u] \tag{10}$$

$[A] = \left[\frac{\partial h}{\partial a_i}\right]$ represents the partial derivative matrix of the deviation transfer equation for the part size, where $[\delta H_o]$ represents the dimension deviation matrix of the assembly function requirements and $[\delta a]$ means the dimension deviation matrix. $[B] = \left[\frac{\partial h}{\partial b_i}\right]$ represents the partial derivative matrix of the deviation transfer equation for the part geometry deviation and $[\delta b]$ means the geometry deviation matrix. $[R] = \left[\frac{\partial h}{\partial r_i}\right]$ means the partial derivative matrix of the deviation transfer equation for assembly deformation, and $[\delta r]$ indicates the assembly deformation deviation matrix. $[U] = \left[\frac{\partial h}{\partial u_i}\right]$ means the partial derivative matrix of the deviation transfer equation for the assembly size, and $[\delta u]$ represents the assembly dimension deviation matrix.

The above formula is the linear equation for assembly deviation transmission. Based on this model, the Monte Carlo algorithm can be further used to analyze the assembly accuracy of WRCs.

6. Assembly Accuracy Analysis Algorithm Process Design Based on Monte Carlo Algorithm

The input data for the assembly accuracy analysis includes the product design tolerance, tooling design tolerance, assembly tolerance, flexible substructure FEA data, assembly process and other information. Among this information, the design tolerance, assembly tolerance and so on are tolerance zones, and the distribution states of individual random variables exhibit diversity. How to randomly extract a set of deviation values that meet the tolerance domain requirements from the tolerance band is the premise of assembly accuracy analysis, and the Monte Carlo method can effectively solve this problem.

The assembly accuracy simulation algorithm flow based on the Monte Carlo method is established and shown in Figure 8, which mainly includes sample quantity definition, tolerance information extraction and definition, random deviation generation, random sampling, assembly accuracy analysis value solution, simulation statistical parameter calculation, and simulation result output. The specific steps are as follows.

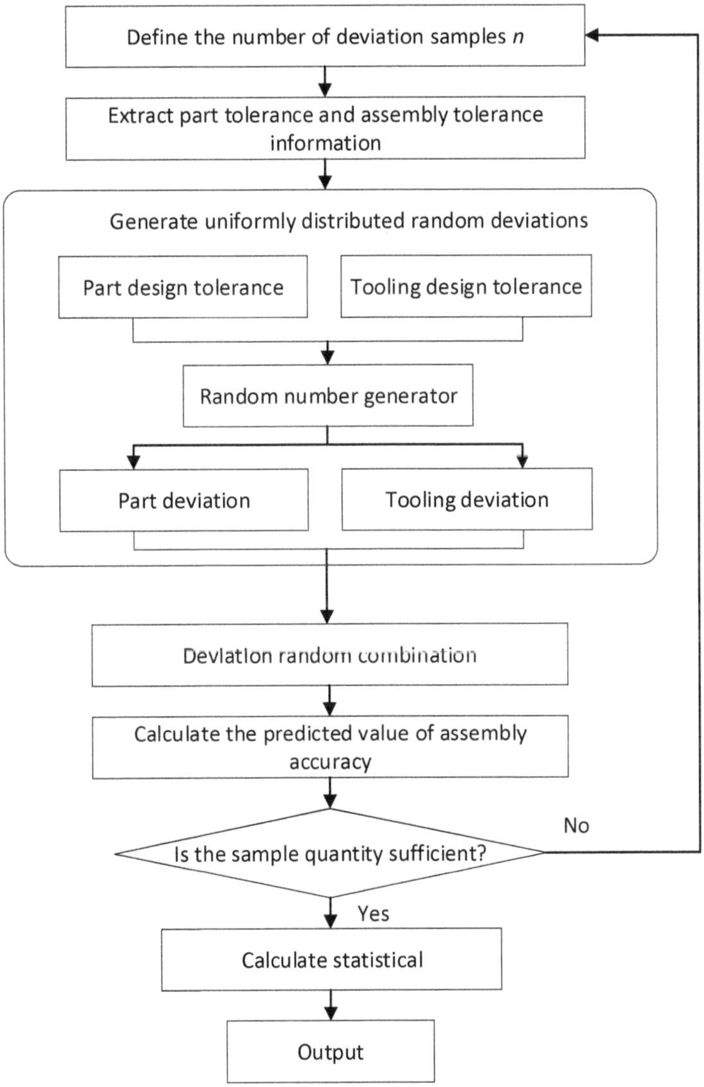

Figure 8. Assembly accuracy analysis algorithm flow based on the Monte Carlo algorithm.

- Step 1: Define the deviation sample size n. A larger sample size increases the accuracy of assembly accuracy simulations. However, excessively large samples proportionally increase the computational time and cost. Since product assembly qualification rates do not require 100% perfection, a statistically representative Gaussian distribution can be approximated within limited computational time. Thus, this study sets the simulation frequency to 5000.
- Step 2: Extract and define tolerance information. Extract the part design tolerance in the design model and the assembly tolerance in the process model. Define the equivalent deviation information, tooling positioning tolerance and other information in the assembly accuracy information model as input information for the assembly accuracy simulation.
- Step 3: Generate random deviation. By using a random number generator, the deviation values in the tolerance domain are randomly extracted from the various tolerance information input in step 2 to generate random deviations.
- Step 4: Calculate the predicted value of assembly accuracy. Solve the assembly accuracy analysis function and calculate the assembly accuracy analysis value. If the sample quantity requirement is met, go to step 5; otherwise, go to step 1 to redefine the sample quantity n.
- Step 5: Statistical analysis of assembly accuracy prediction values. Perform statistical analysis on the calculated values of multiple samples, compare them with the predetermined assembly accuracy requirements (design requirements), and provide the predicted evaluation value and the average value and qualification rate of the calculation analysis function.

7. Case Study and Discussion

To validate the effectiveness of the proposed method, this section presents the assembly of a thin-walled frame (one of the WRCs) as an example and constructs the deviation transfer vector loop and vector equation of rigid–flexible coupling. The Monte Carlo method is applied to analyze the assembly accuracy, and the validity of the analysis results is explained.

The design model of the thin-walled frame assembly is shown in Figure 9a, including part A, part B, and part C. Part C is regarded as the basis, and parts A and B are assembled at the same time during assembly. In ideal assembly, there is no gap between the fitting surfaces of parts A and B. However, due to the deformation of thin-walled parts, manufacturing and process deviations, there is an assembly gap between the fitting surfaces of parts A and B in actual assembly, as shown by G_y in the Figure 9. The allowable tolerance of G_y is ± 0.5 mm, which is the functional requirement for this assembly. Since G_y cannot be guaranteed through single point measurement, it is decomposed into multiple measurement points G_{yi} on the mating surface. By predicting the assembly accuracy of each measurement point G_{yi} on the mating surface, the assembly functional requirements of G_y can be ensured.

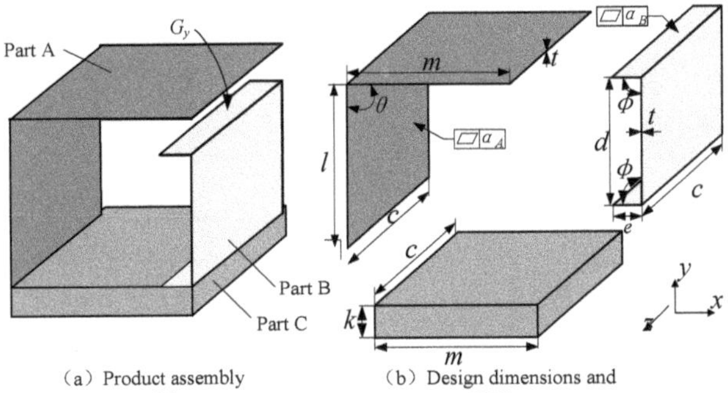

(a) Product assembly model (b) Design dimensions and DRFs

Figure 9. Thin-walled frame assembly.

The design dimension parameters of the assembly are shown in Figure 9b, and the design dimensions and deviations are shown in Table 1.

Table 1. Design dimensions and deviations of thin-walled frame assembly.

Dimension Type	Design Dimensions											
Dimension parameters	l	c	t	θ	α_A	α	d	e	φ	α_B	m	k
Nominal value/mm	250	350	2	90	0	3	225	38	90	0	250	25
Deviation value/±mm	1	1	0.2	5 *	0.15 *	1 *	1	0.2	5	0.25 *	1	0.2
Dimension Type	**Assembling Dimensions**											
Dimension parameters	u_1	u_2	u_3	u_4	f_i	g_i	$^{**}f_1$	$^{**}f_2$	$^{**}f_3$	$^{**}g_1$	$^{**}g_2$	$^{**}g_3$
Nominal value/mm	225	25	25	225	**	**	25	100	250	25	100	250
Deviation value/±mm	1	0.2	0.2	1	**	**	0.2	0.8	1	0.2	0.8	1

Notes, * represents the equivalent deviation of part assembly deformation or geometric tolerance, and this example only considers the deformation of part A. Due to the deformation of part A under the action of gravity, which affects the angle θ, another angle deviation α is introduced; ** represents the position, size, and deviation of the measurement point corresponding to the functional requirements.

Firstly, based on the part design benchmark and assembly positioning benchmark, the DRFs (DRF_A, DRF_B, and DRF_C) of the parts and assembly connection points Joints J (P_B, P_A), J (P_C, P_A), and J (P_C, P_B) are defined based on the assembly constraint relationships. The defined assembly DRFs and Joints are shown in Figure 10a. On this basis, the assembly connection form is analyzed, and according to the method and principles for constructing a deviation transfer multi-dimensional vector loop proposed in Section 5.2, a deviation transfer multi-dimensional vector loop for the thin-walled frame assembly is established, as shown in Figure 10b.

In this multi-dimensional vector loop model of assembly deviation transmission for the thin-walled frame, G_{yi} is the assembly clearance deviation vector of the i-th measurement point on the mating surface. φ_1, d, k are angle or size vectors, while θ and α are the angle deviation introduced by the deformation of the part. b_i is the geometric deviation vector, and u_1, u_2, u_3, u_4 are assembly dimension vectors. f_i and g_i are the measurement point position size vectors required for functional requirements.

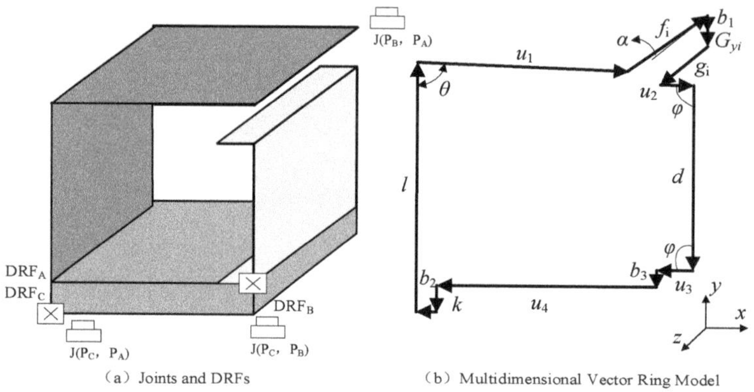

Figure 10. Multi-dimensional vector loop model for deviation transmission in thin-walled frame assembly.

Secondly, based on the multi-dimensional vector loop of assembly deviation transmission, the vector equation is established according to Formula (5). Since there is only vector open loop and no vector closed loop in this example, the deviation transfer vector equation is established in the direction of vector open loop, as shown in Equation (11).

$$\begin{aligned} G_{yi} &= u_2\sin(-180+\varphi+\varphi)+d\sin(0-\varphi)+u_3\sin(180) \\ &+b_3\sin(-90)+u_4\sin(180)+k\sin(-90)+b_2\sin(180) \\ &+a\sin(90)+u_1\sin(-90+\theta)+f_i\sin(\alpha)+g_i\sin(0)-b_1\sin(90) \end{aligned} \tag{11}$$

Then, the above equation is linearized according to Equation (9) and the parameters $\delta a, \delta_b, A'$ are solved, as shown in Equation (12):

$$\begin{aligned} [\delta a] &= \begin{bmatrix} \delta l & \delta d & \delta k & \delta\theta & \delta\varphi & \delta\alpha & \delta f_1 & \delta f_2 & \delta f_3 & \delta g_1 & \delta g_2 & \delta g_3 \end{bmatrix}^T \\ &= \begin{bmatrix} 1 & 1 & 0.2 & 5 & 5 & 2 & 0.2 & 0.8 & 1 & 0.2 & 0.8 & 1 \end{bmatrix}^T \end{aligned} \tag{12}$$

$$[\delta b] = \begin{Bmatrix} \delta b_1 \\ \delta b_2 \\ \delta b_3 \end{Bmatrix} = \begin{Bmatrix} 0.15 \\ 0.15 \\ 0.15 \end{Bmatrix} \tag{13}$$

$$\begin{aligned} [A'] &= \begin{bmatrix} \frac{\partial G_{y1}}{\partial l} & \frac{\partial G_{y1}}{\partial d} & \frac{\partial G_{y1}}{\partial k} & \frac{\partial G_{y1}}{\partial \theta} & \frac{\partial G_{y1}}{\partial \varphi} & \frac{\partial G_{y1}}{\partial \alpha} & \frac{\partial G_{y1}}{\partial f_1} & \frac{\partial G_{y1}}{\partial f_2} & \frac{\partial G_{y1}}{\partial f_3} & \frac{\partial G_{y1}}{\partial g_1} & \frac{\partial G_{y1}}{\partial g_2} & \frac{\partial G_{y1}}{\partial g_3} \\ \frac{\partial G_{y2}}{\partial l} & \frac{\partial G_{y2}}{\partial d} & \frac{\partial G_{y2}}{\partial k} & \frac{\partial G_{y2}}{\partial \theta} & \frac{\partial G_{y2}}{\partial \varphi} & \frac{\partial G_{y2}}{\partial \alpha} & \frac{\partial G_{y2}}{\partial f_1} & \frac{\partial G_{y1}}{\partial f_2} & \frac{\partial G_{y1}}{\partial f_3} & \frac{\partial G_{y2}}{\partial g_1} & \frac{\partial G_{y2}}{\partial g_2} & \frac{\partial G_{y2}}{\partial g_3} \\ \frac{\partial G_{y3}}{\partial l} & \frac{\partial G_{y3}}{\partial d} & \frac{\partial G_{y3}}{\partial k} & \frac{\partial G_{y3}}{\partial \theta} & \frac{\partial G_{y3}}{\partial \varphi} & \frac{\partial G_{y3}}{\partial \alpha} & \frac{\partial G_{y3}}{\partial f_1} & \frac{\partial G_{y1}}{\partial f_2} & \frac{\partial G_{y1}}{\partial f_3} & \frac{\partial G_{y3}}{\partial g_1} & \frac{\partial G_{y3}}{\partial g_2} & \frac{\partial G_{y3}}{\partial g_3} \end{bmatrix} \\ &= \begin{bmatrix} 1 & -1 & -1 & 225 & 0 & 25 & 0.05 & 0 & 0 & 0 & 0 & 0 \\ 1 & -1 & -1 & 225 & 0 & 100 & 0.05 & 0 & 0 & 0 & 0 & 0 \\ 1 & -1 & -1 & 225 & 0 & 250 & 0.05 & 0 & 0 & 0 & 0 & 0 \end{bmatrix} \end{aligned} \tag{14}$$

$$[B'] = \begin{bmatrix} \frac{\partial G_{y1}}{\partial b_1} & \frac{\partial G_{y1}}{\partial b_2} & \frac{\partial G_{y1}}{\partial b_3} \\ \frac{\partial G_{y2}}{\partial b_1} & \frac{\partial G_{y2}}{\partial b_2} & \frac{\partial G_{y2}}{\partial b_3} \\ \frac{\partial G_{y3}}{\partial b_1} & \frac{\partial G_{y3}}{\partial b_2} & \frac{\partial G_{y3}}{\partial b_3} \end{bmatrix} = \begin{bmatrix} -1 & 0 & -1 \\ -1 & 0 & -1 \\ -1 & 0 & -1 \end{bmatrix} \tag{15}$$

Since this example only has open-loop vector equations, there are no parameters A, B, R, and U. According to Formula (10), the accuracy analysis function for assembly clearance δG_{yi} is defined as:

$$\delta G_{yi} = \sqrt{[A']^2 \delta a_i^2 + [B']^2 \delta b_i^2} \tag{16}$$

Finally, assembly accuracy simulation is conducted based on the Monte Carlo method, with ±0.5 mm specified as the upper and lower limits allowed for assembly clearance. The simulation times are set to n = 5000, and the obtained simulation results are shown in Figure 11a.

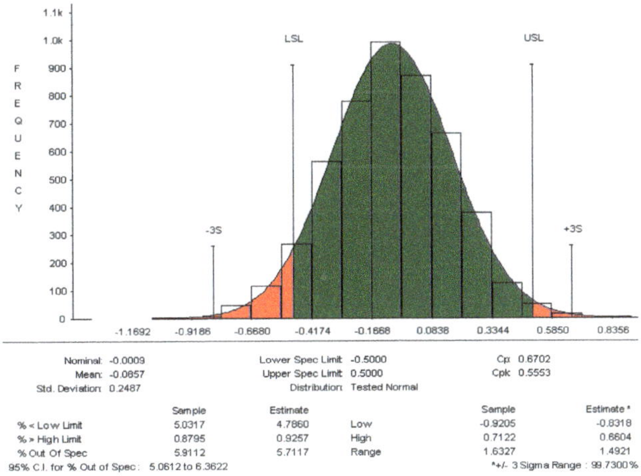

(a) Simulation result of WRC

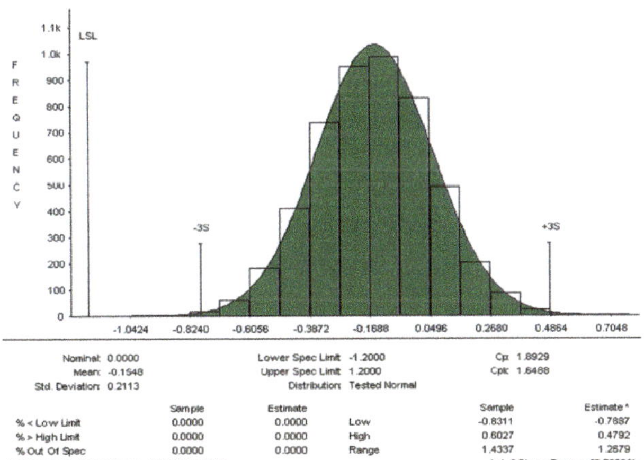

(b) Simulation result of rigid component

Figure 11. Accuracy simulation results of assembly gap G_{yi} in a thin-walled frame.

According to the simulation results, the probability of the G_{yi} deviation exceeding ± 0.5 mm is 5.91%, and the predicted average assembly accuracy of the gap is 0.248 mm. To demonstrate the effectiveness of the assembly deviation transfer model and accuracy analysis for WRCs, the accuracy of the assembly gap G_{yi} of the thin-walled frame is simulated and predicted using the rigid component deviation transfer model and accuracy analysis method. The simulation results show no deviation, and the predicted average assembly accuracy of the gap is 0.211 mm. A comparison of the results obtained from the two accuracy analysis methods is shown in Table 2.

Table 2. Comparison of different assembly accuracy analysis methods with measured values.

Control Object		Design Tolerance/±mm	Simulation Results	
			Accuracy Analysis Method for Rigid Components	Accuracy Analysis Method for WRC
G_{yi}	Assembly accuracy/mm	0.5	0.211	0.248
	Qualification rate/%		100	94.09

When the rigid component accuracy analysis method is applied, the influence of assembly deformation deviation is not considered, and the obtained assembly qualification rate is higher. From the simulation results, the qualification rate is 100%, which meets the design requirements. However, the assembly accuracy and assembly qualification rate obtained by the WRC assembly accuracy analysis method proposed in this paper are relatively low (94.09%). The probability that the assembly gap between the mating surfaces of parts A and B is within the design range is 94.09%, which cannot meet the assembly accuracy requirements under the 3σ quality criterion, It is necessary to further optimize the assembly deviations and the assembly process.

This indicates that accounting for weak rigidity issues enables comprehensive coverage of the assembly deviation sources, thereby ensuring safer assembly processes and more reliable assembly accuracy analysis results.

8. Conclusions

To address the impact of assembly deformation on the accuracy of assembly accuracy analysis for WRCs, this study proposes a novel assembly accuracy analysis method. Through equivalent analysis of the deformation deviation, a unified deviation vector representation model that conforms to the actual WRC assembly is established. Based on the multi-dimensional vector loop, the assembly deviation transmission mechanism is investigated, leading to the development of a rigid–flexible coupled deviation propagation model that reveals the influence patterns of the mating configurations and datum references on the assembly accuracy in WRCs. Furthermore, a computational method for accumulated deviation propagation in rigid–flexible coupled WRCs is developed, with case studies validating the reliability of the proposed approach. This method provides a theoretical basis and implementation method for assembly accuracy simulation of WRCs, and it can improve the reliability of assembly accuracy analysis results for WRCs.

Author Contributions: Conceptualization, D.Z.; methodology, D.Z.; software, X.Z.; validation, X.Z. and Z.Y.; writing—original draft preparation, D.Z.; writing—review and editing, G.W. All authors have read and agreed to the published version of the manuscript.

Funding: This research was funded by the Natural Science Basic Research Project of Shaanxi Province, China, grant numbers 2025JC-YBMS-584 and 2025JC-YBMS-565.

Data Availability Statement: The original contributions presented in this study are included in the article. Further inquiries can be directed to the corresponding author.

Conflicts of Interest: The authors declare no conflict of interest.

References

1. Maltauro, M.; Passarotto, G.; Concheri, G.; Meneghello, R. Bridging the gap between design and manufacturing specifications for non-rigid parts using the influence coefficient method. *Int. J. Adv. Manuf. Technol.* **2023**, *127*, 579–597. [CrossRef]
2. Atik, H.; Chahbouni, M.; Amegouz, D.; Boutahari, S. Optimization tolerancing of surface in flexible parts and assembly: Influence Coefficient Method with shape defects. *Int. J. Eng. Technol.* **2018**, *7*, 90. [CrossRef]
3. Polini, W.; Corrado, A. Methods of influence coefficients to evaluate stress and deviation distribution of flexible assemblies—A review. *Int. J. Adv. Manuf. Technol.* **2020**, *107*, 2901–2915. [CrossRef]
4. Hillyard, R.C.; Braid, I.C. Characterizing non-ideal shapes in terms of dimensions and tolerances. *Acm Siggraph Comput. Graph.* **1978**, *12*, 234–238. [CrossRef]
5. Zhang, B.C. Geometric Modeling of Dimensioning and Tolerancing. Ph.D. Thesis, Arizona State University, Tempe, AZ, USA, 1992.
6. Shen, T.H.; Lu, C. An assembly accuracy analysis approach of mechanical assembly involving parallel and serial connections considering form defects and local surface deformations. *Precis. Eng.* **2024**, *88*, 44–64. [CrossRef]
7. Zhao, B.; Wang, Y.; Sun, Q.; Zhang, Y.; Liang, X.; Liu, X. Monomer model: An integrated characterization method of geometrical deviations for assembly accuracy analysis. *Assem. Autom.* **2021**, *41*, 514–523. [CrossRef]
8. Cui, L.; Sun, M.; Cao, Y.; Zhao, Q.J.; Zeng, W.-H.; Guo, S.-R. A novel tolerance geometric method based on machine learning. *J. Intell. Manuf.* **2021**, *32*, 799–821. [CrossRef]
9. Laifa, M.; Sai, W.B.; Hbaieb, M. Evaluation of machining process by integrating 3D manufacturing dispersions, functional constraints, and the concept of small displacement torsors. *Int. J. Adv. Manuf. Technol.* **2014**, *71*, 1327–1336. [CrossRef]
10. Jin, S.; Chen, H.; Li, Z.; Lai, X. A small displacement torsor model for 3D tolerance analysis of conical structures. *Proc. Inst. Mech. Eng. Part C J. Mech. Eng. Sci.* **2015**, *229*, 2514–2523. [CrossRef]
11. Wu, Y.G. Tolerance mathematical model based on the variation of control points of geometric elemen. *J. Mech. Eng.* **2013**, *5*, 138. [CrossRef]
12. Guo, F.; Bao, Q.; Liu, J.; Sha, X. Assembly quality control technologies in forced clam and compensation processes for large and integrated aeronautical composite structures. *Machines* **2025**, *13*, 159. [CrossRef]
13. Cao, Y.L.; Srinivasan, V. Special Issue: Geometric tolerancing. *J. Comput. Inf. Sci. Eng.* **2015**, *15*, 44–57. [CrossRef]
14. Anwer, N.; Schleich, B.; Mathieu, L.; Wartzack, S. From solid modelling to skin model shapes: Shifting paradigms in computer-aided tolerancing. *CIRP Ann.-Manuf. Technol.* **2014**, *63*, 137–140. [CrossRef]
15. ASME. Dimensioning and Tolerancing Principles. In *ANSI Standard Y14.5M*; ASME: New York, NY, USA, 1994.
16. Johnson, R.H. *Dimensioning and Tolerancing Final Report*; Report R-84GM-02.2; Computer Aided Manufacturing International: Arlington, TX, USA, 1985.
17. Roy, U.; Liu, C.R. Feature-based representational scheme of a solid modeler for providing dimensioning and tolerancing information. *Robot. Comput.-Integr. Manuf.* **1988**, *4*, 335–345. [CrossRef]
18. Mujezinovic, A.; Davidson, J.K.; Shah, J.J. A new mathematical model for geometric tolerances as applied to polygonal faces. *Int. J. Mech. Des.* **2004**, *126*, 504–518. [CrossRef]
19. Desrochers, A.; Desrochers, A. A CAD/CAM representation model applied to tolerance transfer methods. *Int. J. Mech. Des.* **2003**, *125*, 22–31. [CrossRef]
20. Liu, Y.; Gao, S.; Wu, Z.; Yang, J. Feature-based hierarchical tolerance information representation model and its implementation. *Int. J. Mech. Eng.* **2003**, *39*, 1–7.
21. Glancy, C.G.; Chase, K.W. *A Second-Order method for assembly tolerance analysis*; American Society of Mechanical Engineers: New York, NY, USA, 1999; pp. 977–984.
22. Chase, K.W.; Gao, J.; Magleby, S.P.; Spencer, P.M. General 2-D tolerance analysis of mechanical assemblies with small kinematic adjustments. *J. Des. Manuf.* **1995**, *5*, 263–274.
23. Tang, W.; Li, Y.; Yan, T.; Zhang, M. Assembly accuracy analysis method based on multi-stage linearized contact. *Eng. Rep.* **2025**, *7*, e13118. [CrossRef]
24. Liu, S.G.; Wang, P.; Li, Z.G. Non-normal statistical tolerance analysis using analytical convolution method. *Int. J. Adv. Manuf. Technol.* **2011**, *7*, 127–130. [CrossRef]
25. Marler, J.D. Nonlinear tolerance analysis using the direct linearization method. Ph.D. Thesis, Brigham Young University Department of Mechanical Engineering, Provo, UT, USA, 1988.
26. Tsai, J.C.; Kuo, C.H. A novel statistical tolerance analysis method for assembled parts. *Int. J. Prod. Res.* **2012**, *50*, 3498–3513. [CrossRef]

27. Cao, Y. Dynamic prediction and compensation of aerocraft assembly variation based on state space model. *Assem. Autom.* **2015**, *35*, 183–189. [CrossRef]
28. Hu, W.; Liu, J.H.; Jiang, K.; Guo, C.Y. Assembly precision prediction method for spacecraft based on 3D model. *Comput. Integr. Manuf. Syst.* **2013**, *19*, 990–999.
29. Polini, W.; Corrado, A. Geometric tolerance analysis through Jacobian model for rigid assemblies with translational deviations. *Assem. Autom.* **2016**, *36*, 72–79.
30. Guo, F.; Hou, Y.; Xiao, Q.; Zhang, X.; Xiao, S.; Wang, Z. Reliability improvement on assembly accuracy with maximum out-of-tolerance probability analysis and prior precise repair optimization. *Adv. Eng. Inform.* **2023**, *55*, 101866.
31. Yang, J.; Wang, J.; Wu, Z.; Anwer, N. Statistical tolerancing based on variation of point-set. *Procedia Cirp* **2013**, *10*, 9–16. [CrossRef]
32. Liu, S.C.; Hu, S.J. Variation Simulation for Deformable Sheet metal assemblies using finite element methods. *J. Manuf. Sci. Eng.* **1997**, *119*, 368–374. [CrossRef]
33. Spathopoulos, S.C.; Stavroulakis, G.E. Springback prediction in sheet metal forming based on finite element analysis and artificial neural Network Approach. *Appl. Mech.* **2020**, *1*, 97–110. [CrossRef]
34. Liu, Y.; Zhao, Y.; Lin, Q.; Pan, W.; Wang, W.; Ge, E. Deviation GAN: A generative end-to-end approach for the deviation prediction of sheet metal assembly. *Mech. Syst. Signal Process.* **2023**, *204*, 110822. [CrossRef]
35. Shi, S.; Liu, J.; Gong, H.; Shao, N.; Anwer, N. Assembly accuracy analysis and phase optimization of aero-engine multistage rotors considering surface morphology and non-uniform contact deformation. *Precis. Eng.* **2024**, *88*, 595–610. [CrossRef]
36. Dupac, M.; Beale, D.G. Dynamic analysis of a flexible linkage mechanism with cracks and clearance. *Mech. Mach. Theory.* **2010**, *45*, 1909–1923. [CrossRef]
37. Yan, J.; Ding, H.; Liang, H. Application of Digital Twin technology in high-precision assembly: Structural accuracy analysis and rework optimization for large-scale equipment. *Int. J. Adv. Manuf. Technol.* **2025**, *138*, 1229–1250. [CrossRef]
38. Imani, B.M.; Pour, M. Tolerance analysis of flexible kinematic mechanism using DLM method. *Mech. Mach. Theory.* **2009**, *44*, 445–456. [CrossRef]
39. Schleich, B.; Anwer, N.; Mathieu, L.; Wartzack, S. Skin Model Shapes: A new paradigm shift for geometric variations modelling in mechanical engineering. *Comput. Aided Des.* **2014**, *50*, 1–15. [CrossRef]
40. Franciosa, P.; Gerbino, S.; Patalano, S. Simulation of variational compliant assemblies with shape errors based on morphing mesh approach. *Int. J. Adv. Manuf. Technol.* **2011**, *53*, 47–61. [CrossRef]

machines

MDPI

Article

Six-Dimensional Spatial Dimension Chain Modeling via Transfer Matrix Method with Coupled Form Error Distributions

Lu Liu [1], Xin Jin [1,*], Huan Guo [2] and Chaojiang Li [1,*]

[1] Department of Mechanical Engineering, Beijing Institute of Technology, No. 5 Zhongguancun South Street, Haidian District, Beijing 100081, China; liulubit@163.com

[2] Beijing Institute of Computer Application, No. 10 Chedaogou, Haidian District, Beijing 100089, China; gh2005gh2005@126.com

* Correspondence: goldking@bit.edu.cn (X.J.); mecjli@bit.edu.cn (C.L.)

Abstract

In tolerance design for complex mechanical systems, 3D dimension chain analyses are crucial for assembly accuracy. The current methods (e.g., worst-case analysis, statistical tolerance analysis) face limitations from oversimplified assumptions—treating datum features as ideal geometries while ignoring manufacturing-induced spatial distribution of form errors and failing to characterize 3D coupled error constraints. This study proposes a six-dimensional spatial dimension chain (SDC) model based on transfer matrix theory. The key innovations include (1) a six-dimensional model integrating position and orientation vectors, incorporating geometric error propagation constraints for high-fidelity error prediction and tolerance optimization, (2) the characterization of spatially distributed form errors and 3D coupled errors of spatial dimension chain-based multiple mating-surface constraint (SDC-MMSC) using six-degree-of-freedom (6-DoF) geometric error components, reducing the assembly topology complexity while improving the efficiency, and (3) a 6-DoF error characterization method for non-mating-constrained data, providing the theoretical basis for SDC modeling. The experimental validation on an aero-engine casing assembly shows that the SDC model captures multidimensional closed-loop spatial errors, with absolute errors of max–min closed-loop distances below 9.3 μm and coaxiality prediction errors under 8.3%. The SDC-MMSC method demonstrates superiority, yielding normal vector angular errors <0.008° and envelope surface RMSE values <0.006 mm. This method overcomes traditional simplified assumptions, establishing a high-precision, multidimensional distributed-form-error-driven SDC model for complex mechanical systems.

Keywords: assembly; spatial dimension chain; spatial distribution of form errors; multiconstraint

1. Introduction

During the design phase, a dimensional chain analysis serves as a critical tool for the rational tolerance allocation of components [1], while providing essential references for subsequent control of the machining and assembly accuracy. With increasing product complexity and higher integration of component structures, the conventional dimensional chain analysis methods face significant challenges. In high-complexity and large-dimension scenarios, such as aero-engines, the effects of error propagation and accumulation substantially complicate dimensional chain analyses. This study focused on the influence of geometric errors, particularly the spatial distribution of form errors, on spatial dimensional chains. Spatially distributed form errors refer to the widely spaced deviations of the real surface from the nominal surface [2]. Such errors result in inconsistent clearances. As illustrated in Figure 1, the impact of spatially distributed form errors on the final assembly accuracy is demonstrated. The figure shows that when a plane with spatially distributed form errors is assembled with an ideal plane, the non-uniform distribution of form errors within the error area induces a rotational displacement of the ideal plane. Furthermore, as shown in Figure 2, the casing assembly consists of multiple components. The combined effects of dimensional errors in individual parts, geometric errors of mating surfaces, and multidirectional constraints on composite mating surfaces of the flange interfaces (including both axial constraints such as end-face contacts and radial constraints such as spigot fits) may lead to out-of-spec closed-loop dimensions (e.g., coaxiality errors in the casing assembly), potentially requiring assembly process adjustments that extend the production lead time and increase manufacturing costs. As shown in Figure 3, the axial assembly distance of the rotor is a critical parameter in ensuring aero-engine assembly quality. However, due to geometric errors at component mating surfaces, the axial relative positioning between the high-pressure compressor rotor and turbine rotor requires multiple trial assemblies to select appropriate shims for dimensional compensation. When treating either the shim thickness or post-assembly axial clearance as the closing loop, the accurate determination of this loop becomes key to reducing aero-engine manufacturing costs. The complexity of assembly analyses stems from multiple factors, including the large dimensions of component mating surfaces, cumulative errors during multicomponent assembly, and complex multiconstraint mating relationships at assembly interfaces. A conventional dimensional chain analysis is inadequate in addressing these challenges, creating an urgent need to develop distributed-form-error-based spatial dimensional chain analysis methods.

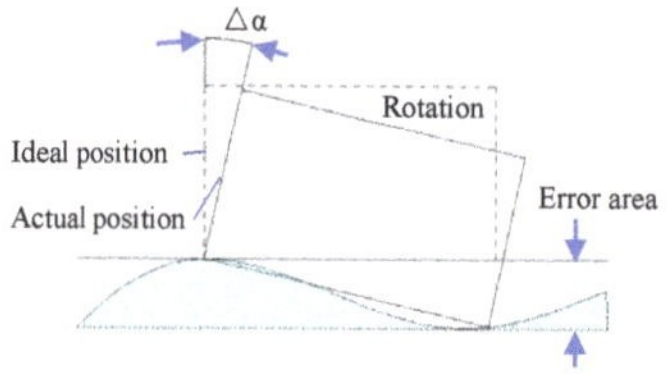

Figure 1. Effects of spatially distributed form errors.

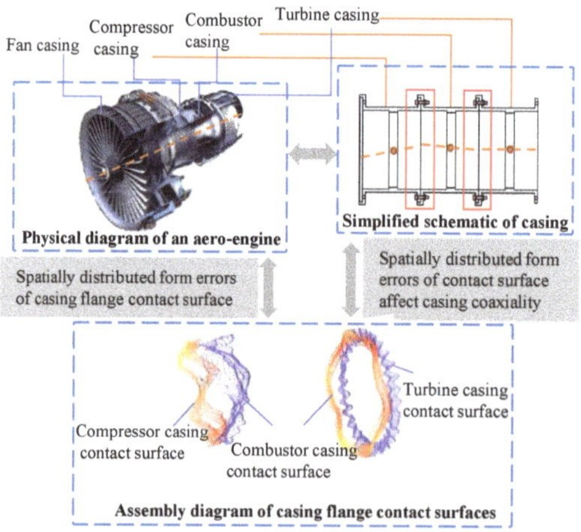

Figure 2. Influence of spatially distributed form errors on the functional dimension (casing assembly coaxiality).

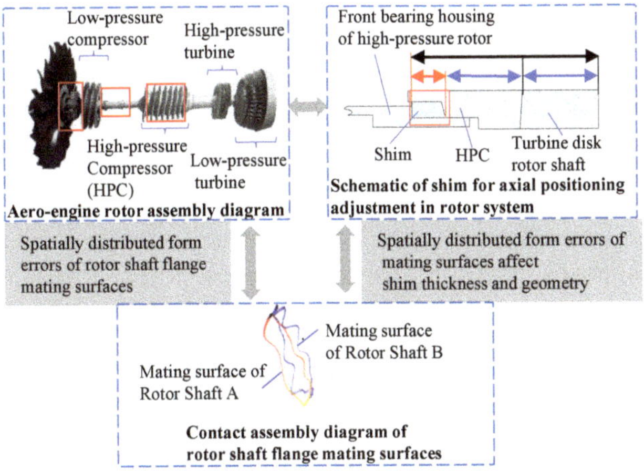

Figure 3. Influence of spatially distributed form errors on the functional dimension (shim thickness).

Previous studies have developed various types of dimensional chain models. For instance, one-dimensional or two-dimensional assembly dimensional chains were established based on 2D drawings, followed by traditional numerical analysis methods such as the worst-case method [3–5] and statistical method [6] to calculate assembly errors in the dimensional chain. Subsequently, manual tolerance allocation and optimization were performed based on engineering experience. However, these methods fail to fully account for (1) geometric errors of mating surfaces in 3D space, (2) coupling effects between assembly features, and (3) multidimensional characteristics of closed loops. Consequently, when applied to dimensional chains involving large-scale mating surfaces (which cannot be simplified as points) where geometric errors are non negligible, they produce inaccurate tolerance design results. Several studies have developed three-dimensional spatial dimensional chain models, such as the tolerance map (T-MAP) model [7–9]. Chase et al. [10–13] employed two-dimensional and three-dimensional direct linearization methods, respectively, to convert deviations caused by geometric tolerances into dimensional deviations. Clément et al. applied the small displacement torsor (SDT) model [14–20], which utilizes deviation torsors and clearance torsors to describe geometric deviations and mating clearances between parts, and solves functional requirements through algebraic composition operations. Other researchers have approached dimensional chain modeling from the perspective of geometric relationship propagation. Desrochers and Rivière [21] proposed a matrix model, arguing that feature variations relative to a datum can be decomposed into six independent components—three translational displacements along the coordinate axes and three rotational displacements about the axes. Wang et al. [22] introduced a CAD-model-based automated generation method for 3D assembly dimensional chains using graph theory. El Mouden et al. [23] developed a 3D tolerance analysis method combining the Jacobian matrix with the torsor model, improving the computational efficiency for complex assemblies. Additionally, Zhang et al. [24] proposed a 3D tolerance zone computation method based on SDT, robotic kinematics, and convex set theory models, addressing challenges in process dimensioning and tolerance design for 3D manufacturing. Jain [25] proposed a method by fitting an orthotope inside an ellipsoid, employing the system Jacobian matrix to address challenges in tolerance analyses, tolerance allocation, and compensator selection. Tang et al. [26] introduced an improved Monte Carlo simulation approach utilizing surrogate models and low-discrepancy sequences, significantly enhancing the computational efficiency in dimensional variation analyses while maintaining accuracy.

In another study [27], the authors developed a tolerance analysis tool based on Monte Carlo simulation to mitigate assembly rework caused by dimensional variability in prefabricated and offsite construction, subsequently reducing the rework risks through process optimization. Additionally, McGregor [28] applied a support vector regression (SVR) machine learning approach to develop a geometric shape prediction and qualification tool based on manufacturing parameters and feature descriptors, effectively predicting and diagnosing geometric defects in additive manufacturing while improving the part classification accuracy. Maltauro et al. [29] introduced the concept of variable-dependent admissible limits for tolerance stack-up analyses, where the limits are adaptively adjusted based on both geometrical and non-geometrical variables. While the aforementioned studies have made positive contributions to tolerance accumulation in geometric relationship transmission, these spatial dimension chain models face two primary limitations. First, most models are tolerance-based, where the tolerances merely describe the permissible variation range of geometric errors. Such approaches oversimplify the situation by reducing the geometric errors to 'extreme boundaries' within tolerance zones, treating post-machining surface form errors as scalar quantities (as shown in Figure 3). In reality, form errors exhibit spatial distributions. Second, in conventional dimension chains, the closed loop is typically represented by a scalar quantity (such as Δh in Figure 4) as the accumulation result. However, due to the stacking of spatially distributed form errors and other geometric errors, the closed-loop clearance actually demonstrates multidimensional spatial distribution characteristics, leading to discrepancies between design values and actual assembly outcomes. Under high-dimensional and large-scale conditions with limited part quantities, assembly process adjustments through part replacement become unfeasible. Consequently, accurate dimension chain prediction during the design phase can fundamentally reduce the subsequent manufacturing costs, time expenditures, and labor requirements. Moreover, precise dimension chain prediction at the design stage can provide optimized feedback for more accurate tolerance allocation. Thus, developing methods to obtain more precise dimension chain prediction results, thereby achieving more accurate tolerance allocation outcomes, represents an urgent and significant challenge.

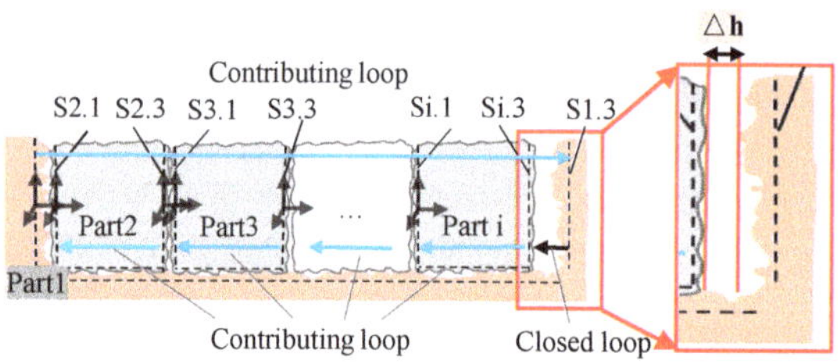

Figure 4. Spatial multidimensional distribution characteristics of the closed loop.

Meanwhile, in practical engineering, geometric errors propagate through the geometric relationships in the dimensional chain. The constrained mating surfaces of parts often involve more than one pair of critical interfaces (as shown in Figure 4, where the key mating surface of part 2 serves as a key datum plane in the dimensional chain, with its position and orientation jointly constrained by both the horizontal plane and the left vertical plane of part 1). The resulting three-dimensional error coupling from these constraints ultimately directly affects the product performance. Zhang et al. [30,31] proposed a method for solving contact problems of mating surfaces by considering spatially distributed form errors. While these studies account for the influence of spatially distributed form errors on assembly posture, they primarily focus on single-reference assembly problems (i.e., assemblies constrained by only one pair of mating surfaces), with limited research on posture constraints imposed by dual mating surfaces. Liu et al. [32] introduced a general analysis method incorporating the parallel effects of geometric structures, utilizing constrained registration to compute the optimal mating results. Shi et al. [33] comprehensively considered the influence of the surface morphology and non-uniform contact deformation on the error propagation process, further adopting a phase optimization method to improve the assembly accuracy of multistage rotors. However, these methods primarily focus on error propagation mechanisms, while the systematic modeling of geometric constraints and assembly relationships within dimensional chains remains unresolved.

In response to the aforementioned challenges, the main contributions of this work are as follows: (1) We developed a distributed-form-error-driven spatial dimension chain model establishing mapping relationships between geometric features and functional dimensions through six-dimensional vector functions. The model incorporates spatially distributed form errors in datum definitions and employs the transfer matrix method to represent closed loops in both positional and orientational dimensions, enabling high-fidelity error prediction for large-scale assemblies. (2) We addressed multiple mating surface constraints by characterizing form errors and 3D coupled errors using 6-DoF geometric error components, significantly reducing the computational complexity while improving the prediction accuracy. (3) Non-mating-constrained datum errors were characterized using 6-DoF geometric error components based on spatially distributed form errors and modified envelope conditions, ensuring interchangeability and assembly quality. (4) We validated the model through a casing assembly case, demonstrating its effectiveness in analyzing six-dimensional closed-loop characteristics. (5) We applied the proposed six-dimensional spatial dimension chain model to the tolerance optimization design of aero-engine casing components, where its multidimensional error prediction capability facilitates precision tolerance allocation.

2. Six-Dimensional Spatial Dimension Chain Modeling

The proposed six-dimensional spatial dimension chain (SDC) model accounts for spatially distributed form errors during the design stage. The processing of the spatially distributed form errors follows the methodology in [34]—an error modeling approach that considers the distribution patterns of form errors induced by existing manufacturing processes at the design phase. This approach is predicated on the assumption that the machining errors remain statistically stable during a machine tool's precision stability period. The machining errors on part surfaces comprise systematic errors and random errors. Systematic errors, as deterministic components, exhibit repeatability and predictability, constituting the dominant portion of the total errors; random errors lack regularity and typically account for a minor proportion. To establish manufacturing-process-referenced spatially distributed form errors, we inspect historically machined workpieces and employ NURBS (non-uniform rational B-spline)-based surface regression [35] to extract and separate systematic and random errors, subsequently utilizing the systematic error morphology (including both magnitude and spatial distribution patterns) as the primary error model component. Anwer et al. [36] made significant contributions to the field of tolerance analysis [37,38] by developing skin-model-based approaches, which utilize statistical distributions and idealized geometric deviation assumptions. In contrast to these methods, the present work establishes a fundamentally distinct manufacturing-feature-driven modeling process based on actual machining process data.

2.1. Definition of Modeling Data

The error modeling method based on existing manufacturing processes and the NURBS surface regression method for extracting systematic errors serve as the foundational error data for constructing the SDC model. The core of SDC modeling lies in establishing constraint relationships between geometric elements (e.g., planes, axes) and analyzing error accumulation through a datum reference system. In this study, the term 'datum' refers to a theoretically exact feature (as defined in ISO GPS [39]). Its closed-loop characteristic relies on well-defined geometric data and constraint propagation paths. Thus, defining a datum coordinate system via geometric elements (e.g., mating surfaces) to constrain the error directions constitutes the fundamental requirement for establishing an SDC. Subsequently, the modeling data for mating surfaces are selected and classified. Since most parts demand high assembly precision and exhibit substantial rigidity, deformation is not considered in this model. Therefore, this study operated under the assumption of rigid body conditions.

As shown in Figure 5, when considering the influence of spatially distributed form errors [31], the data are categorized based on the constraint properties of mating surfaces:

- Mating-constrained datum planes: The coordinate systems are established at both the ideal mating positions (e.g., surface b or c) and the actual mating positions (e.g., surface b′), referred to as the ideal mating-constrained datum and actual mating-constrained datum, respectively;
- Non-mating-constrained datum planes: The coordinate systems are defined at the ideal locations of critical surfaces (e.g., surfaces a and d) and their actual locations (e.g., surfaces a′ and d′), designated as the ideal non-mating-constrained datum and actual non-mating-constrained datum, respectively.

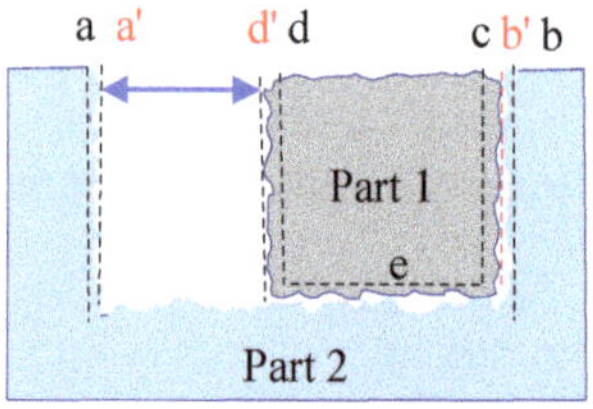

Figure 5. Datum for SDC modeling.

2.2. Establishment of SDC

Next, the SDC model with planar mating surfaces is constructed to characterize the mapping relationships between geometric features and functional dimensions.

The definition of the starting point in the SDC model affects its mathematical formulation. Since this section focuses on methodological research, the following fundamental rules are adopted to ensure the generality of the SDC model and to align with conventional measurement or design datum selection practices:

1. The assembly datum plane of the component is selected as the primary reference for the global coordinate system of the SDC model;
2. Local coordinate systems are then established on the critical surfaces of the part, with their relationship to the global coordinate system illustrated in Figure 6;
3. The SDC model is then decomposed into two branches:
   - First branch: Starting from the assembly datum (i.e., the origin of the global coordinate system), local coordinate systems are sequentially constructed along the transmission path of the contributing loops until reaching the datum plane at one end of the closed loop.
   - Second branch: Similarly, starting from the assembly datum, the process continues until reaching the datum plane at the other end of the closed loop.

- Figure 7 illustrates the implementation of this principle via a distributed-form-error-based multidimensional SDC framework, covering both its construction and application (specific variables are detailed in the following two paragraphs).

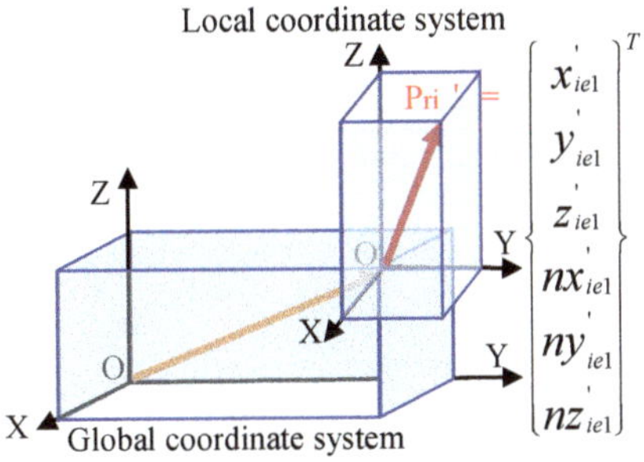

Figure 6. Relationship between global and local coordinate systems of critical surfaces.

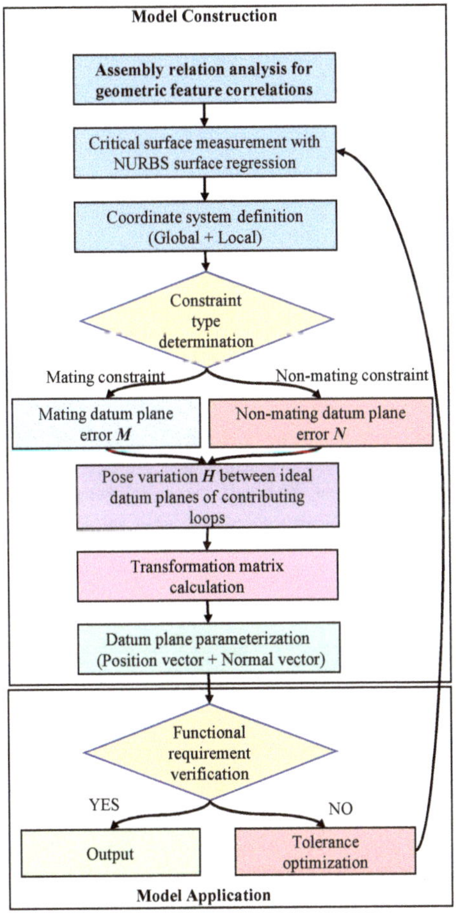

Figure 7. Construction and application of SDC model.

The mechanistic implementation of this workflow is demonstrated through a representative case study, as shown in Figure 8. Based on the fundamental rules of modeling, mating surface S1.1 of part 1 serves as the assembly datum for the system. Therefore, the global coordinate system is established at the geometric center of mating surface S1.1/2.1. Subsequently, part 2 is assembled relative to part 1, and local coordinate systems are defined at their key mating surfaces—the ideal mating-constrained data (S1.1 and S2.1) and the actual mating-constrained datum (S1.1/2.1). The geometric error of the actual mating-constrained datum S1.1/2.1 from the ideal mating-constrained datum S2.1 is defined as the mating-constrained datum error $M_{2.1}$. (In this study, the mating-constrained datum error refers to the error between the ideal and actual mating-constrained data caused by spatially distributed form errors of critical surfaces.) This geometric error can be treated as a contributing loop with a nominal dimension of zero and incorporated into the contributing loops of part 2.

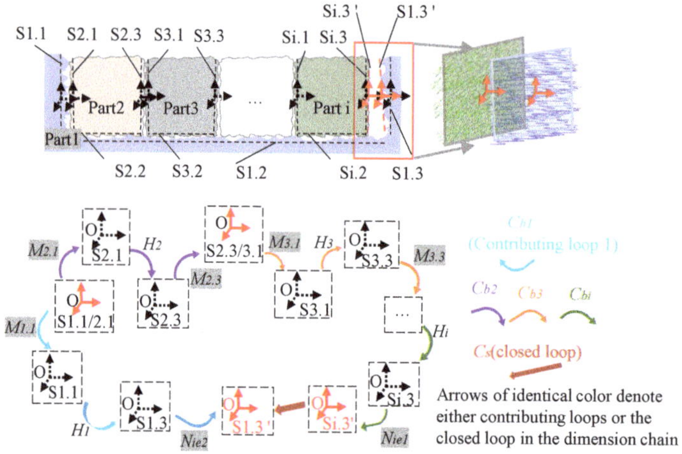

Figure 8. Modeling mechanism for six-dimensional SDC.

Similarly, when assembling part 3 relative to part 2, the local coordinate systems are defined at their key mating surfaces—the ideal mating-constrained data (S2.3 and S3.1) and the actual mating-constrained datum (S2.3/3.1). The geometric error $M_{2.3}$ between S2.3/3.1 and S2.3 is also treated as a zero-nominal-dimension contributing loop and merged into part 2's contributing loops. The homogeneous transformation matrix (HTM) H_i characterizes the pose variation between ideal datum planes within the loop (e.g., H_2 represents the relationship between datum planes S2.1 and S2.3), where the translation matrix T_i describes positional offsets and the rotation matrix R_i accounts for angular changes in Equation (1). In Figure 8, elements connected by arrows of the same color belong to the same contributing loop. The set of contributing loops is denoted as $\{C_{b1}, C_b2, \ldots, C_{bi}\}$.

$$H_i = \begin{bmatrix} R_i & T_i \\ 0 & 1 \end{bmatrix} \tag{1}$$

This assembly logic applies recursively to part i (where i denotes the generic part index). At the non-mating-constrained critical surfaces, the ideal non-mating-constrained datum (Si.3) and actual non-mating-constrained datum (Si.3′) are established. The error between them is defined as the non-mating-constrained datum error $N_{i.3}$. (Here, the non-mating-constrained datum error refers to the error between the ideal and actual non-mating-constrained data due to spatially distributed form errors.) The spatial vector gap between part i's actual non-mating-constrained datum (Si.3′) and part 1's actual non-mating-constrained datum (S1.3′) forms the closed loop (C_s).

The relationship between the geometric features (the mating-constrained datum error (M), the non-mating-constrained datum error (N), and SDC contributing loops) and the functional dimension (SDC closed loop C_s) is illustrated in Figure 8.

As demonstrated above, the integration of the error modeling method based on the current machining process with the NURBS surface regression method provides an effective solution for extracting, generating, and characterizing spatially distributed form errors. The transfer matrix method (TMM) employs rigid body kinematics to model pose errors (position + orientation) caused by geometric error propagation, and can also analyze coupled errors between geometric features (e.g., coupling errors caused by multimating surface constraints). However, it does not directly provide geometric error constraint relationships (e.g., assembly functional requirements). The conventional dimension chain models, based on tolerance boundaries or deviation analysis, disregard the influence of spatially distributed form errors. Consequently, they fail to characterize the non-uniform distribution of form errors in actual machining processes, inadequately describe 3D spatial error coupling, and lack precision in representing spatially distributed geometric features and critical functional dimensions. To address these limitations, this study integrates spatially distributed form errors, derived from the geometric error modeling method based on the current machining process and the NURBS surface regression method, with the transfer matrix into an SDC model. Furthermore, the SDC model is constructed using a six-dimensional vector function, establishing the mapping relationship between the geometric features and functional dimensions, as illustrated in Figure 9.

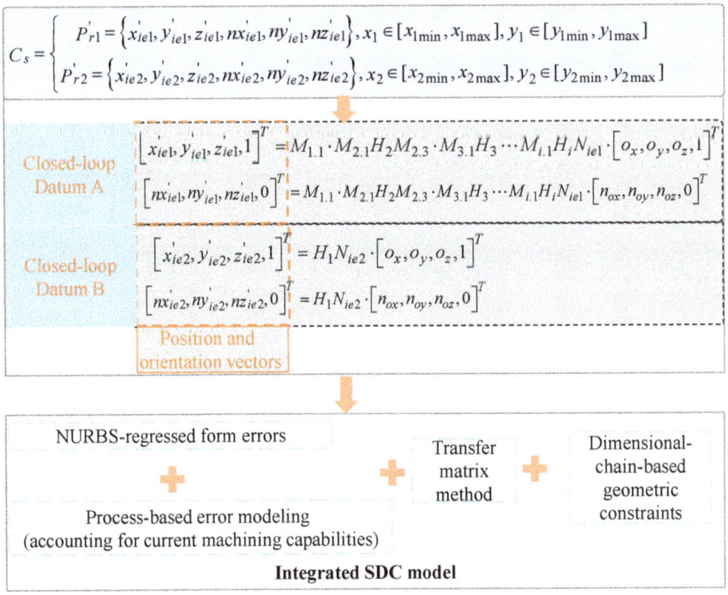

Figure 9. SDC model.

Here, P_{ri}' (i = 1,2) denotes the pose of the closed-loop datum in the global coordinate system. The spatial gap of the closed loop is defined as the relationship between a pair of six-dimensional closed-loop datum coordinate systems P_{ri}'. Based on this relationship, additional multidimensional closed-loop parameters can be derived, such as the maximum and minimum distances of closed loops and the parallelism of one datum plane relative to another. In Figure 9, (x_1,y_1) and (x_2,y_2) represent the boundary ranges of the closed-loop data.

The closed-loop datum plane P_{ri}' is defined in the local coordinate system with the origin position vector $[o_x, o_y, o_z, 1]^T$ and unit normal vector $[n_{ox}, n_{oy}, n_{oz}, 0]^T$ (e.g., $[0, 0, 1, 0]^T$ for Z-normal planes, respectively. The mating-constrained and non-mating-constrained datum plane errors are formulated as small displacement screws M and N, respectively (Equation (2)). Each component of M and N represents an error along six degrees of freedom, including rotations about the x-, y-, and z-axes and translations along these axes (Figure 10).

$$M = (\Delta\alpha_i, \Delta\beta_i, \Delta\gamma_i, \Delta x_i, \Delta y_i, \Delta z_i);\ N_{ie} = (\Delta\alpha_{ie}, \Delta\beta_{ie}, \Delta\gamma_{ie}, \Delta x_{ie}, \Delta y_{ie}, \Delta z_{ie}) \quad (2)$$

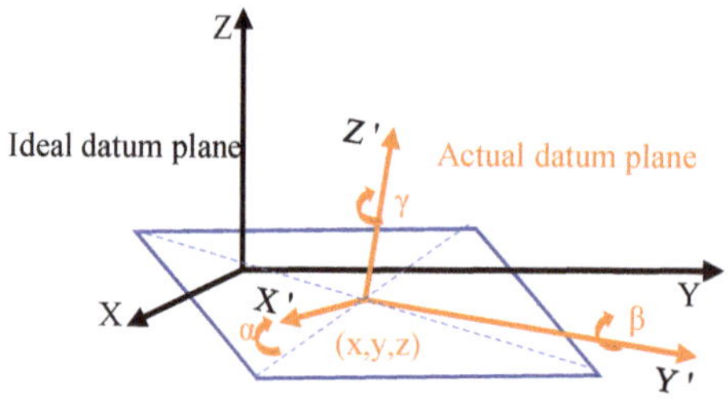

Figure 10. Schematic of mating-constrained and non-mating-constrained datum errors.

Next, the key parameters of the SDC model are characterized. The actual mating-constrained datum and its error (M) are derived in Section 2.3, while the non-mating-constrained datum and its error (N) are addressed in Section 2.4. Finally, an aero-engine casing assembly case study (Section 2.5) demonstrates the model's application, with the validation results discussed in Section 3.

2.3. *Characterization for Mating-Constrained Datum Error*

As illustrated in Figure 11, the middle panel serves as the primary schematic. A global coordinate system is established at the assembly datum, followed by the sequential assembly of part A and part B. The distance between the top surface of part B and the bottom surface of part A constitutes the closed loop, while the spatial distances between the top and bottom surfaces of part A and part B act as the contributing loops. The mating surfaces of both parts belong to the mating-constrained datum plane category. Based on the quantity of mating surfaces, the mating-constrained datum planes are classified into the single mating surface constraint (SMSC, left panel of Figure 11) and the multiple mating surface constraint (MMSC, right panel of Figure 11). The subsequent analyses focus on characterizing the mating-constrained datum plane and solving its associated errors under both scenarios, with the key procedural steps detailed in Figure 12.

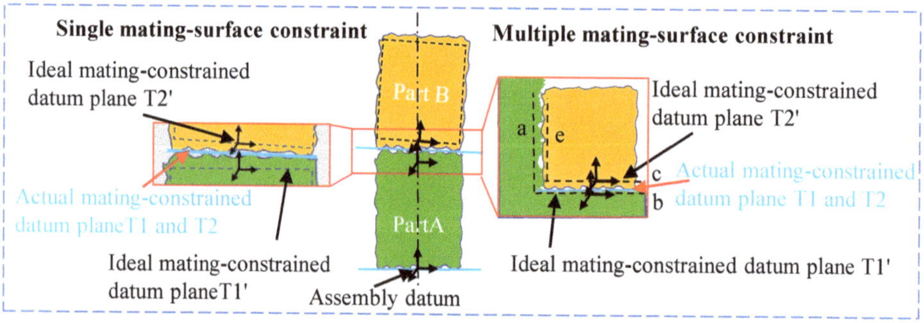

Figure 11. Types of mating-constrained datum planes.

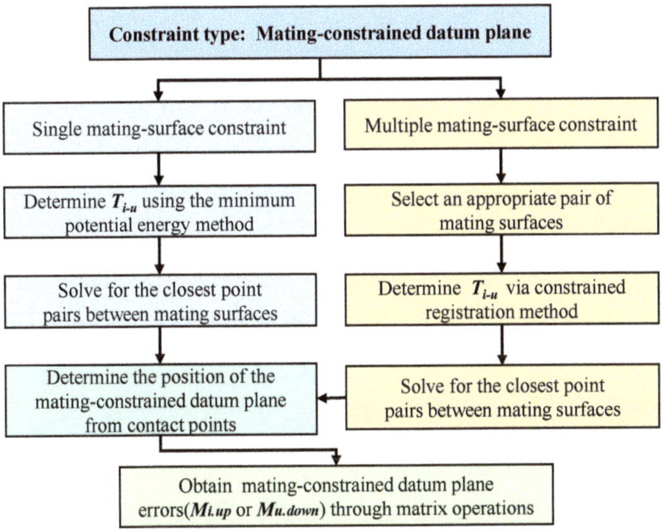

Figure 12. Characterization flowchart for mating-constrained datum errors.

(1) The SMSC: As shown in the left panel of Figure 11, the coordinate system of the mating-constrained datum plane is established at both the ideal mating position (critical surfaces T1′ and T2′) and the actual mating position (critical surfaces T1 or T2). The mating-constrained datum error, denoted as $M = (\Delta\alpha_i, \Delta\beta_i, \Delta\gamma_i, \Delta x_i, \Delta y_i, \Delta z_i)$, corresponds to the error between the ideal mating-constrained datum plane T1′ and the actual mating-constrained datum plane T1, or between T2′ and T2. This error is characterized by six geometric error components, including rotations about and translations along the *x*-, *y*-, and *z*-axes.

The mating-constrained datum error is solved using the minimum potential energy method, with the following steps.

When the assembly reaches a stable state [30], the potential energy of part B relative to part A is minimized. The optimization objective is to minimize the potential energy of part B relative to part A, with the constraint that the mating surfaces of parts A and B do not interfere, as specified in Equation (3). The bare bones particle swarm optimization (BBPSO) algorithm is employed to determine the optimal solution for matrix $T_{i\text{-}u}$ and to identify three pairs of contact points on the two mating surfaces (three non-collinear contact points ensure a stable assembly).

The surface containing these three contact points on the same part defines the actual mating-constrained datum plane (T1 or T2). Through matrix operations, the six-degree-of-freedom (6-DoF) geometric error components of the ideal mating-constrained datum plane T1′ relative to the actual mating-constrained datum plane T1 (denoted as $M_{u.down}$) and those of T2′ relative to T2 (denoted as $M_{i.up}$) can be obtained. The matrix relationship among $T_{i\text{-}u}$, $M_{i.up}$, and $M_{u.down}$ is then given by Equation (4).

$$\min\sum(T_{i-u}\cdot D^1 - D^2)mg \quad S.T. \quad T_{i-u}\cdot D^1 - D^2 \geq 0 \tag{3}$$

Here, D^1 and D^2 represent the point cloud data of the mating surfaces of part A and part B in the assembly coordinate system, respectively. The term ($T_{i\text{-}u}\cdot D^1 - D^2$) denotes the distance between part A and part B, where m is the mass of each point and g is the gravitational acceleration (a constant).

$$
\begin{aligned}
T_{i-u} &= ER_{i.up} \cdot ER_{u.down} \\
&= \begin{bmatrix} 1 & -\Delta\gamma_{i.up} & \Delta\beta_{i.up} & \Delta x_{i.up} \\ \Delta\gamma_{i.up} & 1 & -\Delta\alpha_{i.up} & \Delta y_{i.up} \\ -\Delta\beta_{i.up} & \Delta\alpha_{i.up} & 1 & \Delta z_{i.up} \\ 0 & 0 & 0 & 1 \end{bmatrix} \begin{bmatrix} 1 & -\Delta\gamma_{u.down} & \Delta\beta_{u.down} & \Delta x_{u.down} \\ \Delta\gamma_{u.down} & 1 & -\Delta\alpha_{u.down} & \Delta y_{u.down} \\ -\Delta\beta_{u.down} & \Delta\alpha_{u.down} & 1 & \Delta z_{u.down} \\ 0 & 0 & 0 & 1 \end{bmatrix} \\
&= \begin{bmatrix} 1+(-\Delta\gamma_{i.up}).\Delta\gamma_{u.down}+(\Delta\beta_{i.up})(-\Delta\beta_{u.down}) & (-\Delta\gamma_{u.down})+(-\Delta\gamma_{i.up})+\Delta\beta_{i.up}.\Delta\alpha_{u.down} & \Delta\beta_{u.down}+(-\Delta\gamma_{i.up})(-\Delta\alpha_{u.down})+\Delta\beta_{i.up} & \Delta x_{u.down}+(-\Delta\gamma_{i.up})\Delta y_{u.down}+\Delta\beta_{i.up}.\Delta z_{u.down}+\Delta x_{i.up} \\ \Delta\gamma_{i.up}+\Delta\gamma_{u.down}+(-\Delta\alpha_{i.up})(-\Delta\beta_{u.down}) & \Delta\gamma_{i.up}.(-\Delta\gamma_{u.down})+1+(-\Delta\alpha_{i.up}).(\Delta\alpha_{u.down}) & \Delta\gamma_{i.up}.\Delta\beta_{u.down}+(-\Delta\alpha_{u.down})+(-\Delta\alpha_{i.up}) & \Delta\gamma_{i.up}.\Delta x_{u.down}+\Delta y_{u.down}+(-\Delta\alpha_{i.up}).\Delta z_{u.down}+\Delta y_{i.up} \\ -\Delta\beta_{i.up}+\Delta\alpha_{i.up}.\Delta\gamma_{u.down}+(-\Delta\beta_{u.down}) & (-\Delta\beta_{i.up})(-\Delta\gamma_{u.down})+\Delta\alpha_{i.up}+\Delta\alpha_{u.down} & (-\Delta\beta_{i.up}).\Delta\beta_{u.down}+\Delta\alpha_{i.up}.(-\Delta\alpha_{u.down})+1 & (-\Delta\beta_{i.up}).\Delta x_{u.down}+\Delta\alpha_{i.up}.\Delta y_{u.down}+\Delta z_{u.down}+\Delta z_{i.up} \\ 0 & 0 & 0 & 1 \end{bmatrix}
\end{aligned}
\quad (4)
$$

Here, $T_{i\text{-}u}$ describes the pose error of the ideal mating-constrained datum plane of the u-th part relative to that of the i-th part, caused by mating surface errors; Δx, Δy, and Δz denote the infinitesimal translations along the x-, y-, and z-axes, respectively; $\Delta\alpha$, $\Delta\beta$, and $\Delta\gamma$ represent the infinitesimal rotations about the x-, y-, and z-axes, respectively.

(2) The MMSC: The conventional dimension chain methods consider only a single pair of mating surfaces along a specified direction while neglecting others, with a primary focus on dimensional magnitudes and tolerances. These methods fail to account for the coupled effects of geometric errors from two pairs of mating surfaces in different directions on the SDC.

To address these limitations, we first selected an optimal pair of mating surfaces (the pair of mating surfaces that maximally constrains the part's degrees of freedom) to obtain a datum plane as the actual mating-constrained datum plane in the SDC model (Figure 11, right panel), based on the assembly dimension chain direction. Subsequently, based on the selected mating-constrained datum plane, the constrained registration method [40] was employed to characterize the actual mating-constrained datum plane and solve its associated 6-DoF error M.

Whether for SMSC or MMSC, the mating-constrained datum plane error is inherently linked to the part's manufacturing process. Under different processes, the magnitude and range of the geometric error components vary, typically following a probability distribution such as a normal distribution. These variations in geometric error components ultimately influence the characterization and distribution of the SDC error.

The mating-constrained datum plane error is solved using the constrained registration method, with the following procedure.

A constrained registration optimization model is constructed to obtain the optimal solution for matrix $T_{i\text{-}u}$ between the ideal mating-constrained datum planes T2′ and T1′ in a multimating surface assembly structure. When the two parts are stably assembled, three pairs of the closest points (approaching infinitesimal proximity to zero) are selected on the designated mating surfaces c and b as contact points. These three contact point pairs can be recorded as $C1(x_{1i}, y_{1i}, z_{1i}, 1)$ and $C2(x_{2i}, y_{2i}, z_{2i}, 1)$ $(i = 1, 2, 3)$. The surfaces containing these three contact points on each part defines the actual mating-constrained datum planes T1 or T2. Subsequently, through matrix operations, the mating-constrained datum plane errors $M_{i.up}$ and $M_{u.down}$ can be obtained. Here, $M_{i.up}$ represents the geometric error of the ideal mating-constrained datum plane T2′ relative to the actual mating-constrained datum plane T2, while $M_{u.down}$ represents T1′ versus T1. The matrix relationship among $T_{i\text{-}u}$, $M_{i.up}$, and $M_{u.down}$ is given by Equation (4).

This methodology, which incorporates geometric errors (particularly spatially distributed form errors) and employs constrained registration to solve both the actual mating-constrained datum planes in SDC and their associated error (M) under the MMSC, effectively reduces the topological complexity in assembled component surfaces. Consequently, it achieves computational simplification while enhancing both the efficiency and accuracy of SDC predictions.

2.4. Characterization for Non-Mating-Constrained Datum Error

To ensure part interchangeability and assembly quality, this subsection introduces an improved envelope condition (also referred to as the envelope requirement in some standards) and proposes new definitions of the envelope surface and envelope area. The envelope surface is then used to characterize the actual non-mating-constrained datum plane, from which the non-mating-constrained datum plane error $N_{ie} = (\Delta\alpha_{ie}, \Delta\beta_{ie}, \Delta\gamma_{ie}, \Delta x_{ie}, \Delta y_{ie}, \Delta z_{ie})$ is derived. This error is characterized by six geometric error components, including rotations about and translations along the *x*-, *y*-, and *z*-axes. As illustrated in Figure 13, taking a simplified cylindrical casing as an example, a global coordinate system is established at the assembly datum O, followed by the sequential assembly of part A and part B. The distance between the top surface of part B and the bottom surface of part A forms the closed loop. As established in Section 2.1, the top surface of Part B is a non-mating-constrained datum plane. To characterize this plane using the envelope surface, the envelope surface must first be determined based on the envelope condition—a constraint ensuring that the non-mating-constrained datum plane error maintains dimensional and form compliance with assembly requirements under error conditions.

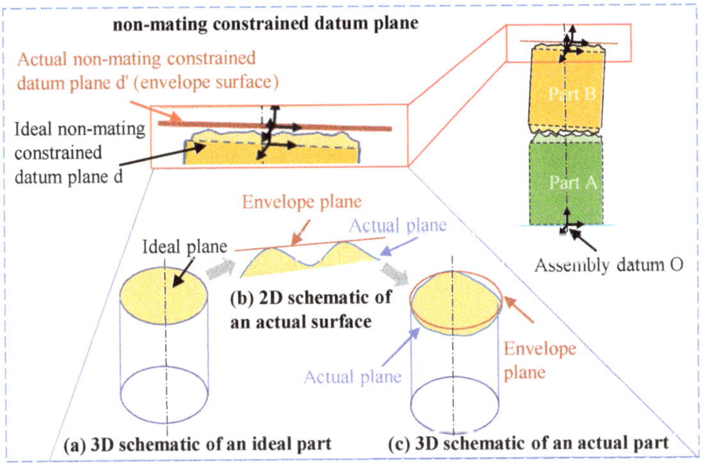

Figure 13. Schematic of envelope plane characterization for actual non-mating-constrained datum planes.

The improved envelope condition further specifies four criteria:

1. Envelope condition definition: The envelope condition is an error analysis method. The envelope condition bounds all elements of the spatially distributed form errors within the envelope surface, controlling the surface's position, orientation, and form to meet interchangeability and assembly quality requirements.
2. Envelope surface definition: In a given direction, an ideal surface approximates the actual surface, with its position and form satisfying the envelope condition (criterion 1 above).
3. Envelope area definition: The envelope surface and actual plane must intersect at $\geq$3 points, forming a triangular area termed the envelope area.
4. Maximization requirement: The envelope area must be maximized to closely conform to the actual error surface.

The envelope surface with the maximum envelope area is identified as the actual non-mating-constrained datum plane. Subsequently, the six-degree-of-freedom (6-DoF) non-mating-constrained datum plane error (N) of the envelope surface d′ relative to the ideal non-mating-constrained datum plane d is then derived. The characterization of the envelope surface and the solution for N are detailed below in Figure 14.

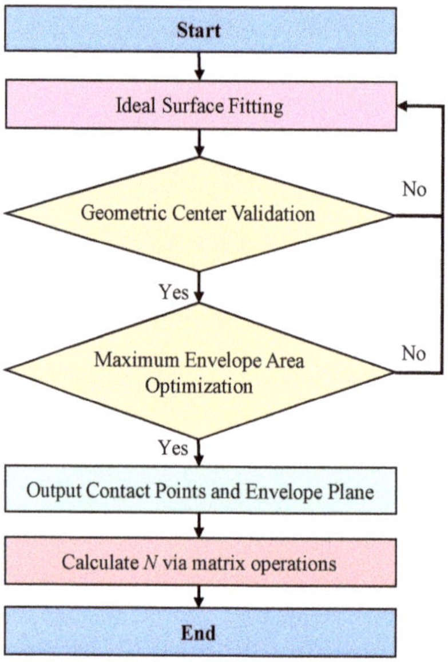

Figure 14. Flowchart for solving envelope surface and non-mating-constrained datum errors.

1. Ideal Surface Fitting

An ideal datum plane is fitted to the actual error surface S through asymptotic convergence, subject to the following constraints. The ideal plane does not interfere with the actual error surface S, and there exist at least three contact points between the ideal plane and S, $P_1(x_1{}', y_1{}', z_1{}')$, $P_2(x_2{}', y_2{}', z_2{}')$, and $P_3(x_3{}', y_3{}', z_3{}')$, as formalized in Equation (5):

$$V \cdot (x_j, y_j, z_j) \leq 0 \quad \forall (x_j, y_j, z_j) \in S \tag{5}$$

where V is the normal vector of the envelope plane and (x_j, y_j, z_j) represents all measured points on S.

2. Geometric Center Validation

Since the casing part is a hollow cylinder, its centroid must lie within the cavity. To ensure the envelope surface better conforms to the actual condition, the geometric center of the casing's critical surface must lie within the triangle formed by contact points P_1, P_2, and P_3. If this condition fails, the algorithm returns to step 1.

3. Maximum Envelope Area Optimization

To ensure the envelope surface maximally encompasses the actual surface for optimal conformity and improved model accuracy, the maximum envelope area (max S_{area}) must be determined. The envelope area S_{area}, defined by the contact points P_1, P_2, and P_3

(Equation (6)), is evaluated. If the condition is satisfied, the procedure proceeds; otherwise, it returns to step 1.

$$S_{area} = \frac{1}{2} \times \|(P2 - P1) \times (P3 - P1)\| \quad (6)$$

where $\times$ denotes the cross product and $\|\cdot\|$ the vector norm

4. Output Contact Points and Envelope Plane

The coordinates of P_1, P_2, and P_3 are recorded, and the position of the envelope plane d' is determined.

5. Non-Mating-Constrained Datum Error Calculation

The 6-DoF geometric error components of d' relative to the ideal datum plane d (denoted as N) are computed and output.

This subsection accounts for spatially distributed form errors and characterizes the actual non-mating-constrained datum planes error using more precise 6-DoF parameters. The N is correlated with the part's manufacturing processes, as different processes yield varying error magnitudes and ranges that typically following probability distributions such as normal distributions. Variations in error ultimately affect the spatial distribution of the closed loop.

2.5. Six-Dimensional Modeling for Aero-Engine Casing Assembly

Taking the assembly of a simulated aero-engine casing as an example, we analyzed two typical assembly issues caused by geometric error accumulation: (1) an end-face closed-loop SDC, where the spatial distance of the closed loop is referenced to the flange end face as the datum plane; (2) an axis-based closed-loop SDC, where the coaxiality spatial error is referenced to the casing housing axis as the datum axis. Additionally, Table 1 summarizes the advantages of the SDC model.

Table 1. Comparative advantages of the SDC model for aero engine casing assembly

Key Issue	Limitation of Conventional Methods	SDC-Based Solution	Application Value
Flange spatially distributed form errors	Tolerance-based dimension chain modeling	SDC model incorporating spatially distributed form errors	Solves high-precision prediction of mating surfaces
Multiconstrained flange structure	Single mating surface constraint assumption (end face only)	Multimating surface constraints for 3D coupled errors	Enables 3D coupled error analyses in SDCs
Functional dimension representation	Scalar-based expression	Six-dimension parameters (position + orientation vectors)	Supports multidimensional spatial characterization

(1) SDC propagation law with the end face as the closed-loop datum.

The contributing loops are C_{b11} and C_{b12}, as shown in Figure 15a, with the closed loop C_s representing the spatial dimension from the upper end face O_{ie1} of the upper casing to the assembly datum face O_{ldown}. Following the aforementioned computational process, the coordinate systems for critical surfaces are established and the influencing factors of SDC errors are analyzed, as depicted in Figure 15b.

The predominant error sources in SDC are:

1. The mating-constrained datum plane error $_{M1up}$ of the lower casing top surface O_{lup} (comprising two constrained surfaces—the flange end face and flange spigot cylindrical face);

2. The mating-constrained datum plane error M_{2down} of the upper casing bottom surface O_{2down} (also including two constrained surfaces—the flange end face and flange spigot cylindrical face);
3. The non-mating-constrained datum plane error N_{ie1} at the end face of the upper casing's top surface;
4. The mating-constrained datum plane error M_{1down} at the end face of the lower casing's bottom surface.

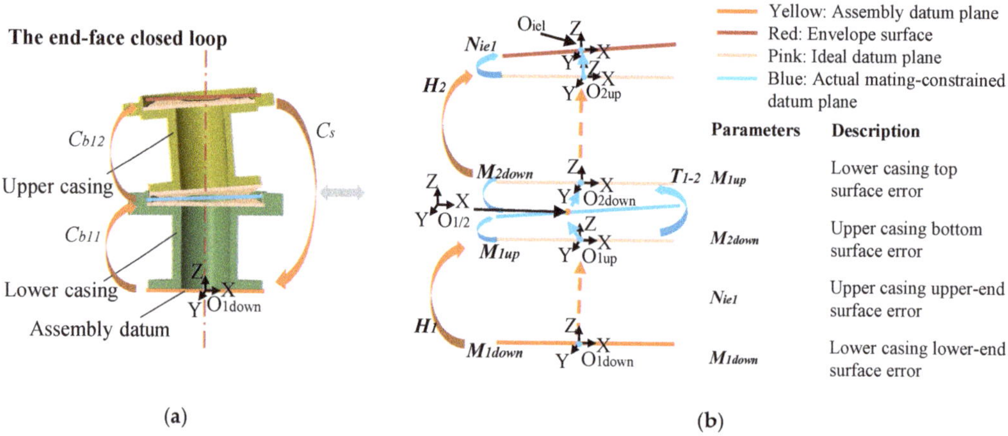

Figure 15. End face closed-loop SDC in simulated casing assembly: (**a**) schematic of simulated casing assembly; (**b**) datum coordinate system construction and error contributors.

Here, the lower casing bottom end face and the assembly datum plane O_{1down} are considered nominally ideal surfaces; therefore, the error component M_{1down} can be disregarded.

The initial pose of the closed-loop datum plane P_{r1}' is defined by the position vector of the coordinate system origin: $O = [o_x, o_y, o_z, 1]^T$; the unit normal vector of the datum plane: $n = [n_{ox}, n_{oy}, n_{oz}, 0]^T$ (e.g., $[0, 0, 1, 0]^T$ for Z-axis alignment).

The influence of geometric errors on the closed loop is analyzed using the transfer matrix method, where the pose of the closed-loop datum in the global coordinate system is characterized by a six-dimensional parameter P_r1', as expressed in Equations (7) and (8). Finally, the closed-loop metrics are evaluated against assembly or design requirements. If the results are unsatisfactory, the process iterates for redesign optimization.

$$
\begin{aligned}
O_{ie1} &= (x_{ie1}', y_{ie1}', z_{ie1}', 1)^T \\
&= H_1 \cdot M_{1up} \cdot M_{2down} \cdot H_2 \cdot N_{ie1} \cdot [o_x, o_y, o_z, 1]^T \\
&= H_1 \cdot T_{1-2} \cdot H_2 \cdot N_{ie1} \cdot [o_x, o_y, o_z, 1]^T
\end{aligned}
$$

$$
= \begin{bmatrix} \cos\beta_1 \cos\gamma_1 & -\cos\beta_1 \sin\gamma_1 & \sin\beta_1 & x_1 \\ \cos\beta_1 \sin\gamma_1 + \sin\alpha_1 \sin\beta_1 \cos\gamma_1 & \cos\alpha_1 \cos\gamma_1 - \sin\alpha_1 \sin\beta_1 \sin\gamma_1 & -\sin\alpha_1 \cos\beta_1 & y_1 \\ \sin\beta_1 \sin\gamma_1 - \cos\alpha_1 \sin\beta_1 \cos\gamma_1 & \cos\alpha_1 \cos\gamma_1 + \cos\alpha_1 \sin\beta_1 \sin\gamma_1 & \cos\alpha_1 \cos\beta_1 & z_1 \\ 0 & 0 & 0 & 1 \end{bmatrix} \cdot \begin{bmatrix} 1 & -\Delta\gamma_{1-2} & \Delta\beta_{1-2} & \Delta x_{1-2} \\ \Delta\gamma_{1-2} & 1 & -\Delta\alpha_{1-2} & \Delta y_{1-2} \\ -\Delta\beta_{1-2} & \Delta\alpha_{1-2} & 1 & \Delta z_{1-2} \\ 0 & 0 & 0 & 1 \end{bmatrix} \tag{7}
$$

$$
\cdots \begin{bmatrix} 1 & -\Delta\gamma_{ie1} & \Delta\beta_{ie1} & \Delta x_{ie1} \\ \Delta\gamma_{ie1} & 1 & -\Delta\alpha_{ie1} & \Delta y_{ie1} \\ -\Delta\beta_{ie1} & \Delta\alpha_{ie1} & 1 & \Delta z_{ie1} \\ 0 & 0 & 0 & 1 \end{bmatrix} \cdot \begin{bmatrix} o_x \\ o_y \\ o_z \\ 1 \end{bmatrix}
$$

$$
n_{ie1} = (nx_{ie1}', ny_{ie1}', nz_{ie1}', 0)^T = H_1 \cdot M_{1up} \cdot M_{2down} \cdot H_2 \cdot N_{ie1} \cdot [n_{0x}, n_{0y}, n_{0z}, 0]^T = H_1 \cdot T_{1-2} \cdot H_2 \cdot N_{ie1} \cdot [n_{0x}, n_{0y}, n_{0z}, 0]^T \tag{8}
$$

Here, M_{1up}, M_{2down}, and $T_{1\text{-}2}$ share the same definitions as discussed in Section 2.3. Similarly, N_{ie1} retains the meaning introduced in Section 2.4:

$$H_i = H_{kj} = \begin{bmatrix} \cos\beta_{kj}\cos\gamma_{kj} & -\cos\beta_{kj}\sin\gamma_{kj} & \sin\beta_{kj} & x \\ \cos\beta_{kj}\sin\gamma_{kj} + \sin\alpha_{kj}\sin\beta_{kj}\cos\gamma_{kj} & \cos\alpha_{kj}\cos\gamma_{kj} - \sin\alpha_{kj}\sin\beta_{kj}\sin\gamma_{kj} & -\sin\alpha_{kj}\cos\beta_{kj} & y \\ \sin\beta_{kj}\sin\gamma_{kj} - \cos\alpha_{kj}\sin\beta_{kj}\cos\gamma_{kj} & \cos\alpha_{kj}\cos\gamma_{kj} + \cos\alpha_{kj}\sin\beta_{kj}\sin\gamma_{kj} & \cos\alpha_{kj}\cos\beta_{kj} & z \\ 0 & 0 & 0 & 1 \end{bmatrix} \quad (9)$$

where $H_i = H_{kj}$ denotes the matrix representation of positional and orientational variations between multiple ideal datum planes on the same component within an SDC (e.g., the pose relationship between ideal datum planes O_{2down} and O_{2up} on the upper casing in Figure 15). It specifically describes the coordinate system pose relationship between the k-th and j-th critical surfaces on the i-th component. In the matrix, x, y, and z represent the displacements of coordinate system O_{ij} relative to O_{ik} along the x-, y-, and z-axes, respectively; α_{kj}, β_{kj}, and γ_{kj} denote the rotation angles of O_{ij} relative to O_{ik} about the x-, y-, and z-axes. In this case, $i = 1, 2 \ldots$, $k = 1$ (constant), and $j = 2$ (constant).

From Equations (9) and (10), the position and orientation vector expression $P_{r1}' = \{x_{ie1}', y_{ie1}', z_{ie1}', nx_{ie1}', ny_{ie1}', nz_{ie1}'\}$ is derived for datum plane O_{2up} at one end of the closed loop in the global coordinate system. Similarly, the position and orientation vector expression $P_{r2}' = \{x_{ie2}', y_{ie2}', z_{ie2}', nx_{ie2}', ny_{ie2}', nz_{ie2}'\}$ is obtained for the other datum plane O_{1down} in the closed loop. In this example, to simplify the model description and avoid redundant terms, O_{1down} is treated as an ideal plane; hence, $P_{r2}' = \{0, 0, 0, 0, 0, 1\}$. Consequently, the closed-loop characterization is given by Equation (12).

$$C_{sl}(close\ loop) = \begin{cases} P_{r1}' = \left\{x_{ie1}', y_{ie1}', z_{ie1}', nx_{ie1}', ny_{ie1}', nz_{ie1}'\right\}, x_1 \in [x_{1c} - r, x_{1c} + r], y_1 \in [y_{1c} - \sqrt{(r^2 - (x_1 - x_{1c})^2)}] \\ P_{r2}' = \left\{x_{ie2}', y_{ie2}', z_{ie2}', nx_{ie2}', ny_{ie2}', nz_{ie2}'\right\} = \{0\}, x_2 \in [x_{2c} - r, x_{2c} + r], y_2 \in [y_{2c} - \sqrt{(r^2 - (x_2 - x_{2c})^2)}] \end{cases} \quad (10)$$

Here, (x_{1c}, y_{1c}) and (x_{2c}, y_{2c}) represent the coordinates of the center points of the closed-loop end faces along the x- and y-axes, respectively, and r denotes the radius of the circle.

Based on Equation (10), the spatial six-dimensional distribution (position and orientation vectors) of a pair of datum planes for the end-face closed loop in the global coordinate system can be obtained. This further enables the derivation of parameters related to the end-face closed loop, such as the maximum and minimum spatial distances, parallelism, or relative normal vector relationship between the evaluated datum plane P_{r1}' and the other datum plane.

(2) SDC propagation law with the axis as the closed-loop datum.

Based on the aforementioned computational process, the coordinate systems of critical surfaces are established (Figure 16). The primary error contributors in the SDC are:

- The mating-constrained datum plane error M_{1up} of the lower casing's top surface (including two constraint surfaces—the flange end face and the flange spigot cylindrical surface);
- The mating-constrained datum plane error M_{2down} of the upper casing's bottom surface (similarly comprising the flange end face and spigot cylindrical surface).

This model primarily analyzes spatially distributed form errors, while radial errors (e.g., cylindricity of the casing's inner surface) and spatial relative positions (e.g., axial misalignment between upper and lower casings) remain uncharacterized.

For the closed loop, one datum axis L_{down} is defined by the equation $L_{down} = \{lx_{ie}', ly_{ie}', lz_{ie}', nlx_{ie}', nly_{ie}', nlz_{ie}'\}$, where L_{down} (the assembly datum axis) is treated as a known value.

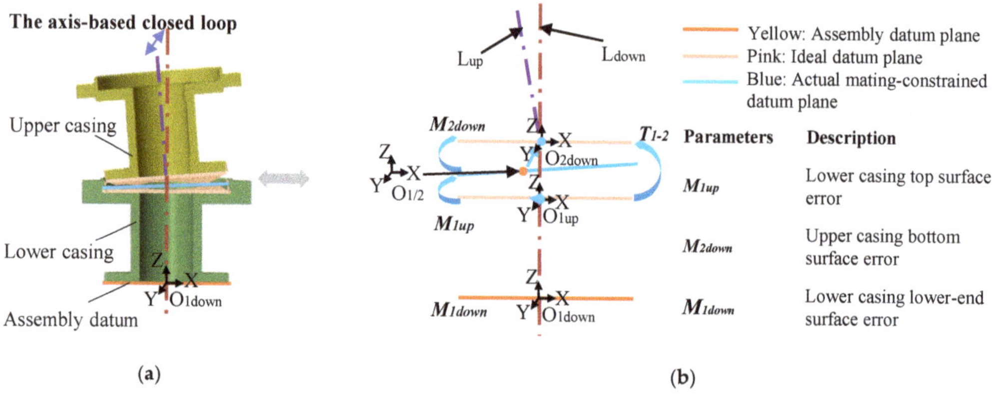

Figure 16. Axis-based closed-loop SDC in simulated casing assembly: (**a**) schematic of simulated casing assembly; (**b**) datum coordinate system construction and error contributors.

The initial pose of the other datum axis L_{up} is specified as:

- Position vector of the coordinate origin: $[o_{lx}, o_{ly}, o_{lz}, 1]^T$;
- Unit normal vector of the datum axis: $[nl_{ox}, nl_{ox}, nl_{ox}, 0]^T$ (e.g., $[0, 0, 1, 0]^T$ for Z-axis alignment).

After error propagation through the SDC, the pose of the closed-loop datum in the global coordinate system is described using a six-dimensional parameter L_{up} (position vector + orientation vector), as defined in Equations (11) and (12).

$$L_{ie1} = (lx_{ie1}{}', ly_{ie1}{}', lz_{ie1}{}', 1)^T = H_1 \cdot M_{1up} \cdot M_{2down} \cdot \left[o_{lx}, o_{ly}, o_{lz}, 1\right]^T = H_1 \cdot T_{1-2} \cdot \left[o_{lx}, o_{ly}, o_{lz}, 1\right]^T \quad (11)$$

$$nL_{ie1} = (nlx_{ie1}{}', nly_{ie1}{}', nlz_{ie1}{}', 0)^T = H_1 \cdot M_{1up} \cdot M_{2down} \cdot \left[nl_{0x}, nl_{0y}, nl_{0z}, 0\right]^T = H_1 \cdot T_{1-2} \cdot \left[nl_{0x}, nl_{0y}, nl_{0z}, 0\right]^T \quad (12)$$

Here, M_{1up}, M_{2down}, H_1, and T_{1-2} share the same definitions as in Equations (7) and (8).

Theorem-type environments (including propositions, lemmas, corollaries, etc.) can be formatted as follows. Consequently, the closed-loop expression (Equation (13)) is derived from Equations (11) and (12).

$$C_{sl}(close\ loop) = \begin{cases} L_{up} = \{lx_{ie1}{}', ly_{ie1}{}', lz_{ie1}{}', nlx_{ie1}{}', nly_{ie1}{}', nlz_{ie1}{}'\}, x_1 \in [x_{1\min}, x_{1\max}], y_1 \in [y_{1\min}, y_{1\max}] \\ L_{down} = \{lx_{ie}{}', ly_{ie}{}', lz_{ie}{}', nlx_{ie}{}', nly_{ie}{}', nlz_{ie}{}'\} = \{0,0,0,0,0,1,0\}, x_2 \in [x_{2\min}, x_{2\max}], y_2 \in [y_{2\min}, y_{2\max}] \end{cases} \quad (13)$$

Here, (x_1, y_1) and (x_2, y_2) represent the interval range of the datum axis for the closed loop. Based on Equation (13), the spatial six-dimensional distribution (position and orientation vectors) of the datum axis pair for the closed loop in the global coordinate system can be obtained. This further enables the derivation of coaxiality-related parameters, such as the spatial pose normal vector of the evaluated axis L_{up} and the coaxiality error itself.

3. Experimental Validation of the SDC Model for Casing Assembly

3.1. Experimental Design

The SDC model in this study is an error modeling method based on existing manufacturing processes, using the error distribution patterns induced by current machining techniques as design references. Machined casings from previous productions were inspected, and the corresponding systematic errors were extracted and separated using a

NURBS-based surface regression method to serve as the primary errors in the SDC model. The validity of the SDC model was verified through casing assembly experiments.

To enable multiple tests with a limited number of parts, several bolt holes were arranged on the flange end face, allowing the flange to be rotated at specific angles to obtain multiple sets of assembly results. The casing part samples are shown in Figure 17a, and the assembled casing structure is illustrated in Figure 17b. The upper surface of the base was ground to serve as an ideal assembly reference plane. To more clearly validate the multidimensional characteristics of the closed loop, the design tolerance band for the components was intentionally widened.

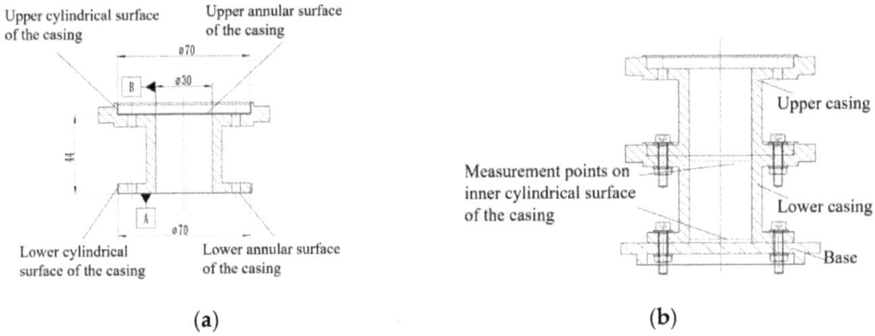

Figure 17. (**a**) Casing component diagram and (**b**) casing assembly diagram.

Using datum A, the parallelism of the upper annular surface of the lower casing was 0.121 mm, while that of the upper casing was 0.07 mm. The perpendicularity of the upper annular surface of the lower casing, relative to datum B, was 0.08 mm. Two upper casings (casing 1 and casing 2) and one lower casing were machined. During assembly, the cylindrical mating surfaces at the flange interface were designed as transition fits.

3.2. Experimental Procedure

3.2.1. Component Measurement

This study employed a coordinate measuring machine (CMM) to collect data points from the critical surfaces of the parts. The CMM model used was a Hexagon SCIROCCO 09.10.07, with a measurement uncertainty of (2.0 + 3L/1000) μm, and the software system was PC-DMIS 2020. The measured critical surfaces included the lower annular surface, lower cylindrical surface, upper annular surface, upper cylindrical surface, and inner cylindrical surface of the casing, as shown in Figure 17a. A total of 240 points were collected from five circular paths on both the upper and lower annular surfaces of the casing. For the cylindrical surfaces at the upper end of the lower casing and the lower end of the upper casing, 96 points were collected from two circular paths each (the number of points depended on the sampling interval, which was set to 3–5 mm in this study). The inner cylindrical surface of the casing was measured along two circular paths, totaling 16 points (the line connecting the centers of these circles was used as the part axis). All measurements were performed after CMM calibration, with the errors primarily being dependent on the equipment's inherent accuracy. The orientation of the measurement coordinate system is illustrated in Figure 15. Taking the lower annular surface as an example, the coordinate system origin is located at the center of the lower annular surface, with the positive X-axis pointing to the right and the positive Z-axis pointing upward. The measurement of an individual part is shown in Figure 18a. To ensure experimental repeatability, two sets of experiments were conducted using casing 1 and casing 2 for the upper casing.

(**a**) (**b**)

Figure 18. (**a**) Component measurement and (**b**) assembly measurement.

3.2.2. Assembly Measurement

(1) One objective of assembly measurement is to obtain the measured pose of the upper end face of the upper casing in the end face closed-loop model, as shown in Figure 18b. This allows the characterization of key parameters such as the envelope surface (one of the closed-loop datum planes), the distance from the envelope surface to the assembly datum (another closed-loop datum plane), the perpendicularity of the envelope surface relative to the lower casing axis, and its parallelism relative to the assembly datum plane—thereby validating the multidimensional characteristics and accuracy of the closed loop. (2) To further validate the model's effectiveness, the second objective is to acquire spatial closed-loop parameters formed by the axes of the upper and lower casings in the axis-based closed-loop model, thereby deriving optimization-relevant metrics such as the coaxiality error.

As illustrated in Figure 15, an assembly coordinate system was established on the base, coinciding with the coordinate system of the lower casing's bottom annular face. The lower casing served as the assembly datum, to which casing 1 was mounted and adjusted to a stabilized configuration before being secured with M3 bolts. Post-assembly, measurements were taken at the upper casing's top annular face (240 points across 5 concentric circles) and the inner cylindrical surfaces of both upper and lower casings (16 points per cylinder, sampled in 2 separate circles), as shown in Figure 17b.

To ensure experimental repeatability and validate the impact of assembly phase differences between the upper and lower casings on the SDC model predictions—particularly when the same set of components exhibits identical spatially distributed form errors and geometric errors—the following procedure was implemented. Casing 1 was rotated clockwise around the z-axis by random angles to vary the bolt hole engagement sequence during assembly. Five experimental trials were conducted under these conditions. Subsequently, casing 1 was replaced with casing 2 while repeating the same assembly and measurement steps. This process yielded 10 comparative experimental sets (5 sets for the casing 1 assembly test group and 5 sets for the casing 2 assembly test group).

3.3. Results and Analysis

First, in the local coordinate system, the mating-constrained datum error was calculated using the method described in Section 2.3, while the non-mating-constrained datum error was determined using the approach from Section 2.4. Next, the surface data of the upper casing's top face were measured in the assembly coordinate system, and an envelope surface was characterized based on the method in Section 2.4. Then, in the global coordinate system, the characterized parameters were substituted into Equations (7)–(13) to predict the pose of the envelope surface (Equation (14)). Finally, comparative experiments demonstrated the effectiveness of the SDC modeling method, thereby achieving the goal of guiding the tolerance design optimization.

3.3.1. Prediction Results of the Datum Plane for the End Face SDC Closed Loop

To demonstrate the multidimensional advantages of the SDC model, this subsection analyzes the prediction results using the following parameters. The predicted and actual envelope surfaces are shown in Figure 19, the maximum and minimum distances of the closed loop are presented in Figure 20, the parallelism in Figure 21, and the perpendicularity in Figure 22.

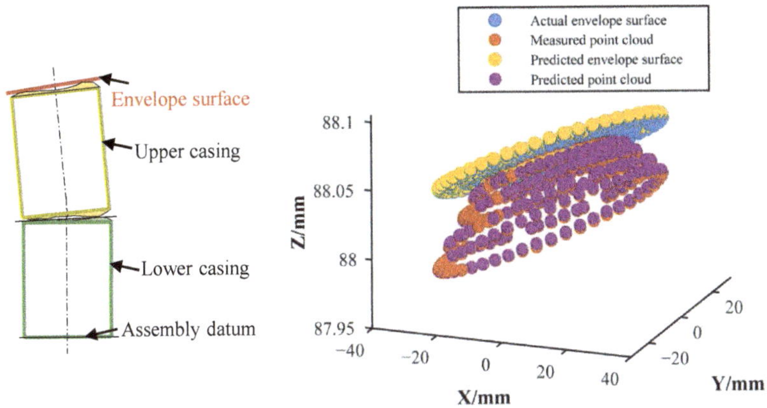

(a) (b)

Figure 19. (**a**) Schematic diagram of parameter physical locations and (**b**) comparison of measured and predicted data (point clouds and envelope surfaces).

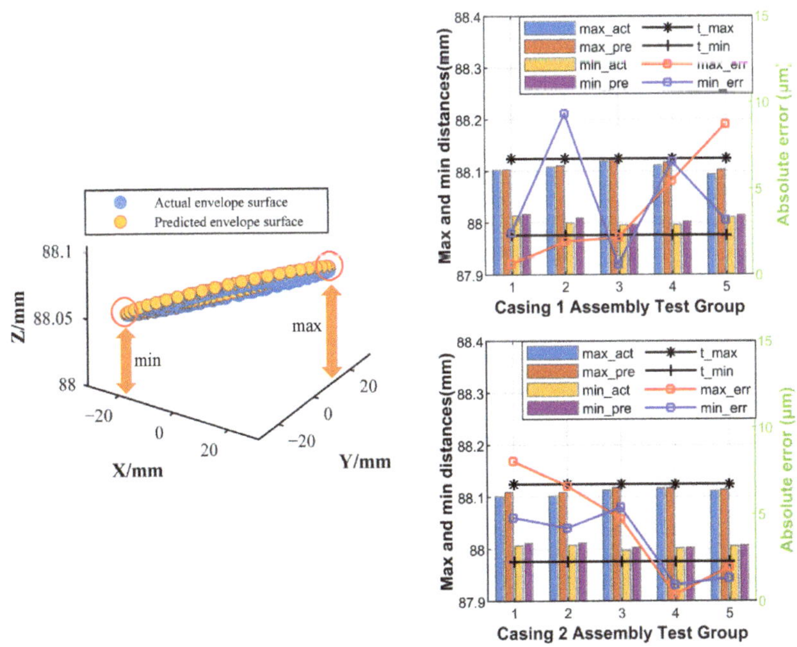

(a) (b)

Figure 20. (**a**) Definition schematic and (**b**) comparison of actual vs. predicted maximum and minimum distances of the closed loop.

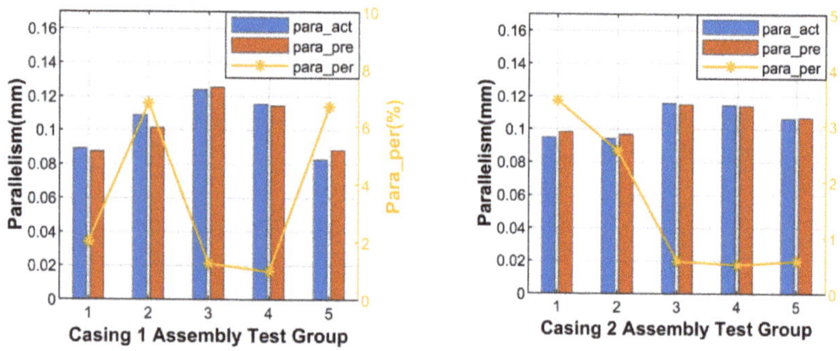

Figure 21. Comparison of actual vs. predicted parallelism of the envelope surface.

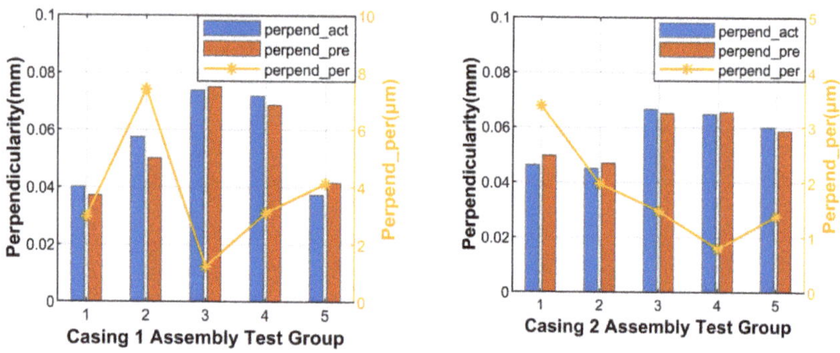

Figure 22. Comparison of actual vs. predicted perpendicularity of the envelope surface.

1. Characterization results of the closed-loop datum plane

Figure 19 shows the state of the upper end face of casing 1 when rotated 0° clockwise around the z-axis in the assembly coordinate system (dataset 1 from the casing 1 assembly test group). The red point cloud represents the measured point cloud data, the blue point cloud represents the actual envelope surface obtained after characterization of the measured point cloud, the purple point cloud represents the predicted point cloud data obtained from the SDC model, and the yellow point cloud represents the predicted envelope surface after characterization of the predicted point cloud.

Characterizing parameters of the envelope surface: $P_{r1}{}' = \{-0.0818, -0.0143, 88.0601, -0.0012, 0.0008, 1\}$

The characteristic equation is expressed as:

$$\begin{cases} (-0.0012) \times (x - (-0.0818)) + 0.0008 \times (y - (-0.0143)) + (z - 88.0601) = 0 \\ x \in [-0.0818 - 31, -0.0818 + 31] \\ y \in \left[-0.0143 - \sqrt{31^2 - (x + 0.0818)^2}, -0.0143 + \sqrt{31^2 - (x + 0.0818)^2}\right] \end{cases} \tag{14}$$

2. Maximum and minimum distances of the closed loop

$$\begin{cases} max\_pre = max|Qp_i|; max\_act = max|Qa_j|; \\ max\_err = \left|max|Qp_i| - max|Qa_j|\right|; \\ min\_pre = min|Qp_i|; min\_act = min|Qa_j|; \\ min\_err = \left|min|Qp_i| - min|Qa_j|\right|, \end{cases} (i, j = 1, 2, \ldots, n) \tag{15}$$

where Qp_i—the z-coordinates of the predicted envelope point cloud; Qa_j—the z-coordinates of the actual envelope point cloud; *max_pre*—maximum value of Qp_i; *max_act*—maximum value of Qa_j; *max_err*—absolute error between *max_pre* and *max_act*; *min_pre*—minimum value of Qp_i; *min_act*—minimum value of Qa_j; *min_err*—absolute error between *min_pre* and *min_act*; *n*—number of verification points.

A comparison of the actual results of the envelope surface, the predicted results of the envelope surface, and the predicted values from the traditional worst-case dimension chain method (closed-loop dimensional errors range between t_min = 87.975 mm and t_max = 88.124 mm) is shown in Figure 20. The results obtained by the traditional worst-case dimension chain method remain unchanged regardless of the phase rotation of the upper casing. This is because the worst-case method yields a fixed interval range (a scalar value) that does not account for the distribution of form errors. In contrast, the spatial gap of the closed loop varies with the assembly phase shift between the upper and lower casings, demonstrating the advantage of the SDC model in providing multidimensional closed-loop results. The absolute errors between the predicted and actual values are all below 9.3 μm, further confirming the accuracy of the predictions. These findings provide a theoretical basis for optimizing tolerance design.

3. Parallelism of the envelope surface

$$para\_per = \left| \frac{para\_act - para\_pre}{para\_act} \right| \times 100\% \tag{16}$$

where *para_act* denotes the parallelism of the actual envelope surface, *par_pre* denotes the parallelism of the predicted envelope surface, and *para_per* denotes the relative error between *para_act* and *par_pre*.

As shown in Figure 21, as the upper casing rotates relative to the lower casing, the parallelism of the closed loop varies, further demonstrating the advantage of the SDC model in dimensional characterization. Moreover, the relative error between the actual and predicted parallelism values remains below 6.8%, validating that the model's predictions closely match the real results.

4. Perpendicularity of the envelope surface

$$\text{perpend\_per} = |\text{perpend\_act} - \text{perpend\_pre}| \tag{17}$$

where *perpend_act* denotes the perpendicularity of the actual envelope surface, *perpend_pre* denotes the perpendicularity of the predicted envelope surface, and *perpend_per* denotes the absolute error between *perpend_act* and *perpend_pre*.

As shown in Figure 22, the perpendicularity of the envelope surface relative to the datum axis changes with the assembly phase variation between the upper and lower casings, further highlighting the SDC model's advantage in dimensional characterization. Additionally, the relative error between the actual and predicted perpendicularity values remains below 7.4%, confirming the model's prediction accuracy.

3.3.2. Effect of SMSC and MMSC on End Face Closed-Loop Accuracy

To validate the accuracy improvement of the spatial dimension chain-based multiple mating-surface constraint (SDC-MMSC) method for assembled structures with multiinterface constraints (Figure 15, showing the flange fitting between the lower casing's top surface and the upper casing's bottom surface), this subsection compares three parameters among the SDC-MMSC method, the spatial dimension chain-based single mating surface constraint (SDC-SMSC) method (simplifying multiple constraints to a single mating surface constraint), and experimental measurements:

1. The normal vector of the envelope surface (Figure 23);
2. The angular error θ between the predicted and actual normal vectors (Figure 24);
3. The root mean square error (*RMSE*) between the predicted and actual envelope surfaces (Figure 25).

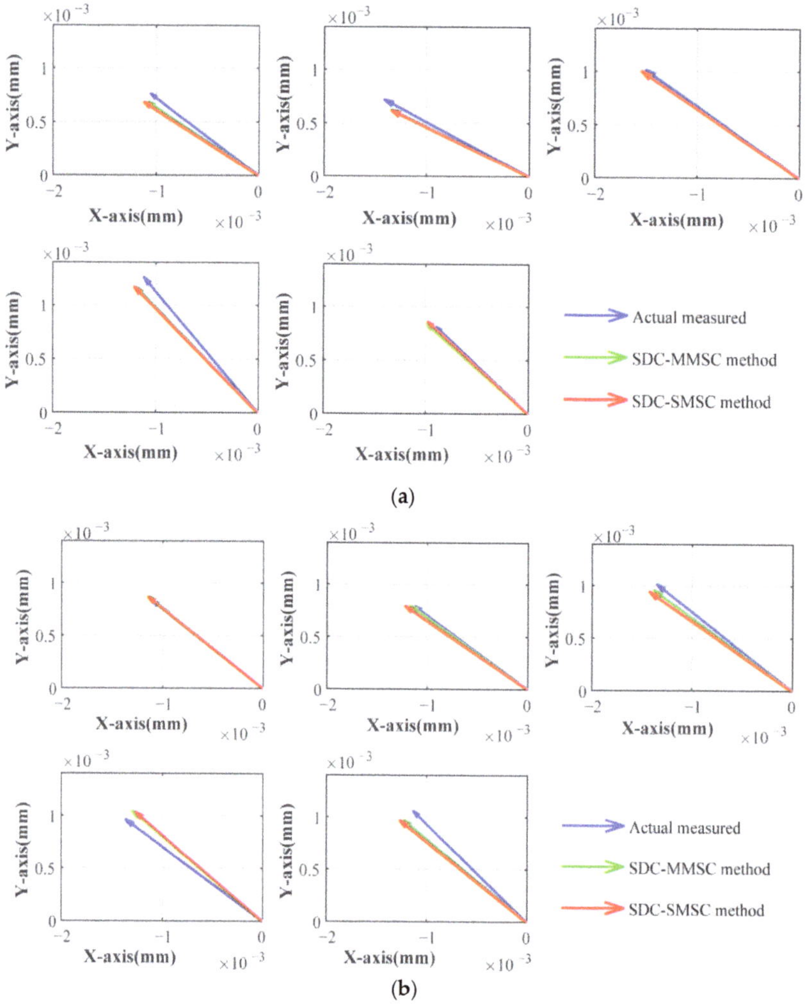

Figure 23. Comparison between predicted normal vectors (SDC-MMSC/SDC-SMSC methods) and actual measurements on envelope surfaces: (**a**) casing 1 test group; (**b**) casing 2 test group.

Figure 23 shows that the influence of the two pairs of mating surfaces on dimensional chain accuracy cannot be neglected. From Figure 24b, the angular error θ obtained via the SDC-MMSC method remains below 0.008°. Among the 10 datasets, the SDC-MMSC method exhibits smaller relative errors than the SDC-SMSC method in 7 cases, while the two methods achieve comparable accuracy in the remaining 3 cases. Figure 25 indicates that the RMSE of the SDC-MMSC method remains below 0.006 mm, demonstrating its superior accuracy over the SDC-SMSC method. The experimental results confirm the method's effectiveness and demonstrate that multiconstraint consideration significantly enhances the accuracy over existing techniques.

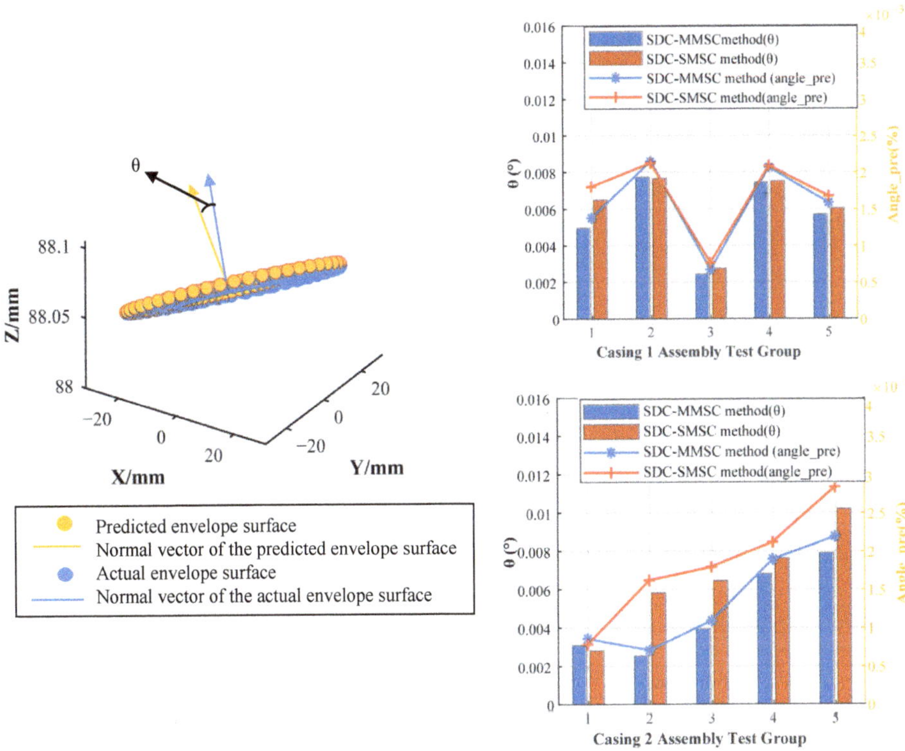

(a) (b)

Figure 24. Angular error (θ) between predicted and actual normal vectors: (**a**) definition schematic; (**b**) comparison between SDC-MMSC and SDC-SMSC methods.

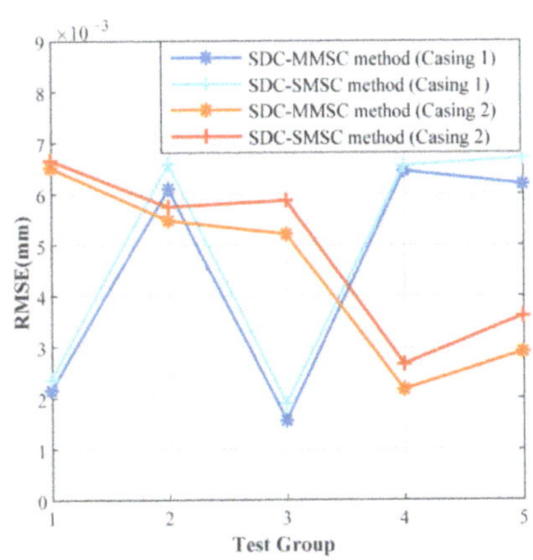

Figure 25. RMSE between predicted and actual envelope surfaces.

$$\begin{cases} \text{Normal vector extraction from envelope surfaces :min} \sum_{i=1}^{K} \frac{(ax_i + by_i + cz_i - d)^2}{2} \\ \theta = \arccos(\frac{\vec{m} \cdot \vec{n}}{\left\|\vec{m}\right\| \cdot \left\|\vec{n}\right\|}) \\ angle\_per = \left|\frac{\theta}{360}\right| * 100\% \end{cases} \tag{18}$$

where $\vec{m} = (a, b, c)$ is the normal vector of the plane, d is the constant term in the plane equation composed of K points, (x_i, y_i, z_i) represents the point cloud coordinates of the envelope surface, K is the number of point clouds, θ is the angle between two vectors, $\vec{m}, \vec{n}$ denote two vectors, $||\cdot||$ represents the magnitude (norm) of a vector, and *angle_pre* is the relative error (normalized to 360°) between the predicted surface normal vector and the actual surface normal vector.

$$RMSE = \sqrt{\sum_{s=1}^{k} \frac{(P^E - P')^2}{k}} = \sqrt{\sum_{s=1}^{k} \frac{(x_s - x_s')^2 + (y_s - y_s')^2 + (z_s - z_s')^2}{k}} \tag{19}$$

where k denotes the number of measurement points, $P^E = (x_s, y_s, z_s)$ represents the actual envelope surface, and $P' = (x_s', y_s', z_s')$ denotes the predicted envelope surface.

3.3.3. Coaxiality Prediction in Axis-Based Closed Loop

In practical engineering, secondary factors are often empirically neglected to simplify calculations. This study focuses on geometric errors of casing mating surfaces (e.g., flatness, parallelism) that significantly affect the coaxiality. By substituting the characteristic parameters into Equations (11) and (12), Equation (13) is derived, representing the spatial six-dimensional (position and orientation vectors) distribution of a pair of datum axes in the *axis-based* closed loop. This subsection validates the accuracy of the SDC model through the coaxiality metric (Equation (20)), comparing the dimension chain worst-case method, the proposed model's predictions, and experimental results, as shown in Figure 26.

$$coax\_per = \left|\frac{coax\_act - coax\_pre}{coax\_act}\right| \times 100\% \tag{20}$$

where *coax_act* denotes the actual coaxiality, *coax_pre* represents the predicted coaxiality, and *coax_per* is the relative error between the predicted and actual coaxiality.

A comparative analysis was conducted between the predicted values from the *axis-based closed-loop* model and actual measured values, along with results from the worst-case assembly dimension chain method. As illustrated, due to differences in the assembly phase between the upper and lower casings, the coaxiality also varies. This further demonstrates that spatially distributed form errors cannot be neglected in the SDC model. Moreover, the relative error between the predicted and actual coaxiality is less than 8.3% in all cases and consistently smaller than the results obtained using the traditional worst-case dimension chain method (coaxiality errors of 0.185 mm for casing 1 test group and 0.165 mm for casing 2 test group). These results validate that the predictions of the proposed model closely align with the actual measurements.

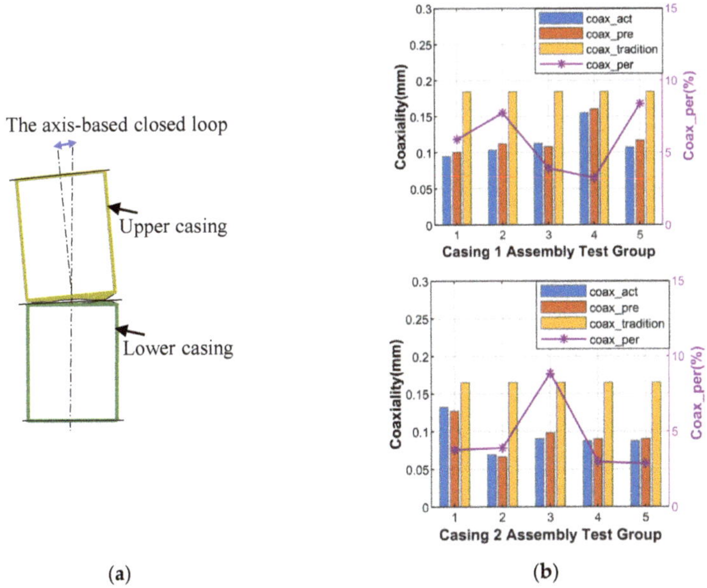

Figure 26. Coaxiality comparison in axis-based closed loop: (**a**) schematic of axis-based closed loop; (**b**) predicted vs. measured coaxiality.

4. Application of SDC Model for Tolerance Allocation Optimization

The effectiveness of the SDC model has been experimentally validated in the previous section using an engine casing case study. The ultimate goal of constructing this model is to enable more precise tolerance allocation in design. Taking the engine casing from Section 3 as an example, the application steps of the SDC model are as follows:

1. Define the assembly's functional requirements: The parallelism of the envelope surface derived from the upper casing's top face must be $\leq$0.03 mm.
2. Identify contributing loops: The specific parameters are illustrated in Figure 15.
3. Assign tolerances: Based on design experience and process capability, tolerances are allocated to each contributing loop (see case 1).
4. Establish the SDC model: Based on the tolerance range defined in step 3, previously machined casings within this tolerance are inspected. Systematic errors are extracted and separated using NURBS-based surface regression [35], serving as the error component in the SDC model. The actual mating-constrained datum surface and non-mating-constrained datum surface are characterized, and their respective errors (M and N) are solved. The SDC model is then established, followed by Monte Carlo simulation.
   A critical point is that form tolerances must comply with orientation tolerances; for instance, in design and simulation, the generated form errors (e.g., flatness) must be smaller than the corresponding orientation errors (e.g., parallelism).
   To elucidate the connection between steps 3 and 4, we provide illustrative examples that demonstrate their interrelationship in the geometric error generation process.
   (1) Form Error (e.g., Flatness)
   Evaluation method: Two parallel planes minimally enclose the actual surface (minimum zone method).
   (2) Orientation Error (e.g., Parallelism)
   Evaluation method: First, an ideal reference plane is fitted from the measured

datum surface using a mathematical method such as least squares regression. Second, the maximum vertical deviation of all measured surface points from this reference plane is calculated, which defines the parallelism error value.

(3) Generating Form Errors (e.g., Flatness)

(a) Assign a flatness error magnitude V.
(b) Incorporate the error distribution pattern from existing manufacturing processes.
(c) Scale the Z-coordinates of measured point clouds:

$$Z_{new} = Z_{ori} \times \frac{V}{V_{ori}} \tag{21}$$

where Z_{ori} denotes the original coordinate; V_{ori} denotes the measured flatness error.

(4) Generating Orientation Errors (e.g., Parallelism)

(a) Assign a parallelism error magnitude.
(b) Rotate the form-corrected point cloud (about the x-axis, y-axis, or both axes) to control the quadrant of maximum Z-deviation.

(5) Monte Carlo Simulation Workflow

(a) Set tolerances for all relevant part surfaces.
(b) Randomly sample error magnitudes within tolerance limits and generate corresponding error point clouds.
(c) Map geometric features to functional dimensions using the 6-DOF vector function.
(d) Calculate the functional parameters (e.g., assembly clearance)
(e) Determine the batch assembly yield.

5. Validate the tolerance design: For example, the rationality can be evaluated by assessing whether the compliance rate of positional errors (e.g., parallelism or perpendicularity) in the datum plane of the closed loop meets the design requirements. As shown in Figure 27, the parallelism of the envelope surface falls within [0, 0.025] mm with a 93% compliance rate, confirming compliance with the $\leq$0.03 mm requirement.
6. Optimize the tolerances: Adjust the tolerances with lower process costs to improve the product accuracy and reduce manufacturing expenses. For instance, if the dimensional or parallelism tolerances of the upper or lower casings are overly stringent (leading to high machining costs), they can be relaxed per case 2. Repeating step 4 yields the results in Figure 28, where the parallelism within [0, 0.029] mm achieves a 78% compliance rate. If acceptable, this relaxation maintains precision while lowering costs.

The advantages of using the SDC model for tolerance allocation include:

1. Cost reductions: Loosening the tolerances where possible without compromising assembly requirements reduces the machining costs.
2. Design optimization: Refining the tolerance allocation process via the distributed-form-error-based SDC model enhances the part design accuracy.

The simulation parameters used in the above steps were as follows.

The error magnitudes were sampled based on a normal distribution within the tolerance range, and the spatially distributed form errors were characterized based on experimental measurement data from existing manufacturing processes. The perpendicularity datum axis was constructed based on measured data from the inner cylinder of the casing, and the basic dimensional parameters of the contributing loops are shown in Figure 17.

After 1000 simulation runs, the pose errors [dz, σx, σy] of the envelope surface at different tolerance grades were obtained, along with the parallelism and perpendicularity.

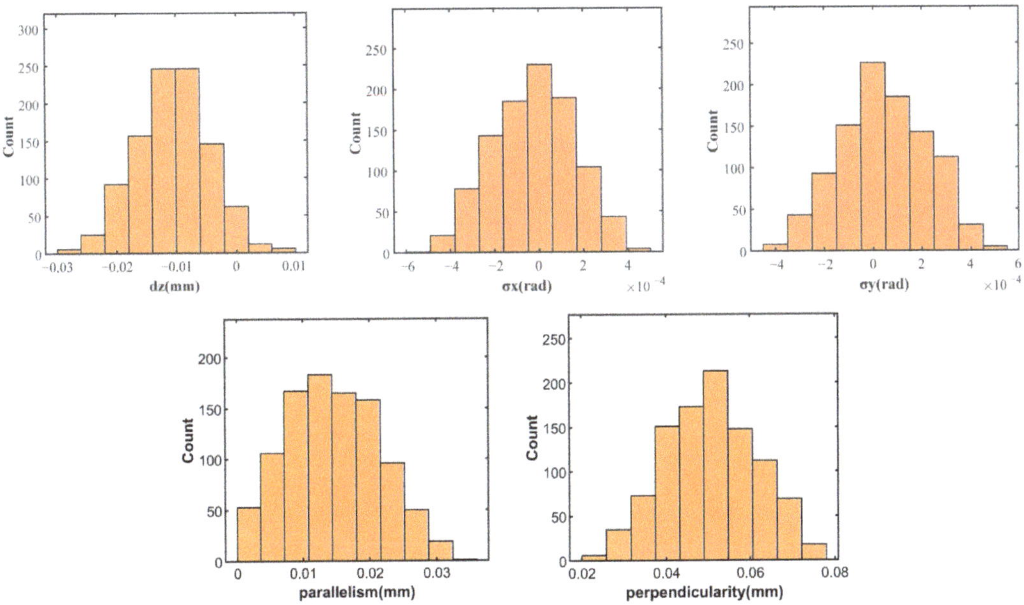

Figure 27. Distributions of the rotational errors (X/Y), translational error (Z), parallelism, and perpendicularity for the envelope surface in case 1.

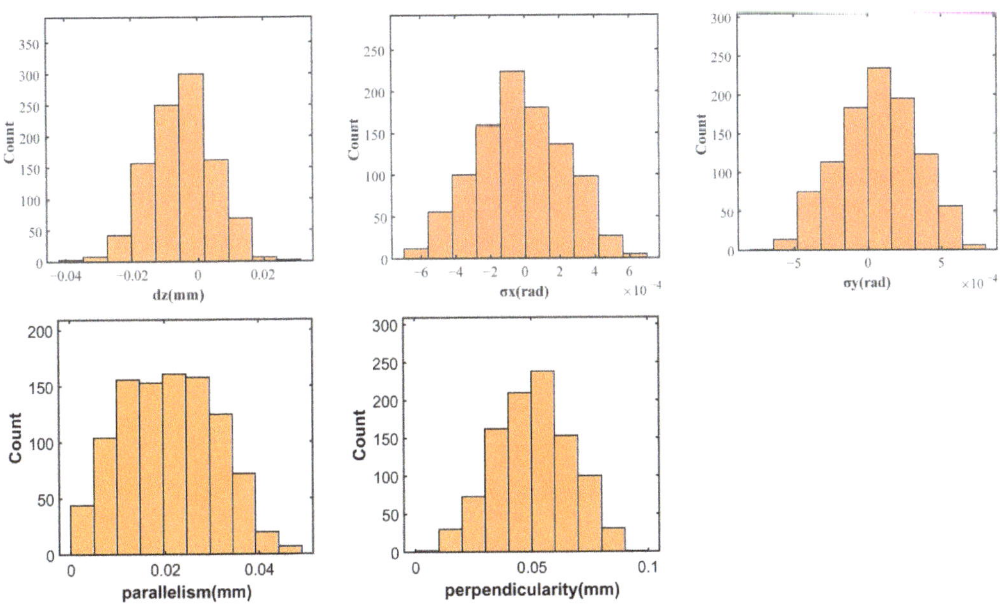

Figure 28. Distributions of the rotational errors (X/Y), translational Error (Z), parallelism, and perpendicularity for the envelope surface in case 2.

Case 1: As previously discussed, the geometric errors in the upper end face of the lower casing and both end faces of the upper casing are critical factors influencing the closed loop. The specific parameters are as follows: dimensional tolerance [−0.011, 0.014] mm, positional tolerance (parallelism) [0, 0.02] mm, and form tolerance (flatness) [0, 0.01] mm. Regarding the simulation results, the displacement in the z-direction is primarily distributed within the range of [−0.014, −0.006] mm, with 492 data points. The rotational error about the x-axis is mainly distributed within [−0.00016, 0.00017] rad, with 604 data points, while the rotational error about the y-axis is mainly distributed within [0.00005, 0.00015] rad, with 411 data points. The parallelism of the envelope surface achieves a compliance rate of 93% within the [0, 0.025] mm range, whereas the perpendicularity compliance rate is 79% within the [0, 0.06] mm range. The result distribution is illustrated in Figure 27.

Case 2: Similarly, the tolerance parameters for the three surfaces, the upper end face of the lower casing and the upper and lower end faces of the upper casing, are specified as follows: dimensional tolerance of [−0.015, 0.024] mm, positional tolerance (parallelism) of [0, 0.03] mm, and form tolerance (flatness) of [0, 0.015] mm. The displacement z is predominantly distributed within [−0.0128, −0.0055] mm, with 550 data points. The rotational error about the x-axis is mainly distributed in [−0.00028, 0.00014] rad, with 565 data points, while the rotational error about the y-axis is primarily within [−0.00016, 0.00032] rad, with 612 data points. The parallelism of the envelope surface yields a 62% compliance rate within [0, 0.0245] mm, 78% within [0, 0.029] mm, and 90% within [0, 0.034] mm. The perpendicularity achieves a 71.5% compliance rate within [0,0.06] mm. The result distribution is illustrated in Figure 28.

5. Conclusions and Future Work

This study established an advanced framework for precision tolerance design, with the core innovation being the development of a six-dimensional SDC model. The effectiveness of the SDC model was validated through an assembly case study using simulated aero-engine casings. The main conclusions are as follows:

(1) A six-dimensional SDC model was developed, which integrates spatially distributed form errors into datum definitions using the transfer matrix theory. By employing six-dimensional vector functions (combining position and orientation parameters), the model achieves high-fidelity mapping between geometric features and functional dimensions, enabling accurate closed-loop predictions for large-scale assemblies.

(2) The six-dimensional SDC model systematically addresses complex constraint scenarios through a hierarchical approach. For the SMSC, the model characterizes spatially distributed form errors by deriving 6-DoF geometric error components through optimized algorithms, thereby enhancing the prediction accuracy. The solution extends to MMSC interactions by characterizing 3D coupled errors through 6-DoF components, effectively reducing the computational complexity while improving the prediction accuracy. Additionally, a novel non-mating-constrained datum plane error characterization method based on modified envelope conditions ensures assembly interchangeability.

(3) Assembly experiments using simulated aero-engine casings demonstrated that all measured values satisfied worst-case dimension chain tolerance limits. Furthermore, the closed-loop spatial distances varied with the assembly phase difference between the upper and lower casings, conclusively validating the necessity of considering spatially distributed form errors in SDC modeling and their multidimensional characteristics. A comparative analysis between the SDC-SMSC and SDC-MMSC methods further confirmed the superiority of the multiconstraint approach.

The model is applicable to planar mating assemblies where surfaces cannot be simplified as points or beams for analyses, enabling error prediction and tolerance allocation during the design phase. Our future work will extend the methodology to curved surface mating assemblies, enhancing its industrial applicability. This research provides both theoretical foundations and practical solutions for the precision assembly and cost-effective manufacturing of complex mechanical systems.

Author Contributions: All authors contributed to the study's conception and design. L.L. designed and conducted the experiments, analyzed the results, and wrote the article. X.J. proposed the basic idea of the article and contributed the materials and measuring instruments. C.L. modified the structure of the article. H.G. helped perform the analysis with constructive discussions. All authors have read and agreed to the published version of the manuscript.

Funding: The work was supported by the National Natural Science Foundation of China (grant number U22B2088).

Institutional Review Board Statement: Not applicable.

Informed Consent Statement: Not applicable.

Data Availability Statement: The original contributions presented in this study are included in the article. Further inquiries can be directed to the corresponding author.

Conflicts of Interest: The authors declare no conflict of interest.

Abbreviations

The following abbreviations are used in this manuscript:

SDC	Spatial dimension chain
6-DoF	Six-degree-of-freedom
NURBS	Non-uniform rational B-spline
SMSC	Single mating surface constraint
MMSC	Multiple mating surface constraint
RMSE	Root mean square error
SDC-MMSC	Spatial dimension chain-based multiple mating surface constraint
SDC-SMSC	Spatial dimension chain-based single mating surface constraint

References

1. Dantan, J.; Homri, L. Overview of mathematical concepts for tolerance analysis. *Procedia CIRP* **2024**, *129*, 73–78. [CrossRef]
2. Jiang, Q.J.Q.; Feng, H.F.H.; Wang, Y.W.Y.; Fang, C.F.C. Batch circular form error characterization and evaluation. *Precis. Eng.* **2017**, *47*, 223–230. [CrossRef]
3. Dantan, J.Y.; Qureshi, A.J. Worst-case and statistical tolerance analysis based on quantified constraint satisfaction problems and Monte Carlo simulation. *Comput. Aided Des.* **2009**, *41*, 1–12. [CrossRef]
4. Mansuy, M.; Giordano, M.; Hernandez, P. A new calculation method for the worst case tolerance analysis and synthesis in stack-type assemblies. *Comput. Aided Des.* **2011**, *43*, 1118–1125. [CrossRef]
5. Mansuy, M.; Giordano, M.; Hernandez, P. A Generic Method for the Worst Case and Statistical Tridimensional Tolerancing Analysis. *Proc. CIRP* **2013**, *10*, 276–282. [CrossRef]
6. Sahani, A.K.; Jain, P.K.; Sharma, S.C.; Bajpai, J.K. Design Verification through Tolerance Stack up Analysis of Mechanical Assembly and Least Cost Tolerance Allocation. *Procedia Mater. Sci.* **2014**, *6*, 284–295. [CrossRef]
7. Bhide, S.; Davidson, J.K.; Shah, J.J. Areal Coordinates: The Basis of a Mathematical Model for Geometric Tolerances. *Geom. Prod. Spec. Verif. Funct. Integr.* **2003**, 35–44. [CrossRef]
8. Davidson, J.K.; Shah, J.J. Geometric tolerances: A new application for line geometry and screws. *Proc. Inst. Mech. Eng. Part C J. Mech. Eng. Sci.* **2020**, *216*, 95–104. [CrossRef]
9. Mujezinovic, A.; Davidson, J.K.; Shah, J.J. A New Mathematical Model for Geometric Tolerances as Applied to Polygonal Faces. *J. Mech. Des.* **2003**, *126*, 504–518. [CrossRef]

10. Chase, K.W.; Gao, J.; Magleby, S.P.; Sorensen, C.D. Including Geometric Feature Variations in Tolerance Analysis of Mechanical Assemblies. *IIE Trans.* **1996**, *28*, 795–807. [CrossRef]
11. Chase, K.; Magleby, S.; Glancy, C.; Chase, K.; Magleby, S.; Glancy, C.; Chase, K.; Magleby, S.; Glancy, C. A Comprehensive System for Computer-Aided Tolerance Analysis of 2-D and 3-D Mechanical Assemblies. In *Geometric Design Tolerancing: Theories, Standards and Applications*; Springer Nature: Berlin/Heidelberg, Germany, 1998; pp. 294–307. [CrossRef]
12. Chase, K.W.; Gao, J.; Magleby, S.P. General 2-D Tolerance Analysis of Mechanical Assemblies with Small Kinematic Adjustments. *J. Design Manuf.* **1995**, *5*, 263–274.
13. Gao, J.; Chase, K.W.; Magleby, S.P. Generalized 3-D tolerance analysis of mechanical assemblies with small kinematic adjustments. *IIE Trans.* **1998**, *30*, 367–377. [CrossRef]
14. Clément, A.; Desrochers, A.; Rivière, A. Theory and Practice of 3-D Tolerancing for Assembly. In *Proceeding of CIRP Seminar on Computer Aided Tolerancing*; Penn State University: University Park, PA, USA, 1991; Volume 25.
15. Clément, A.; Bourdet, P. A Study of Optimal-Criteria Identification Based on the Small-Displacement Screw Model. *CIRP Ann. Manuf. Technol.* **1988**, *37*, 503–506. [CrossRef]
16. Desrochers, A. Modeling Three Dimensional Tolerance Zones Using Screw Parameters. In Proceedings of the ASME 1999 Design Engineering Technical Conferences, Las Vegas, NV, USA, 12–16 September 1999; pp. 895–903.
17. Desrochers, A.; Clément, A. A dimensioning and tolerancing assistance model for CAD/CAM systems. *J. Adv. Manuf. Technol.* **1994**, *9*, 352–361. [CrossRef]
18. Laifa, M.; Sai, W.B.; Hbaieb, M. Evaluation of machining process by integrating 3D manufacturing dispersions, functional constraints, and the concept of small displacement torsors. *Int. J. Adv. Manuf. Technol.* **2014**, *71*, 1327–1336. [CrossRef]
19. Legoff, O.; Villeneuve, F.; Bourdet, P. Geometrical tolerancing in process planning: A tridimensional approach. *Proc. Inst. Mech. Eng. Part B-J. Eng. Manuf.* **1999**, *213*, 635–640. [CrossRef]
20. Teissandier, D.; Couétard, Y.; Gérard, A. A computer aided tolerancing model: Proportioned assembly clearance volume. *Comput. Aided Des.* **1999**, *31*, 805–817. [CrossRef]
21. Desrochers, A.; Rivière, A. A matrix approach to the representation of tolerance zones and clearances. *Int. J. Adv. Manuf. Technol.* **1997**, *13*, 630–636. [CrossRef]
22. Wang, H.; Ning, R.; Tang, C. Automatic generation of 3D as sembly dimension chains. *Chin. J. Mech. Eng.* **2005**, *41*, 7. [CrossRef]
23. El Mouden, M.; Chahbouni, M.; Boutahari, S. A New Approach for Synthesis of 3D Tolerances with Consideration of Flatness Defects in Mechanical Assembly Using Jacobian Torsor Model. *Int. J. Eng. Appl.* **2024**, *12*, 169–179. [CrossRef]
24. Zhang, J.; Qiao, L. Three Dimensional Manufacturing Tolerance Design Using Convex Sets. *Procedia CIRP* **2013**, *10*, 259–266. [CrossRef]
25. Jain, P. Unified approach for tolerance analysis, synthesis, and compensator selection. *Appl. Optics* **2022**, *61*, 8974–8981. [CrossRef]
26. Tang, W.; Xu, G.; Zhang, S. Dimensional Variation Analysis for Rigid Part Assembly With an Improvement of Monte Carlo Simulation. *IEEE Access* **2020**, *8*, 5862–5872. [CrossRef]
27. Rausch, C.; Nahangi, M.; Haas, C.; Liang, W. Monte Carlo simulation for tolerance analysis in prefabrication and offsite construction. *Autom. Constr.* **2019**, *103*, 300–314. [CrossRef]
28. McGregor, D.J.; Bimrose, M.V.; Shao, C.; Tawfick, S.; King, W.P. Using machine learning to predict dimensions and qualify diverse part designs across multiple additive machines and materials. *Addit. Manuf.* **2022**, *55*, 102848. [CrossRef]
29. Maltauro, M.; Meneghello, R.; Concheri, G. Enhancing tolerance stack-up analysis with variable-dependent admissible limits. *CIRP Ann.* **2025**, in press. [CrossRef]
30. Zhang, Q.; Jin, X.; Liu, Z.; Zhang, Z.; Fang, Y.; Zhang, Z.; Ledoux, Y. A New Approach of Surfaces Registration Considering Form Errors for Precise Assembly. *Assem. Autom.* **2018**, *40*, 789–800. [CrossRef]
31. Zhang, Q.; Jin, X.; Zhang, Z.; Zhang, Z.; Shang, K. Assembly Method Based on Constrained Surface Registration. *J. Mech. Eng.* **2018**, *54*, 70–76. [CrossRef]
32. Liu, T.; Zhao, Q.; Cao, Y.; Yang, J. A generic approach for analysis of mechanical assembly. *Precis Eng.* **2018**, *54*, 361–370. [CrossRef]
33. Shi, S.; Liu, J.; Gong, H.; Shao, N.; Anwer, N. Assembly accuracy analysis and phase optimization of aero-engine multistage rotors considering surface morphology and non-uniform contact deformation. *Precis. Eng.* **2024**, *88*, 595–610. [CrossRef]
34. Guo, H.; Zhang, Z.; Xiao, M.; Liu, H.; Zhang, Q. Tolerance optimization method based on flatness error distribution. *Int. J. Adv. Manuf. Tech.* **2021**, *113*, 1–15. [CrossRef]
35. Zhang, Z.; Zhang, Z.; Jin, X.; Zhang, Q. A novel modelling method of geometric errors for precision assembly. *Int. J. Adv. Manuf. Technol.* **2018**, *94*, 1139–1160. [CrossRef]
36. Anwer, N.; Schleich, B.; Mathieu, L.; Wartzack, S. From solid modelling to skin model shapes: Shifting paradigms in computer-aided tolerancing. *CIRP Ann. Manuf. Technol.* **2014**, *63*, 137–140. [CrossRef]

37. Corrado, A.; Polini, W.; Moroni, G.; Petrò, S. 3D Tolerance Analysis with Manufacturing Signature and Operating Conditions. In Proceedings of the 14th CIRP CAT 2016—CIRP Conference on Computer Aided Tolerancing, Gothenburg, Sweden, 18–20 May 2016; Volume 43, pp. 130–135.
38. Qie, Y.; Srinivasan, V.; Anwer, N. On Skin Model Describability using Laplace-Beltrami Operator for Geometrical Product Specification and Verification. *Procedia CIRP* **2024**, *129*, 67–72. [CrossRef]
39. Humienny, Z. Can ISO GPS and ASME Tolerancing Systems Define the Same Functional Requirements? *Appl. Sci.* **2021**, *11*, 8269. [CrossRef]
40. Shen, T.; Lu, C. An assembly accuracy analysis approach of mechanical assembly involving parallel and serial connections considering form defects and local surface deformations. *Precis Eng.* **2024**, *88*, 44–64. [CrossRef]

machines

MDPI

Article

VR Co-Lab: A Virtual Reality Platform for Human–Robot Disassembly Training and Synthetic Data Generation

Yashwanth Maddipatla [1], Sibo Tian [2], Xiao Liang [3], Minghui Zheng [2] and Beiwen Li [4,*]

[1] Department of Mechanical Engineering, Iowa State University, Ames, IA 50011, USA; yash13@iastate.edu
[2] J. Mike Walker '66 Department of Mechanical Engineering, Texas A&M University, College Station, TX 77843, USA; sibotian@tamu.edu (S.T.); mhzheng@tamu.edu (M.Z.)
[3] Zachry Department of Civil and Environmental Engineering, Texas A&M University, College Station, TX 77843, USA; xliang@tamu.edu
[4] School of Environmental, Civil, Agricultural and Mechanical Engineering, University of Georgia, Athens, GA 30602, USA
* Correspondence: beiwen.li@uga.edu

Abstract: This research introduces a virtual reality (VR) training system for improving human–robot collaboration (HRC) in industrial disassembly tasks, particularly for e-waste recycling. Conventional training approaches frequently fail to provide sufficient adaptability, immediate feedback, or scalable solutions for complex industrial workflows. The implementation leverages Quest Pro's body-tracking capabilities to enable ergonomic, immersive interactions with planned eye-tracking integration for improved interactivity and accuracy. The Niryo One robot aids users in hands-on disassembly while generating synthetic data to refine robot motion planning models. A Robot Operating System (ROS) bridge enables the seamless simulation and control of various robotic platforms using Unified Robotics Description Format (URDF) files, bridging virtual and physical training environments. A Long Short-Term Memory (LSTM) model predicts user interactions and robotic motions, optimizing trajectory planning and minimizing errors. Monte Carlo dropout-based uncertainty estimation enhances prediction reliability, ensuring adaptability to dynamic user behavior. Initial technical validation demonstrates the platform's potential, with preliminary testing showing promising results in task execution efficiency and human–robot motion alignment, though comprehensive user studies remain for future work. Limitations include the lack of multi-user scenarios, potential tracking inaccuracies, and the need for further real-world validation. This system establishes a sandbox training framework for HRC in disassembly, leveraging VR and AI-driven feedback to improve skill acquisition, task efficiency, and training scalability across industrial applications.

Keywords: mechanical engineering; manufacturing technology; virtual reality (VR); human–robot collaboration (HRC); e-waste recycling; disassembly and assembly training; virtual simulation; human–robot interaction; synthetic data generation; robot motion prediction; ROS integration; long short-term memory (LSTM); Monte Carlo dropout; trajectory optimization; machine learning in robotics

1. Introduction

Human–robot collaboration (HRC) has gained increasing traction in industrial settings, particularly in assembly and disassembly tasks. The growing complexity of industrial workflows, such as e-waste recycling and precision manufacturing, demands advanced training methodologies seamlessly integrating human expertise with robotic assistance. Traditional training approaches often fall short in preparing workers for these hybrid environments due to several critical limitations: a lack of realistic interaction scenarios, insufficient adaptability to individual learning curves, the absence of real-time feedback mechanisms, and limited scalability for complex industrial workflows [1–4].

The need for effective training systems is further amplified by the increasing adoption of Industry 4.0 principles, emphasizing smart manufacturing, cyber–physical systems, and digital twins [5]. These principles call for training platforms that simulate real-world tasks, adapt dynamically to user performance, and integrate seamlessly with robotic systems. However, existing training methods often lack the flexibility and scalability to meet these demands. For example, traditional classroom-based or on-the-job training approaches are limited by their inability to replicate complex industrial environments or provide real-time feedback tailored to individual users.

Virtual reality (VR) offers a transformative solution by enabling immersive, high-fidelity simulations replicating real-world disassembly processes while allowing for real-time data collection and adaptive feedback. Augmented reality (AR) and mixed reality (MR) further enhance the training experience by overlaying digital guidance in physical environments, improving task efficiency and cognitive load management [6,7]. Prior studies have demonstrated the effectiveness of VR/AR-based training for procedural learning and industrial applications. For instance, Webel et al. [2] highlighted the advantages of AR-based technician training, integrating multimodal feedback to enhance learning outcomes. Similarly, Schwarz et al. [8] demonstrated how immersive VR can improve cognitive training, leading to better memory retention and procedural accuracy. Lerner et al. [9] show how VR can improve performance in surgical environments, demonstrating that skills acquired in virtual training environments can effectively transfer to real-world applications. Recent work by Dwivedi et al. [10] further established that immersive VR environments can reduce learning time by up to 40% while improving task efficiency in manual assembly training. These studies highlight the potential of XR technologies for industrial training.

Despite these advancements, significant challenges remain in developing integrated systems that combine immersive training environments with real-time robotic motion prediction and task adaptation. Current solutions cannot often anticipate human actions dynamically or adapt robot behavior based on human preferences and task context. Additionally, ergonomic considerations such as minimizing worker strain during collaborative tasks are often overlooked in existing systems [11]. Addressing these limitations requires a cohesive framework integrating advanced motion prediction models, adaptive cognitive architectures, and uncertainty-aware planning.

Synthetic data generation techniques have significantly enhanced VR-based training platforms by enabling iterative learning and the refinement of robotic behaviors. Synthetic datasets can simulate various human–robot interactions under varying conditions, providing valuable insights into task optimization and safety improvements [12]. Moreover, advancements in machine learning algorithms such as Long Short-Term Memory (LSTM) networks have shown promise in predicting dynamic human motion patterns [13]. These developments pave the way for creating intelligent systems capable of adapting to complex industrial workflows.

To address these gaps, this paper presents **VR Co-Lab**, a virtual-reality-based platform designed to enhance HRC in disassembly training. This research extends existing knowledge in three key areas:

- Development of an integrated VR-based training environment that combines full-body tracking using the Quest Pro HMD with a real-time ROS–Unity bridge, enabling precise human motion capture and dynamic control of robotic models.
- Implementation of an LSTM-based motion prediction model with Monte Carlo dropout for uncertainty estimation, which continuously refines robotic motion trajec-

tories, leveraging past interaction data for dynamic, real-time task adaptation. The model runs using Unity's Barracuda engine, eliminating external inference server latency and ensuring responsive interactions. This builds upon the work of Liu et al. [13] but extends it to immersive VR environments with real-time feedback loops.

- Creation of a synthetic data generation pipeline that captures complex human–robot interactions to refine disassembly sequence planning and optimize collaborative task execution. Our approach specifically targets human–robot collaboration in disassembly tasks, addressing a critical gap in the literature identified by Jacob et al. [14].

Preliminary system validation shows a promising technical performance, with the LSTM model demonstrating strong alignment between predicted and actual motion trajectories during initial testing. While these early results demonstrate the system's potential for enhancing human–robot collaboration in disassembly tasks, more extensive user evaluation studies will be conducted in future work to assess training effectiveness across different skill levels and task complexities.

This framework provides immersive, interactive training and generates synthetic datasets for iterative learning, enabling the data-driven refinement of robotic disassembly strategies. Additionally, by integrating uncertainty-aware motion forecasting, the system ensures smoother transitions between human and robot actions in shared workspaces. While existing approaches have explored VR training or motion prediction separately [15], our work uniquely combines these elements into a cohesive system designed explicitly for human–robot collaborative disassembly training.

The remainder of this paper is structured as follows: Section 2 reviews the relevant literature on VR/AR training methodologies, human–robot motion prediction frameworks, and collaborative disassembly systems. Section 3 details the system architecture, followed by an evaluation of performance metrics in Section 4. Finally, Section 5 discusses the implications of this approach and outlines future research directions.

2. Related Work

2.1. VR/AR Training Systems for Industrial Applications

Virtual reality (VR) and augmented reality (AR) have emerged as transformative tools for industrial training, offering immersive environments that replicate real-world tasks while providing real-time feedback. Dwivedi et al. [10] demonstrated that immersive VR environments could enhance manual assembly training by providing controlled simulations, reducing learning time, and improving task efficiency. Schwarz et al. [16] evaluated VR's effectiveness in cognitive training, showing improvements in memory retention and cognitive load management.

Webel et al. [2] explored AR-based technician training, incorporating multimodal interaction such as haptic feedback to enhance learning outcomes. Hořejš'i et al. [7] highlighted AR's role in virtual training for parts assembly, proving its effectiveness in reducing assembly time and errors. Similarly, Rocca et al. [17] investigated integrating VR and digital twins into circular economy practices, demonstrating their role in optimizing disassembly workflows. Boud et al. [18] demonstrated VR's effectiveness as a tool for assembly tasks, highlighting how haptic feedback through instrumented objects can improve object manipulation times compared to conventional techniques. Li et al. [4] developed a desktop VR prototype system (V-REALISM) for maintenance training that integrated disassembly sequence planning with virtual reality, demonstrating cost reduction benefits for industrial applications.

Al-Ahmari et al. [19] developed a virtual manufacturing assembly simulation system that offers visual, auditory, tactile, and force feedback, enhancing the realism of the training environment. Numfu et al. [20] presented a VR platform for maintenance training that is easy to use, understand, and transport, enabling users to train anywhere and anytime according to their needs. Gutiérrez et al. [21] developed a multimodal virtual training system to transfer motor and cognitive skills in industrial maintenance and assembly tasks, providing multimodal aids and learning strategies that adapt to users' needs during training.

Li et al. [22] discussed an interactive virtual training system for assembly and disassembly based on precedence constraints, offering both immersive VR and conventional desktop interfaces to balance training quality and cost. Westerfield et al. [23] highlighted intelligent augmented reality training's ability to improve task performance and knowledge retention in industrial operations. Pan et al. [24] designed a training system for robotic component assembly based on VR technology, aimed at enabling students to quickly understand the assembly methods of robot components and mobilize their enthusiasm for practice and innovation. Seth et al. [25] provided a comprehensive review of VR for assembly methods prototyping, identifying key requirements for effective virtual assembly systems including physics-based modeling and haptic feedback.

Recent advancements emphasize the scalability of VR/AR systems for multi-user scenarios and the integration of AI-based adaptive learning models [5]. For example, Carlson et al. [12] discussed synthetic data generation within AR/VR environments to improve training scenarios. Devagiri et al. [26] provided a comprehensive review of AR and AI integration in industrial applications, emphasizing their role in enhancing workforce training and operational efficiency. However, challenges remain in addressing ergonomic considerations [11], ensuring seamless integration with robotic systems, and overcoming physiological effects such as dizziness during extended VR training sessions [27].

2.2. Human Motion Prediction in HRC

Human motion prediction plays a critical role in enhancing task efficiency and ensuring operator safety in HRC settings. Lee et al. [28] proposed a disassembly task planning algorithm that dynamically assigns tasks between human operators and robots by leveraging human behavior prediction models. Liu et al. [13] introduced a task-constrained motion planning approach incorporating uncertainty-informed human motion prediction to improve adaptability and safety. Vongbunyong et al. [29] presented a process demonstration platform that enables skillful human operators to demonstrate disassembly processes, with the knowledge at planning and operational levels being transferred to robots to achieve automated disassembly.

Jacob et al. [14] conducted a rapid review of human–robot cooperation in disassembly, identifying key challenges including task allocation, safety management, and adaptive collaboration. Their work emphasized that effective HRC systems must balance human cognitive capabilities with robotic precision, particularly for complex disassembly sequences. Hjorth and Chrysostomou [30] examined the landscape of human–robot collaborative disassembly (HRCD) and reviewed progress in the field, analyzing principles and elements of human–robot collaboration in industrial environments, such as safety standards and collaborative operation modes.

Ottogalli et al. [31] used virtual reality to simulate human–robot coexistence for an aircraft final assembly line, evaluating ergonomics and process efficiency to identify issues beforehand and prevent unexpected costs.

Recent advancements include frameworks incorporating Gaussian Process Regression (GPR) models [32], which integrate human joint constraints with scene constraints for accurate motion trajectory prediction. Additionally, Recurrent Neural Networks (RNNs), particularly Long Short-Term Memory (LSTM) models, have been extensively utilized for predicting dynamic human motion [13]. While LSTMs excel in capturing temporal dependencies, alternative approaches like Transformer-based models or Kalman filters may offer an improved performance in certain scenarios. Liu et al. [33] developed a dynamic model-informed human motion prediction approach based on an Unscented Kalman Filter, demonstrating improved accuracy in forecasting human movements during collaborative tasks.

Daneshmand et al. [34] surveyed robotic assembly and disassembly in the context of Industry 4.0 and the circular economy, highlighting the importance of integrating human factors into collaborative robotic systems. Their work emphasized that effective HRC requires not only technical solutions for motion prediction but also consideration of ergonomic factors, cognitive load, and user experience aspects that our VR training system specifically addresses through immersive interaction and real-time feedback mechanisms.

2.3. *Inverse and Forward Kinematics in Robotic Training*

Inverse kinematics (IK) is essential for accurately controlling robotic arms during industrial tasks such as assembly or disassembly. Kuts et al. [35] demonstrated that IK ensures high fidelity in replicating physical movements within VR environments, enabling the precise manipulation of virtual robotic models. Similarly, Xu et al. [36] explored adaptive IK algorithms that dynamically adjust joint angles based on task constraints, improving the precision of robotic operations during collaborative tasks.

Forward kinematics (FK), on the other hand, calculates the position of robotic end-effectors based on given joint parameters. Xu et al. [36] highlighted FK's role in simulating robotic movements within VR environments for operator training. Their robotic kinematics teaching system combined VR simulation and remote control to enhance robotics education while mitigating safety risks. They also explored robotic kinematics teaching systems and applications of VR to facilitate the system's ability to provide hands-on experience without physical risks, creating an effective bridge between theoretical knowledge and practical implementation.

By combining FK with IK algorithms, researchers have developed hybrid models capable of optimizing both trajectory planning and collision avoidance during complex industrial workflows [35].

2.4. *Digital Twins and Industry 4.0 Integration*

Digital twins have gained prominence as a key enabler of smart manufacturing under Industry 4.0 principles [5]. Rocca et al. [17] demonstrated the integration of digital twins with VR to optimize disassembly workflows, enabling real-time monitoring and task execution. Pérez et al. [37] developed a digital-twin-based methodology for multi-robot manufacturing cell commissioning, highlighting its potential for improving system efficiency. Sassanelli et al. [38] conducted a systematic literature review on simulation tools supporting disassembly processes with a focus on printed circuit boards, identifying key roles of simulation in sequence planning, process optimization, and training.

Zhao et al. [39] demonstrated the integration of augmented reality and digital twin technologies in a teleoperated disassembly system, enabling improved operator safety and task efficiency through remote visualization and control. Their work particularly emphasized the benefits of digital twins for hazardous disassembly operations, such as EV battery recycling, an application domain directly relevant to our VR training system.

Zhao and Wang [40] developed a machine tool disassembly and assembly training system based on virtual simulation that establishes bidirectional communication between digital models and physical equipment. Their approach enables learners to observe the dynamic responses of machine tool models through the virtual environment, introducing digital twin technology into assembly/disassembly training.

Pérez et al. [41] used VR interfaces for industrial robot control and operator training, increasing usability and reducing complexity. Their approach demonstrated how VR interfaces can simplify complex robotic operations, making them accessible to operators with varying levels of technical expertise. Zhao et al. [42] demonstrate a robot teleoperation system with integrated digital twins and augmented reality for improved safety. Their system allows operators to enter the control loop when anomalies are detected, enabling them to receive real-time feedback while remotely controlling the robot, thereby improving the safety and efficiency of disassembly operations.

However, challenges remain in integrating digital twins with real-time VR/AR systems due to computational complexities and latency issues [43]. Our work addresses these challenges through optimized data pipelines and efficient model deployment strategies, ensuring responsive interaction while maintaining digital twin fidelity.

2.5. Comparison with Existing Approaches

While previous research has made significant contributions to VR/AR training and human–robot collaboration, VR Co-Lab distinguishes itself in several key aspects:

Integration of full-body tracking with robotic control: Unlike systems by Webel et al. [2] and Schwarz et al. [8] that explored AR/VR for technical training without comprehensive body tracking, our system leverages Quest Pro's Inside-Out Body-Tracking technology to capture upper body movements with high precision. This tracking framework enables more natural interactions and provides richer data for motion prediction models.

Bidirectional Unity–ROS communication: While existing platforms often implement unidirectional communication between VR environments and robotic systems, VR Co-Lab features a bidirectional Unity–ROS bridge that enables seamless data exchange. This allows for the real-time adaptation of both the virtual environment and robotic responses based on user interactions, creating a more responsive and adaptive training experience.

LSTM-based motion prediction with uncertainty quantification: Our implementation extends Liu et al.'s [13] work on motion prediction by integrating Monte Carlo dropout for uncertainty estimation directly within an immersive VR environment. This approach not only predicts human movements but also quantifies confidence in these predictions, enabling safer and more adaptive human–robot collaboration during disassembly tasks.

Synthetic data generation for disassembly tasks: Unlike general-purpose VR training systems, VR Co-Lab specifically addresses the challenges of human–robot collaborative disassembly identified by Jacob et al. [14]. Our synthetic data generation pipeline captures complex interaction patterns during disassembly procedures, creating valuable training datasets that enhance both human learning and robotic adaptation.

On-device neural network inference: By implementing the Barracuda engine for neural network inference directly in Unity, our system eliminates the latency associated with external inference servers. This approach ensures responsive real-time interactions essential for effective training and collaboration in VR environments.

These distinctive features collectively create a comprehensive platform that addresses limitations in existing systems by combining immersive VR training, real-time motion prediction with uncertainty estimation, and synthetic data generation specifically tailored for human–robot collaborative disassembly tasks.

The literature strongly supports the integration of VR and AR technologies in industrial training systems. Our research demonstrates that effective VR-based training depends on four key factors: robust observational learning capabilities, balanced cognitive and physical fidelity, immediate feedback mechanisms, and strategic synthetic data utilization. The immersive nature of collaborative VR environments significantly enhances both team-based training outcomes and individual task performance metrics. These insights have directly informed the development of our VR Co-Lab platform, ensuring it delivers the comprehensive capabilities required for advanced industrial training applications in human–robot collaborative disassembly tasks.

3. VR Training System Implementation

The VR training system integrates selected technologies to enhance immersion and interactivity while ensuring a seamless connection between virtual and physical robotics.

3.1. Technology Stack

The VR training system is built upon a carefully selected technology stack that ensures robust performance and scalability. Unity serves as the core of our virtual environment, providing advanced 3D modeling, real-time graphics rendering, and physics simulation capabilities essential for creating detailed and interactive simulations of industrial tasks. This foundation enables a realistic and engaging training experience that closely mimics real-world conditions.

The Unity Perception Toolkit enhances our approach to capturing and utilizing real-time interaction data, supporting immediate feedback and iterative learning enhancements within the training environment. Communication between Unity and the ROS is facilitated by the Unity ROS TCP Connector, ensuring that user interactions within the virtual environment are accurately mirrored by the physical robot's actions, maintaining a coherent experience across both platforms.

For accurate simulation, the Unity URDF Importer recreates realistic robot models within Unity by importing URDF files. This capability is vital for simulating complex mechanical interactions and ensuring that virtual manipulations directly affect the physical robot's behavior. The Robot Operating System (ROS) serves as the backbone for managing robotic communications and control, integrating various hardware and software components necessary for practical training simulations.

The Quest Pro headset was selected for its superior tracking capabilities, enhancing user interaction by providing accurate tracking of body and hand movements, crucial for tasks requiring fine motor control and precise manipulations. To facilitate the easy deployment and management of the VR and ROS environments, Docker is employed with a custom image created using the ROS Melodic base, including additional required ROS packages and submodules to ensure a consistent and reproducible development environment.

3.2. VR Training System Architecture

The architecture of the VR training system is designed to be modular, facilitating easy updates and scalability.

3.2.1. VR Environment

VR environment (Unity scene): Developed in Unity, this component replicates a disassembly workstation with realistic tools and components with virtual replicas of tools and machinery. This setup engages the user and serves educational purposes by providing a safe, controlled environment for complex training scenarios. Figure 1 shows a real-world setup and Figure 2 shows a virtual environment similar to the real-world setup.

Figure 1. Real-world Setup.

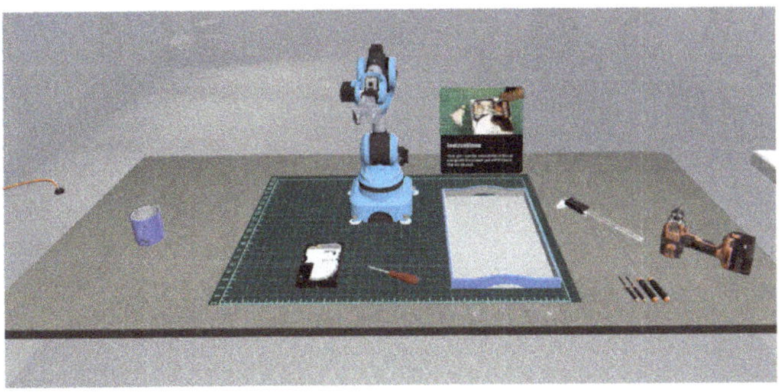

Figure 2. Virtual world training environment.

Interactivity: Users manipulate tools such as Phillips and flathead screwdrivers of various sizes and power tools for hard drive disassembly alongside component parts like screws of different sizes, circuit boards, disk platters, actuator arms, and various electronic components using their hands or controllers. For example, a virtual screwdriver in the environment behaves according to the physics rules defined in Unity, providing tactile feedback if the VR controllers are used.

The system facilitates specific disassembly procedures, including (1) removing the external casing screws using appropriate screwdrivers, (2) detaching and safely removing the disk platter, (3) disconnecting and removing the read/write head and actuator arm, (4) extracting the circuit board and electronic components, and (5) sorting components for recycling or refurbishment. Each procedure requires specific tool selection and proper technique, with the system providing feedback on proper tool usage and procedural accuracy. The system also allows the customization of the setup, type of robot, and different environments to accommodate various training scenarios.

3.2.2. Robotic Control System

Managed by the ROS, this component includes multiple ROS nodes that communicate with each other and with the Unity environment through serialized ROS messages. The nodes handle various tasks such as sensor data processing, control commands, and state monitoring. Communication between the Unity scene and the ROS network is facilitated via a server endpoint that manages the publish/subscribe model of ROS communication.

The Unity–ROS integration leverages the ROS# framework [44], enabling seamless communication between the virtual environment and physical robotic systems. This integration allows for importing various robots as long as there is an associated controller file and a URDF file. The current environment uses the **Niryo One** robot that works alongside the user in disassembling the hard disk.

For neural network inference within Unity, we utilize the Barracuda engine [45], which provides the efficient execution of ONNX models directly in the game engine. This approach eliminates the need for external inference servers, reducing latency and ensuring responsive interaction in the VR environment.

The architecture diagram in Figure 3 illustrates the communication between the Unity scene and the ROS network. The Unity scene contains ROS service, subscriber, and publisher scripts that interact with ROS nodes over a network. Serialized ROS messages facilitate this interaction, ensuring data consistency and real-time communication.

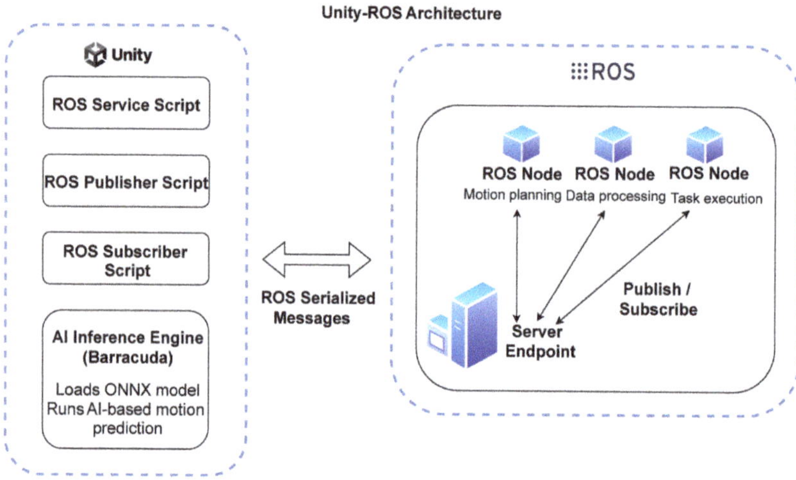

Figure 3. Updated Unity–ROS architecture with AI-based motion prediction using Barracuda.

ROS service script: This script in Unity sends service requests to ROS nodes and processes the responses. It is used for tasks that require a request–response communication pattern, such as querying the state of a robot or commanding it to perform a specific action.

ROS subscriber script: This script subscribes to specific ROS topics and processes incoming data. It allows the Unity environment to receive real-time updates from the ROS nodes.

ROS publisher script: This script publishes data to ROS topics, enabling Unity to send commands and updates to ROS nodes. It is used for controlling the robot and sending state information from the VR environment to the ROS network.

AI inference engine (Barracuda): The Barracuda inference engine is a lightweight, cross-platform solution for running deep learning models in Unity. It supports the Open Neural Network Exchange (ONNX) format, allowing pre-trained models (such as LSTMs and CNNs) to be deployed within Unity environments. In this system, Barracuda facilitates real-time motion prediction to improve human–robot interaction.

- Loads a pre-trained ONNX LSTM model optimized for human motion forecasting.
- Processes real-time user movement data to anticipate the next actions.
- Feeds the predicted motion trajectory into the ROS network for adaptive robot response.
- Enables low-latency inference directly in Unity's runtime, reducing dependence on external servers.
- Supports GPU acceleration when available, improving the efficiency of AI inference in VR.

Integration with ROS: The AI-based motion prediction refines robot control by ensuring that the motion planning and task execution nodes receive real-time human movement predictions. This allows for the following:

- **Adaptive task scheduling**, reducing wait times in collaborative workflows.
- **Enhanced safety mechanisms**, as the robot can proactively adjust to user movements.
- **Increased task efficiency**, with optimized movement paths informed by AI predictions.

The ROS network comprises multiple nodes, each responsible for specific tasks such as sensor data processing, control commands, and state monitoring. The server endpoint manages the publish/subscribe communication model, ensuring messages are delivered reliably and efficiently between nodes and the Unity scene.

Motion planning node: utilizes LSTM-based motion prediction data to plan optimal robot actions.

Data processing node: handles real-time sensor data and refines predictions based on Unity inputs.

Task execution node: executes disassembly actions dynamically based on predicted and real-time user interactions.

This enhancement enables a more intelligent and adaptive robotic training environment, leveraging AI-powered predictive control to improve real-time HRC interactions. The combination of Barracuda, ONNX, and LSTMs allows Unity to act as a robust simulation environment for industrial training.

3.2.3. Data Collection and Feedback Mechanisms

This module integrates advanced data analytics to track performance metrics such as task completion times, error rates, and user engagement. These data are essential for refining training protocols, personalizing the learning experience, and identifying areas for improvement.

Specifically, the task completion times TCT are computed as follows:

$$TCT = \frac{\sum(\text{Time Taken Per Task})}{\text{Number of Tasks}}. \quad (1)$$

This equation calculates the average time users spend to complete each task. A lower average indicates increased proficiency and comfort with the task and the VR environment.

The error rate E is thereby computed as

$$E = \frac{\text{Number of Errors}}{\text{Total Number of Tasks}} \times 100. \quad (2)$$

The user engagement is quantified through parameters $U(D)$ and UEI, where $U(D)$ quantifies the utility of synthetic data and UEI is defined as a user engagement index. Specifically, they are computed as follows:

$$U(D) = 1 - e^{-\lambda D}. \quad (3)$$

where D is the amount of data and λ is a parameter indicating the efficiency of data conversion into learning improvement. This reflects the value added by synthetic data in training machine learning models within the VR system:

$$UEI = \alpha \times \text{Active Time} + \beta \times \text{Interaction Frequency}. \quad (4)$$

This user engagement index UEI combines multiple factors: active time in the system, frequency of interactions, and quality of user feedback. Coefficients α and β weigh the importance of each factor, offering a holistic measure of user engagement.

3.2.4. Task Flow and User Actions

User interaction within the VR environment is carefully designed to ensure that it is intuitive, engaging, and educational. The task flow in the VR training system is designed to guide users through a series of logically sequenced steps, from task initiation to completion, ensuring a comprehensive learning experience.

The flowchart in Figure 4 illustrates the sequence of user interactions within the VR system, incorporating both user actions and real-time robot motion predictions using the LSTM model. The key steps are as follows:

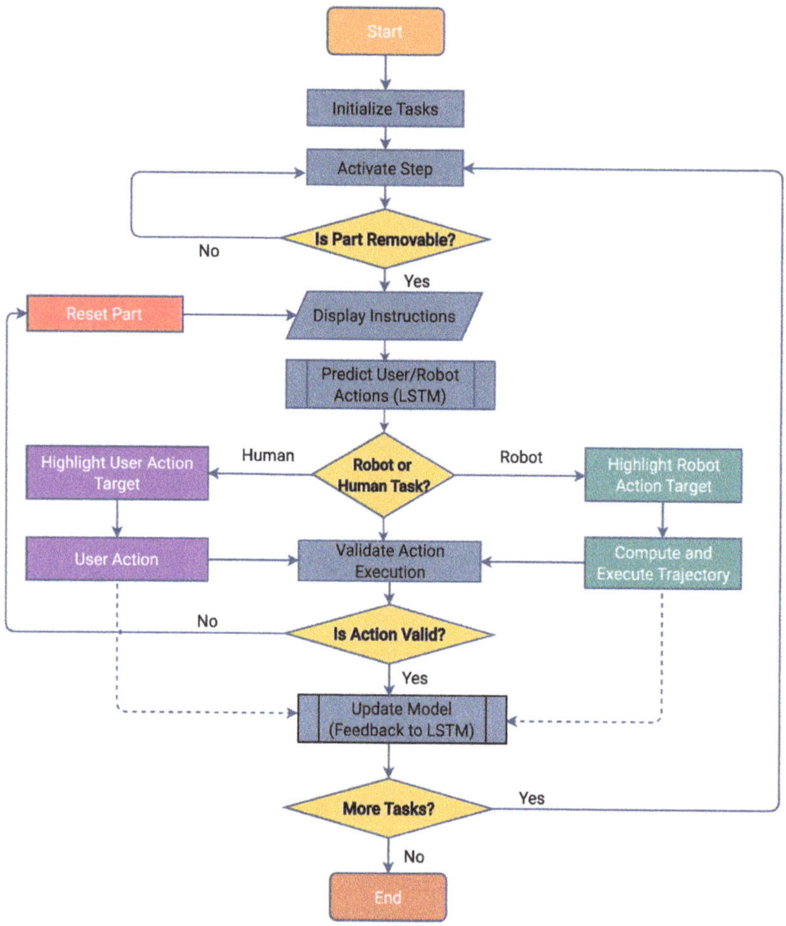

Figure 4. Task flowchart.

1. **Initialize tasks**: the VR environment is set up with all necessary tools, components, and the initial system configuration.
2. **Activate step**: the system prompts the user to begin the task, initiating the first step in the disassembly process.

3. **Is part removable?**: The system checks whether the targeted part can be safely removed. If not, the part is reset, and the process loops back to prompt the user again.
4. **Display instructions**: the VR system provides detailed instructions specific to the current step in the task, guiding the user through the disassembly process.
5. **Predict user/robot actions (LSTM)**: the LSTM model predicts both the user's and the robot's next actions based on the current task state and previous interactions.
6. **Human or robot task?**: the system determines whether the predicted action should be executed by the user (human task) or by the robot (robot task).
7. **Apply shader and execute actions**:
   - **Human task**: a purple shader highlights the part that the user needs to interact with, and the system waits for the user's action.
   - **Robot task**: a green shader is applied to the part to be manipulated by the robot, followed by the computation and execution of the robot's trajectory.
8. **Validate action execution**: the system validates whether the action performed by the user or robot was correct and executed as planned.
9. **Update model (feedback to LSTM)**: the results of the action validation are fed back into the LSTM model to update its predictions and improve future interactions.
10. **Is action valid?**: The system checks if the action taken was valid. If not, it resets the part and loops back to display instructions.
11. **More tasks?**: The system checks if there are additional tasks to be performed. If yes, it returns to the "Activate Step" phase; if not, it proceeds to calculate the final metrics.
12. **Metrics calculation**: at the end of the session, the system computes and displays performance metrics, providing immediate feedback to the user on the effectiveness and efficiency of their interactions.

While real-time updates are displayed via the dashboard, periodic validation metrics ensure consistency between observed and computed performance values. User interactions within the VR environment are designed to be seamless and intuitive, utilizing hand gestures and tool manipulations. Immediate feedback is delivered through visual and auditory signals, aiding users in correcting actions and learning in real time.

The present iteration features a specific training scenario centered on hard drive disassembly that replicates a real-world setup. This focused approach allows us to thoroughly evaluate both the VR training methodology and the LSTM-based motion prediction model within a controlled context. Future implementations will expand to include multiple scenarios with different robotic systems and diverse component types, enabling broader application of the training framework across various industrial contexts.

3.3. Body Tracking and Interaction Through Meta SDKs

Integrating Quest Pro's advanced body and hand-tracking capabilities into the VR training environment allows for replicating real-world movements within the virtual space. The Meta Quest Pro utilizes Inside-Out Body-Tracking (IOBT) technology [46], which leverages the headset's cameras to track upper body movements including wrists, elbows, shoulders, and torso. While this provides accurate tracking for the upper body, the system has inherent limitations for lower body tracking. To address this limitation, Meta has implemented "Generative Legs"—an AI-based solution that creates realistic leg movements by extrapolating from upper body data [47].

Hand tracking on the Quest Pro offers support for multiple gestures including point and pinch, pinch and drag, palm pinch, and direct touch interactions [48]. While this enables a controller-free interaction, the tracking system faces challenges with occlusion when hands overlap or move outside the cameras' field of view, potentially affecting precision during complex manipulation tasks [47].

In the current project phase focused on robotic path planning, eight key upper body tracking points shown in Figure 5 are used to monitor and translate user movements effectively. This integrated tracking framework captures complex user movements and feeds this data back into the system to enhance interaction models and improve the training algorithms continuously. As a result, users can interact with various virtual tools, each behaving according to predefined physics properties. This interaction is crucial for task completion and adds an educational layer by allowing users to experience the physical feel of different tools.

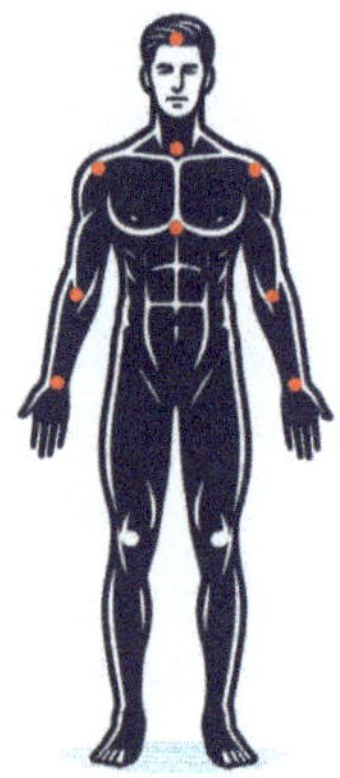

Figure 5. Tracked body points.

The Meta Movement SDK provides developers with tools to implement these tracking capabilities, supporting both controller-based and hand-tracking interactions [47]. For VR training applications, this flexibility allows users to choose the most appropriate interaction method based on specific task requirements and personal preferences.

3.4. *Performance Monitoring and Synthetic Data Generation*

The system monitors performance and generates synthetic data to train machine learning models. These models predict user performance and suggest training adjustments.

3.4.1. Performance Metrics

During the training session, various performance indicators were tracked to assess training effectiveness. Task completion time (TCT) measures the average time taken to complete a full disassembly task, calculated as the sum of time taken per task divided by the number of tasks. Lower TCT values indicate improved task efficiency as users become more proficient with both the task and the VR environment.

Error rate (E) quantifies the percentage of incorrect interactions during the training process, computed using the following formula:

$$E = \frac{\text{Number of Errors}}{\text{Total Number of Tasks}} \times 100.$$

Motion path efficiency analyzes the similarity between predicted and actual movement paths through trajectory deviation analysis. This metric shows improvements across training sessions, illustrating increasing optimization in human–robot movement synchronization.

Preliminary system validation conducted solely by the author indicates a task completion time of 4.2 min during self-testing, with an error rate of 10.8% observed during implementation testing. A motion path efficiency analysis showed that robot trajectories aligned with the author's intended interactions in 82% of cases, suggesting that the proposed system architecture has the potential for modeling human–robot collaboration. These technical validation metrics serve as proof of concept for the implementation rather than formal user evaluation, as this paper focuses on system design and architecture rather than human subjects research.

Synthetic Data for Machine Learning

The VR system continuously monitors performance and generates synthetic data that serve as the foundation for training machine learning models. This data collection process captures the comprehensive movement data of both the user and the robot within the training environment, creating valuable synthetic datasets that replicate complex interaction scenarios. These datasets are then used to train machine learning models that can predict user performance and suggest training adjustments.

The synthetic data enable the development of advanced algorithms capable of predicting user performance and suggesting personalized training modifications. This approach ensures that the training adapts to individual learning styles and speeds, maximizing learning efficiency and personalizing the user experience. Behdad et al. [49] explored leveraging VR experiences with mixed-integer nonlinear programming for the visualization of disassembly sequence planning under uncertainty, demonstrating how virtual environments can enhance decision making in complex disassembly tasks.

Integrating these detailed tracking capabilities and sophisticated data analysis tools ensures that the VR training system is not only immersive and interactive but is also grounded in robust empirical research, enabling continuous improvement and the personalization of training processes. Lerner et al. [9] demonstrated how VR training can improve performance in surgical environments, showing that skills acquired in virtual training environments can effectively transfer to real-world applications.

3.5. Mathematical Foundations and Implementation of Forward Kinematics

Forward kinematics (FK) involves calculating the position and orientation of a robotic manipulator's end-effector based on its joint parameters. The FK process utilizes homogeneous transformation matrices to compute the cumulative transformations from the robot's base to its end-effector. Each joint in the manipulator is described using Denavit–Hartenberg (DH) parameters [50], which provides a standardized method to represent the kinematic parameters of robotic manipulators. The parameters include the following:

- a_i: link length, or the distance between two joint axes along the x-axis.
- α_i: link twist, or the angle between two joint axes around the x-axis.
- d_i: link offset, or the displacement along the z-axis.
- θ_i: joint angle, or the rotation about the z-axis.

The transformation matrix for the i-th joint is given as

$$T_i = \begin{bmatrix} \cos\theta_i & -\sin\theta_i \cos\alpha_i & \sin\theta_i \sin\alpha_i & a_i \cos\theta_i \\ \sin\theta_i & \cos\theta_i \cos\alpha_i & -\cos\theta_i \sin\alpha_i & a_i \sin\theta_i \\ 0 & \sin\alpha_i & \cos\alpha_i & d_i \\ 0 & 0 & 0 & 1 \end{bmatrix}. \quad (5)$$

The overall transformation matrix T from the base to the end-effector is obtained by multiplying the individual transformations for all n joints:

$$T = T_1 \cdot T_2 \cdot \ldots \cdot T_n. \quad (6)$$

The transformation matrix T depends primarily on joint angles (θ_i) rather than time. However, in dynamic scenarios where joint angles change over time, T becomes implicitly time-dependent as $\theta_i(t)$ varies. This temporal dependency is particularly relevant for motion prediction, where the LSTM model forecasts future joint configurations based on previous states, effectively capturing the time evolution of T through sequential joint angle predictions.

This transformation encapsulates the position and orientation of the end-effector in the workspace.

3.5.1. Implementation in Unity

In the Unity environment, forward kinematics (FK) for robotic arms is implemented using the ArticulationBody components, which facilitate the simulation of complex jointed structures essential for modeling robotic arms in virtual environments [51]. The FK process involves several key steps:

1. DH parameter extraction: the joint parameters, such as angles or displacements, are extracted dynamically from Unity's articulation bodies.
2. Transformation matrix calculation: each joint's transformation matrix is computed using the Denavit–Hartenberg (DH) parameters.
3. Composite transformation: the transformation matrices are multiplied sequentially to compute the end-effector's position and orientation relative to the robot's base frame.
4. Simulation output: the computed position is used for visualization and task execution in Unity's 3D environment.

The function 'FK' in the Unity script iteratively calculates the transformation matrices and combines them to yield the final transformation T, which defines the end-effector's pose.

3.5.2. Mathematical Workflow

The workflow for forward kinematics in the Unity script corresponds to the following:

- Input: joint parameters $\boldsymbol{\theta} = (\theta_1, \theta_2, \ldots, \theta_n)$.
- Transformation matrix calculation: for each joint, the transformation matrix T_i is computed using the DH parameter set $[a_i, \alpha_i, \theta_i, d_i]$.
- Cumulative transformation: the transformation matrices are multiplied sequentially to form the composite matrix T, representing the end-effector's pose.
- Output: the position (x, y, z) and orientation (rotation matrix) of the end-effector in the global frame.

Example: Calculating the End-Effector Pose

For a robotic arm with two joints, the transformation matrices T_1 and T_2 are given by

$$T_1 = \begin{bmatrix} \cos\theta_1 & -\sin\theta_1 & 0 & a_1\cos\theta_1 \\ \sin\theta_1 & \cos\theta_1 & 0 & a_1\sin\theta_1 \\ 0 & 0 & 1 & d_1 \\ 0 & 0 & 0 & 1 \end{bmatrix}.$$

$$T_2 = \begin{bmatrix} \cos\theta_2 & -\sin\theta_2 & 0 & a_2\cos\theta_2 \\ \sin\theta_2 & \cos\theta_2 & 0 & a_2\sin\theta_2 \\ 0 & 0 & 1 & d_2 \\ 0 & 0 & 0 & 1 \end{bmatrix}.$$

The final transformation matrix T is computed as

$$T = T_1 \cdot T_2$$

This computation provides the final pose of the end-effector.

3.5.3. Applications in Robotic Training

Forward kinematics enables an accurate simulation of robotic arm movements, crucial in tasks like disassembly training. By incorporating this methodology, Unity-based simulations can achieve high fidelity, aiding real-time task execution and performance evaluation. Yang et al. [52] demonstrated the importance of validating virtual haptic disassembly platforms against real-world performance metrics, particularly when considering the disassembly process and physical attributes of components. Tahriri et al. [53] proposed methods for optimizing robot arm movement time using virtual reality robotic teaching systems, showing increased production rates and decreased cycle times.

3.6. Integrated Pipeline for Predictive Analytics

The integration of the LSTM model into the Unity VR training environment creates a robust system for predictive analytics and real-time robotic control. By leveraging Unity's simulation capabilities [51], Docker containers, the ROS [54], synthetic data generation, and ONNX-based model deployment [55], this pipeline enhances human–robot collaboration for tasks such as disassembly training. Figure 6 illustrates the LSTM architecture used in our implementation.

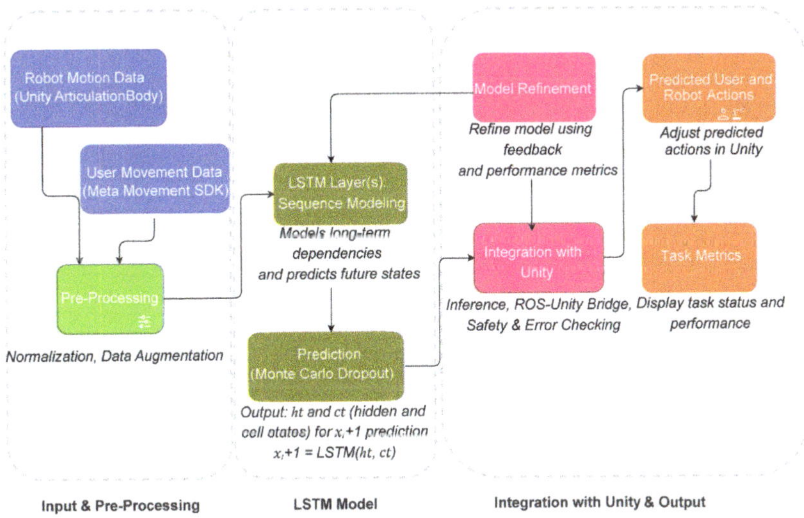

Figure 6. LSTM model.

LSTM was chosen due to its ability to effectively model long-term dependencies in sequential data while maintaining a relatively low computational cost suitable for real-time inference. While architectures such as Transformers or Gaussian Process models could offer alternative solutions, they may introduce latency overheads that impact VR responsiveness. Future work will explore comparative evaluations with alternative architectures.

3.6.1. Synthetic Data Generation and Augmentation

Synthetic data are generated in Unity by simulating user and robotic arm movements, following methodologies similar to those in virtual human modeling for disassembly operations [56]. Unity's `ArticulationBody` components capture joint angles, velocities, and end-effector positions, providing a rich dataset for training machine learning models. Simulated user actions are captured via Meta's Movement SDK, while the ROS# bridge facilitates seamless data transfer between Unity and the ROS environment for preprocessing, storage, and further analysis.

While synthetic data generation enables controlled training scenarios, ensuring its applicability to real-world interactions remains challenging. Future work will include validation with real-world datasets to assess generalization performance and domain adaptation strategies. A key aspect of this validation will be testing the LSTM model's ability to adapt synthetic training data to real-world robotic disassembly tasks, ensuring that movement patterns and trajectory predictions remain applicable outside controlled simulations.

3.6.2. Data Cleaning and Preprocessing

Before training, the collected motion data undergo a cleaning step to remove inconsistencies such as abrupt jumps or unrealistic movements caused by physics simulation artifacts in Unity. This ensures that the dataset accurately represents real-world motion patterns.

3.6.3. Data Augmentation

To enhance model generalization and robustness, data augmentation is applied. The augmentation process involves the following:

- **Motion perturbations:** a combination of Gaussian noise injection (0.005) and a small scaling factor of (0.05) is applied to simulate real-world sensor noise and variations in movement execution.
- **Temporal warping:** slight shifts in the motion sequence timing help simulate variations in human and robotic reaction times.

3.6.4. Augmented Data for Training

The cleaned and augmented dataset is then used to train the LSTM model. The augmentation ensures that the model learns from diverse movement patterns, improving its ability to generalize to unseen user behaviors in the virtual training environment.

The dataset effectively supports the predictive model by integrating synthetic data generation, augmentation, and preprocessing while ensuring realism and variability in human–robot interactions.

3.6.5. LSTM Model Architecture and Training

The LSTM model is trained in PyTorch (version 2.1.0) to predict user movement trajectories and robotic actions [57]. This training approach is inspired by uncertainty-informed human motion prediction in collaborative environments [13]. The training pipeline utilizes sequential data, including the following:

- User movement data captured from the Meta Movement SDK.
- Robotic motion data captured through Unity's `ArticulationBody` components [51].

The model consists of three stacked LSTM layers with 256 hidden units per layer. Each LSTM cell captures sequential dependencies in motion data over time, making it well-suited for human–robot interaction modeling.

The LSTM updates its hidden and cell states at each timestep:

$$(h_t, c_t) = \mathrm{LSTM}(x_t, h_{t-1}, c_{t-1}).$$

where

- x_t is the input motion data at time t.
- h_t is the updated hidden state, encoding learned temporal dependencies.
- c_t is the updated cell state, retaining long-term memory across sequences.

This structure enables the model to predict motion trajectories by leveraging past observations while mitigating issues like vanishing gradients affecting traditional RNNs.

LSTMs were selected over standard RNNs due to their superior ability to retain long-term dependencies and their relatively lower computational overhead compared to attention-based architectures such as Transformers. Future work may explore alternative architectures for improved accuracy and efficiency.

Monte Carlo Dropout for Uncertainty Estimation

To enhance robustness, the model integrates a Monte Carlo dropout (MC dropout) rate of 0.1 during inference. It enables multiple forward passes, producing a distribution of possible predictions. This technique helps quantify uncertainty in motion prediction, ensuring the model does not overfit specific patterns and can adapt to varying user behaviors. Given the dynamic nature of disassembly tasks, incorporating uncertainty estimation allows for more reliable and adaptable robotic trajectory adjustments.

Formally, for a given input sequence X, the final prediction is computed as

$$\hat{x}_t = \frac{1}{M} \sum_{m=1}^{M} f_{\theta_m}(X). \tag{7}$$

where $f_{\theta_m}(X)$ represents a stochastic forward pass and M is the number of MC samples. The standard deviation of these predictions represents the uncertainty estimate.

Loss Function and Training

The model is trained to minimize a loss function L that evaluates the difference between the predicted and actual trajectories:

$$L = \frac{1}{N} \sum_{t=1}^{N} ||\hat{x}_t - x_t||^2. \tag{8}$$

where $\hat{x}_t$ is the predicted trajectory at time t and x_t is the ground truth.

The AdamW optimizer is used with a cosine annealing learning rate to improve convergence. Training is conducted over 100 epochs with a batch size of 128.

3.6.6. Model Training and Evaluation

The LSTM-based model for human–robot motion prediction was trained using a structured pipeline involving data preprocessing, model training, evaluation, and deployment. The training was conducted on Google Colab utilizing a Tesla T4 GPU to accelerate computations. The model predicts human and robotic movement trajectories within the virtual environment, enhancing real-time decision making for collaborative tasks.

Training Setup

The model consists of a three-layer LSTM architecture with 256 hidden units per layer and dropout (0.1) for Monte Carlo uncertainty estimation. The training was performed using the following hyperparameters:

- **Sequence length:** 50 time steps per input sequence.
- **Batch size:** 128.
- **Optimizer:** AdamW with weight decay (10^{-4}).
- **Learning rate:** 0.001, decayed using a cosine annealing schedule.

Evaluation Metrics

The model's performance was assessed using standard regression metrics:

- **Root Mean Squared Error (RMSE):** 0.0179
- **Mean Absolute Error (MAE):** 0.0060
- **R^2 score:** 0.4965

Real-time motion prediction within VR environments exhibits intrinsic noise resulting from dynamic interactions and environmental variables. A lower R^2 score is expected in such applications, and improving it further would require a larger, more diverse dataset encompassing a broader range of user behaviors. These values indicate the model's ability to accurately predict motion sequences. The relatively lower R^2 score is expected due to the inherent variability in human motion data, especially in immersive VR environments where factors such as hand jitter, sensor drift, and user variability contribute to prediction noise. Given that motion prediction in human–robot collaboration is dynamic and influenced by external uncertainties, the primary focus remains on minimizing absolute error rather than achieving a high R^2 score.

Visualization and Results

The predictive performance of the model was analyzed through multiple visualizations:

- **LSTM predictions with Monte Carlo dropout:** a time-series plot comparing predicted and actual motion sequences, including uncertainty estimation using Monte Carlo dropout (Figure 7).
- **True vs. predicted values:** scatter plot comparing predicted and actual values, demonstrating model fit (Figure 8).
- **Performance metrics table:** summarizing model accuracy (Table 1).

The trained model was exported to the Open Neural Network Exchange (ONNX) format for efficient inference and cross-platform compatibility. This evaluation validates the system's capability to improve human–robot interaction through predictive analytics and uncertainty estimation. Future improvements include expanding the dataset and fine-tuning hyperparameters for greater accuracy and reliability.

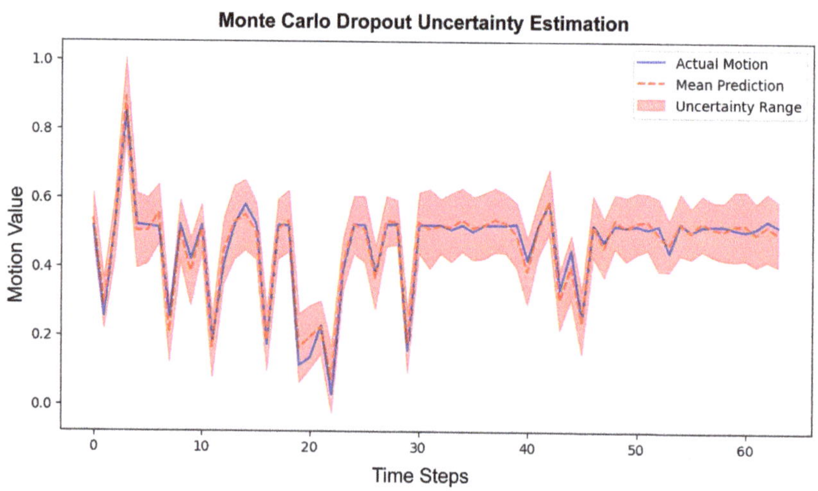

Figure 7. LSTM predictions with Monte Carlo dropout uncertainty estimation.

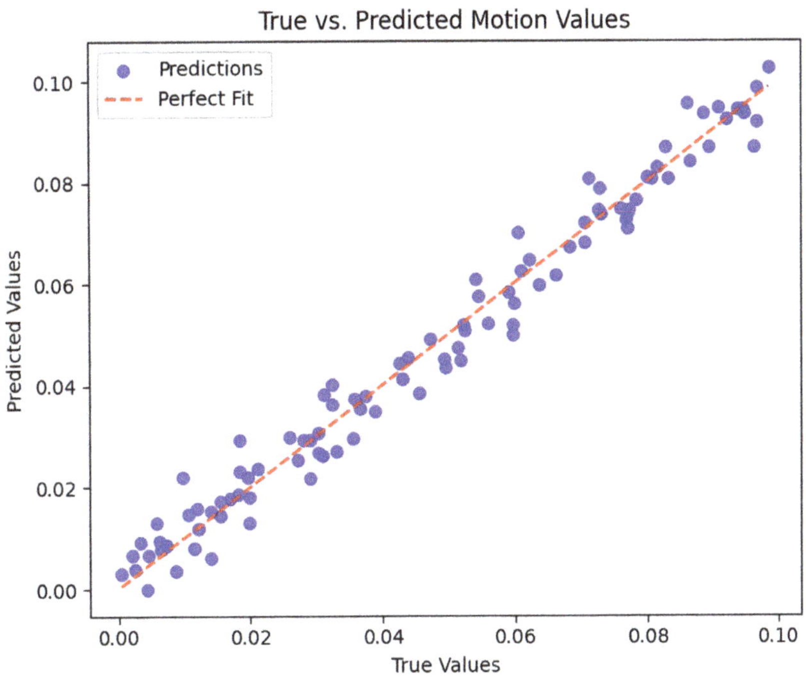

Figure 8. True vs. predicted motion values.

Table 1. Performance metrics for the LSTM Model.

Metric	Value
Root Mean Squared Error (RMSE)	0.0179
Mean Absolute Error (MAE)	0.0060
R^2 score	0.4965

3.6.7. Docker-Based Deployment

To ensure reproducibility and consistency, the system is encapsulated within a Docker container that integrates:

- **ROS Melodic:** facilitates robotic control, sensor integration, and data communication [54].
- **ONNX runtime:** provides efficient execution of the trained LSTM model for real-time inference [55].
- **ROS#:** enables seamless data exchange between Unity and the ROS environment [44].

By encapsulating all dependencies in a containerized environment, the pipeline remains consistent across development and deployment, reducing compatibility issues and simplifying workflow management.

3.6.8. Integration into Unity Environment

The trained LSTM model is deployed in Unity through the following steps:

1. The ONNX model is imported into Unity using the Unity Barracuda library, a lightweight inference engine optimized for neural networks [45].
2. The ROS–Unity bridge is established using ROS#, enabling real-time communication between Unity and the ROS [44].

3. User and robotic motion data are streamed to the ONNX model for inference, generating movement predictions. These predictions guide robotic trajectory adjustments, optimizing collaborative task allocation in disassembly.

3.6.9. Real-Time Inference and Monte Carlo Dropout

The trained ONNX model predicts user movement trajectories and corresponding robotic adjustments during inference. Monte Carlo (MC) dropout is applied to quantify uncertainty, enabling the following:

- More reliable robotic trajectory adjustments by considering uncertainty estimates.
- Enhanced user experience by adapting training difficulty based on confidence levels in motion predictions.
- Identification of high-variance cases where additional data collection or model fine-tuning is needed.

This approach ensures that the system remains adaptable to real-world user behavior and motion variations.

3.6.10. Validation and Refinement

The system undergoes validation using synthetic scenarios generated in Unity, simulating user–robot collaboration for disassembling a hard disk. The validation process includes the following:

- Comparison between predicted and actual motion sequences using RMSE, MAE, and R^2 metrics (Table 1).
- Uncertainty estimation using MC dropout to analyze prediction confidence (Figure 7).
- Error analysis through distribution plots to refine the model (Figure 8).

Initial deployment tests in Unity using ONNX indicate smooth real-time inference, with predictions aligning well with expected motion trajectories. Further optimizations will reduce latency and ensure consistent inference across different hardware configurations.

The insights obtained from these tests iteratively refine the LSTM model, enhancing accuracy and robustness under diverse training conditions.

3.6.11. Benefits of the Integration

- **Efficiency:** ONNX provides fast and lightweight inference suitable for real-time VR applications [55].
- **Cross-platform compatibility:** The ONNX format enables deployment across different devices and hardware configurations [55].
- **Reproducibility:** encapsulation in a Docker container ensures that all components (ROS, Unity, and model dependencies) remain consistent across various development and deployment environments [58].
- **Uncertainty-aware predictions:** MC dropout improves prediction reliability by incorporating uncertainty into robotic motion planning.

Deployment and Future Considerations

This structured integration of deep learning within a virtual training environment effectively enhances human–robot collaboration. Future improvements will explore expanding the dataset with real-world motion capture data. While synthetic data have provided a practical foundation for training, real-world validation remains crucial for assessing the system's generalizability. This will involve collecting user motion data from physical training sessions and evaluating how well the VR-based model translates to real disassembly scenarios.

The integration of synthetic data, Monte Carlo dropout, and ONNX-based real-time inference lays the groundwork for scalable and adaptable motion prediction models in VR-based human–robot collaboration systems.

4. VR Training and Discussion

The VR training system developed integrates cutting-edge virtual reality technology to create an immersive and interactive environment for technical training. This system is designed to simulate real-world tasks within a controlled setting, allowing for detailed monitoring and evaluation of user performance. The primary focus of this system revolves around a VR scenario tailored for a disassembly task, highlighting the system's potential to enhance technical skills and accuracy.

4.1. Hard Disk Disassembly Task

The case study centers on disassembling a hard disk within the VR environment. Users are presented with a virtual hard disk model and a set of tools necessary for the task. The objective is to successfully disassemble it, working alongside the robot using the provided virtual tools and adhering to the standard procedures used in actual hardware maintenance. The user actions are highlighted with a purple material like in Figure 9 and the robot actions are highlighted with a green material like in Figure 10. The corresponding instructions are also shown in the environment to guide the user to use the specific tool for the task.

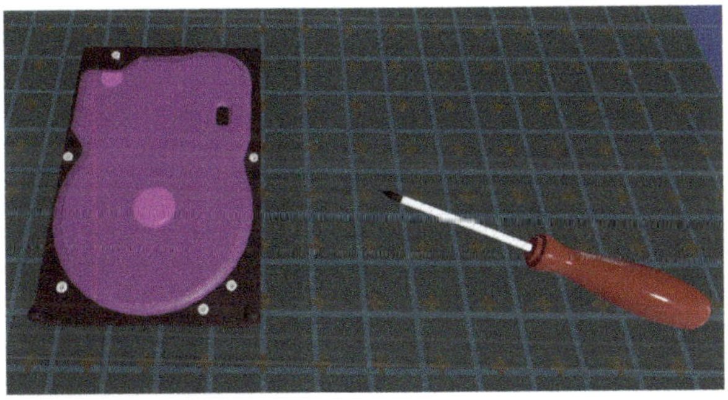

Figure 9. Awaiting user action.

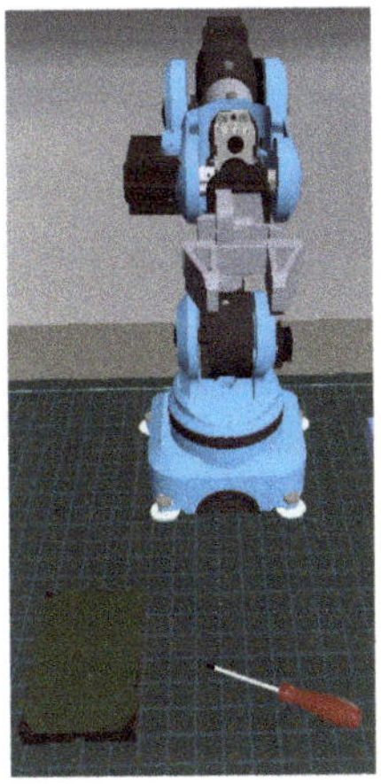

Figure 10. Awaiting robot action.

4.2. Training Setup

The VR setup includes a head-mounted display (HMD)—Quest Pro in this case—hand-tracking, and a space that mimics a typical technician's workstation. The Unity engine and the Meta Presence SDK facilitate realistic interactions with virtual components, such as unscrewing bolts or replacing the disk platter. A demonstration of the VR training environment setup for human–robot collaborative disassembly tasks is shown in Figure 11.

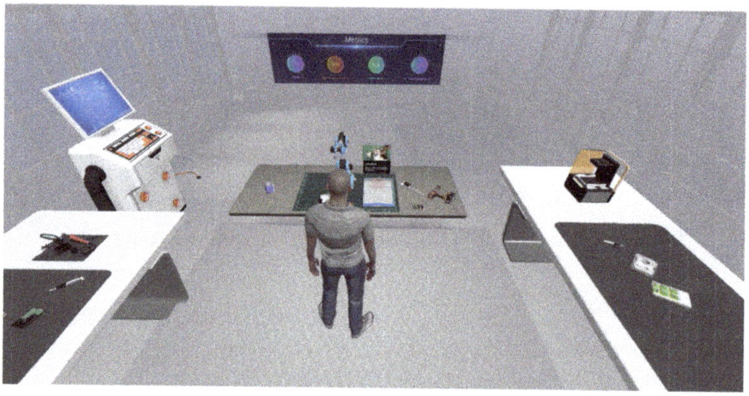

Figure 11. The setup of VR training environment for disassembly task.

4.3. Demonstration of a Short Disassembly Task

We conducted a short disassembly demonstration to evaluate the effectiveness of our VR training environment for human–robot collaborative disassembly tasks. Figure 12 presents snapshots of key steps in the disassembly process, performed collaboratively by a human and a robot (note: the full video is available at https://drive.google.com/file/d/1yc43qBkES99NDzTyj84-tnFHQBmjO3qO/view, (accessed on 16 March 2025).

Safety is a critical component of the disassembly task. As illustrated in Figure 11, the system provides visual and auditory feedback to alert users of errors during disassembly, guiding them in avoiding hazards while handling sensitive components. To further demonstrate the effectiveness of our VR training system, we present a graphical representation of the movement patterns between the human operator and the robot during a typical disassembly task, as shown in Figure 13. The blue line represents the human's movement path while the green line traces the robot's motions. The points where these paths intersect and diverge highlight areas of collaboration and independent action. This synchronization is crucial for ensuring the human and robot can work together seamlessly, minimizing interference and maximizing safety and efficiency. Analyzing these movement patterns allows us to fine-tune the robot's programming to anticipate human actions better, resulting in more intuitive and responsive interactions. This enhances the training experience and prepares both the human and robot for smoother collaboration in real-world tasks.

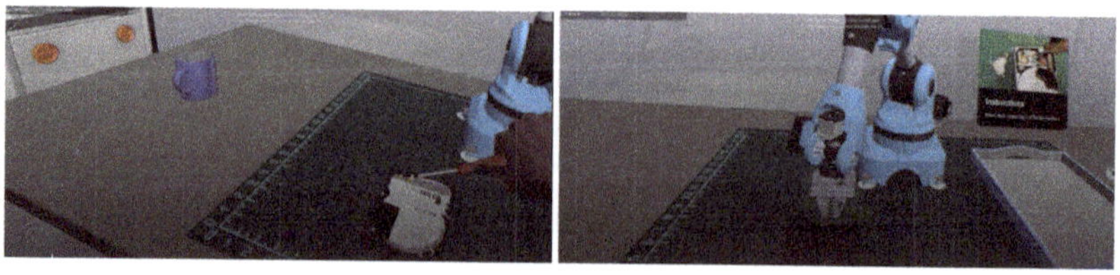

(a) (b)

Figure 12. *Cont.*

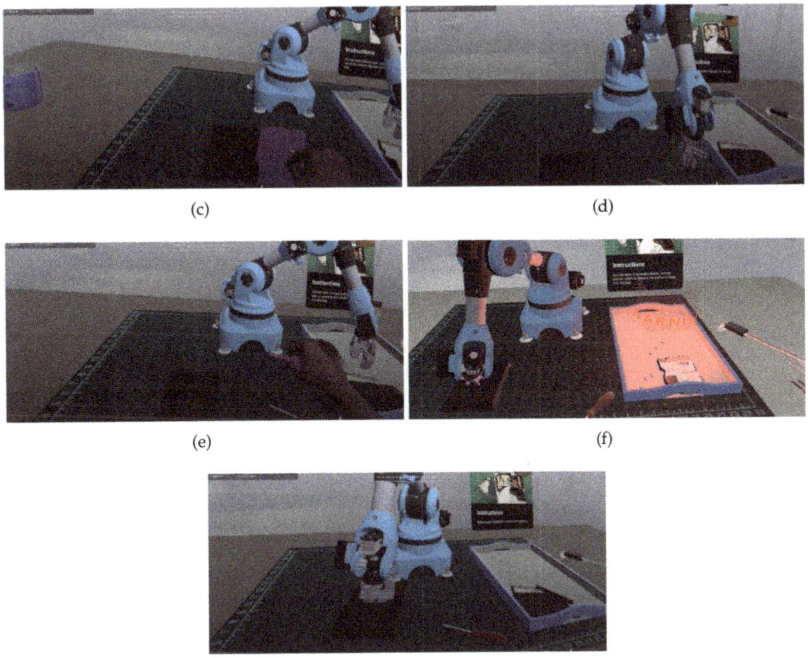

(c) (d)

(e) (f)

(g)

Figure 12. A short demonstration of a set of human-robot collaborative disassembly tasks for a hard disk device: (**a**) Initial setup with the hard disk device positioned for disassembly; (**b**) Robot positioning and alignment for collaborative work; (**c**) Human operator carefully removing a disassembled component and placing it in the designated bin; (**d**) Robot dynamically adjusting its trajectory to avoid collision with human hands during collaborative workspace interaction; (**e**) Robot arm autonomously grasping a separated component and placing it in the bin; (**f**) Error state notification with visual feedback when incorrect action is detected; (**g**) Successful component removal and task completion.

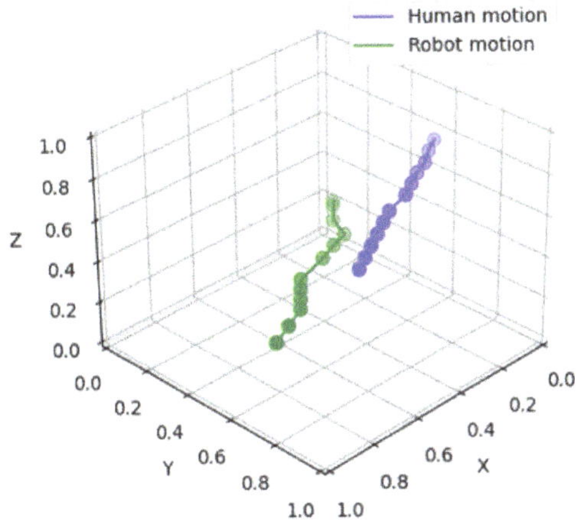

Figure 13. Comparison of human and robot motion paths in VR training.

4.4. Discussion

Analysis of the effectiveness of the VR training indicates significant improvements in task performance due to the immersive and interactive nature of the system. Integrating advanced tracking technologies and realistic interaction scenarios significantly enhances learning outcomes and user engagement. Comparisons with traditional training methods show that VR training increases the retention and applicability of learned skills.

4.4.1. Training Performance Analysis

To evaluate the effectiveness of the VR training environment, we analyzed key performance metrics over multiple training sessions.

Figure 14 presents the trends in task completion time and error rate, demonstrating a consistent improvement with additional training. The decreasing task completion times indicate that users become more efficient over repeated sessions while the declining error rate suggests increased precision in task execution.

Similarly, Figure 15 shows increased motion path efficiency over sessions, reflecting enhanced coordination between human operators and the robot. The trend suggests that VR-based training helps users develop more optimal movement strategies, minimizing unnecessary actions and improving collaboration.

Real-time metrics are displayed within the virtual environment to evaluate user performance during the VR-based human–robot collaborative disassembly task. This allows users to monitor their efficiency and adjust their actions accordingly.

Figure 16 illustrates the in-game **metrics dashboard**, which provides real-time feedback on four key performance indicators:

- **Timer:** displays the session duration in minutes.
- **Task completion time:** shows the time to complete the last disassembly step.
- **Error rate:** tracks the percentage of incorrect actions performed by the user.
- **User engagement:** represents an engagement score based on interaction frequency and consistency.

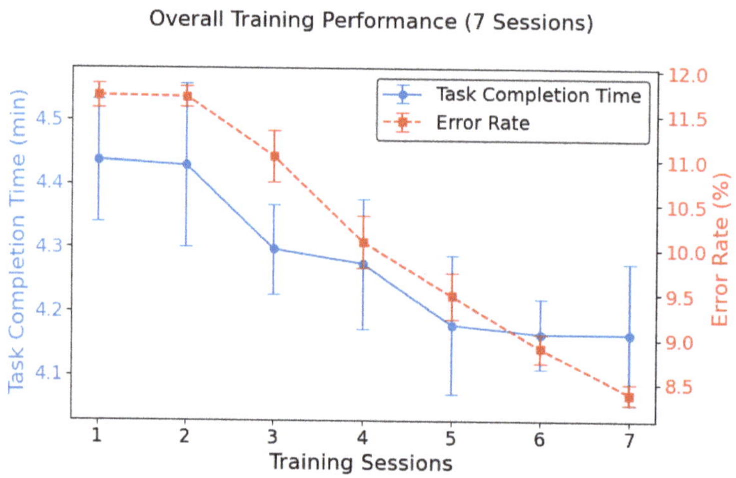

Figure 14. Overall training performance—task completion time and error rate over 7 sessions.

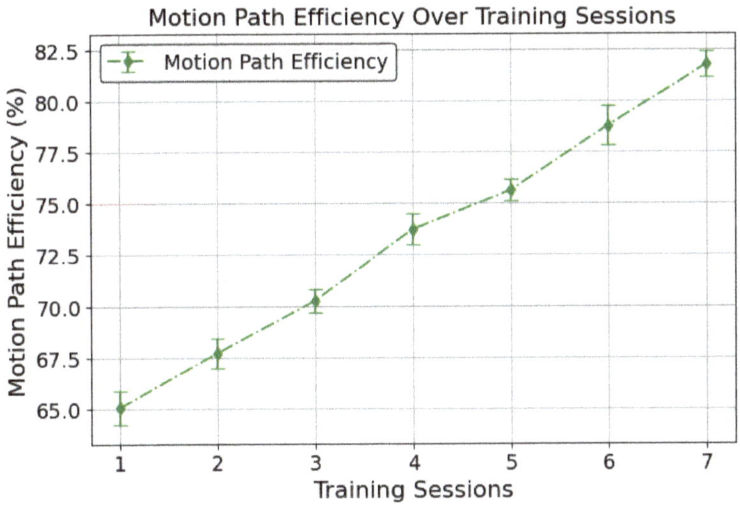

Figure 15. Motion path efficiency over training sessions, showing increased optimal movement patterns.

This dashboard enhances user awareness and encourages iterative improvement by providing instant feedback during training.

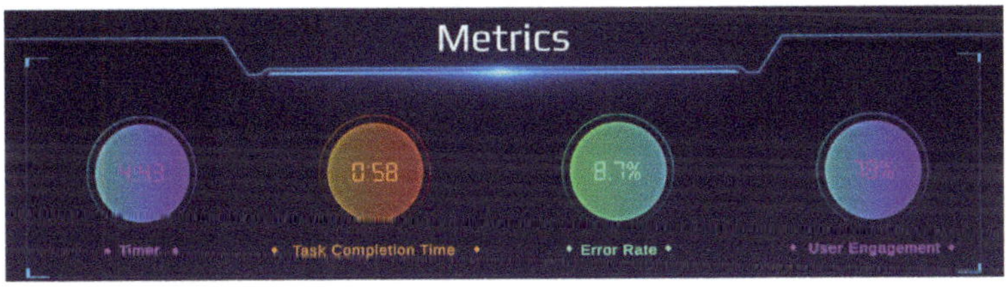

Figure 16. The in-game metrics dashboard displays real-time feedback on session duration, task completion time, error rate, and user engagement, allowing users to assess and optimize their performance.

4.4.2. Effectiveness

Hand- and body-tracking integration proved instrumental in enhancing the training experience by providing real-time feedback and allowing for natural interactions. The data suggest that such immersive training significantly improves precision in tasks that require fine motor skills, as evidenced by the reduced error rates and increased speed of completion.

The VR training system creates a psychologically safe environment where users can practice potentially hazardous disassembly procedures without the risk of damaging expensive components or causing workplace injuries. This safety aspect allows trainees to experiment freely, make mistakes, and learn from them without real-world consequences. Our observations indicate that this psychological safety accelerates confidence building, as users become more willing to attempt complex procedures they might otherwise approach with hesitation. However, the system must balance this safety with appropriate feedback that maintains awareness of real-world risks. Visual and auditory cues during error states (as shown in Figure 12f) help maintain this awareness, ensuring that the confidence gained in VR translates appropriately to physical environments where consequences are tangible.

4.4.3. Comparison with Traditional Training Methods

Compared to conventional training methods, the VR system speeds up the learning process and engages users more effectively, maintaining high attention levels throughout the session. The VR environment's immediate feedback and interactive nature foster enhanced learning outcomes, preparing participants for real-world applications more effectively than traditional theoretical training.

While the VR environment provides a consistent and organized training space compared to potentially cluttered real-world workstations (as seen in the comparison between Figures 1 and 2), this controlled nature presents both advantages and limitations. The consistent environment enables focused learning without distractions and standardized evaluation metrics across users. However, this idealized environment may not fully prepare trainees for the unpredictability of real-world scenarios where tools may be misplaced, components may vary in condition, or environmental factors may interfere with task execution. Future iterations of the system will incorporate controlled variability to better bridge this gap between virtual consistency and real-world unpredictability.

4.5. Implications for VR Training

The findings from this case study highlight the transformative potential of VR training systems in enhancing technical education and training across various sectors. By simulating a diverse array of realistic scenarios, such as complex disk assembly and disassembly, VR facilitates a deeper understanding of user performance and generates synthetic data that can be instrumental in developing more sophisticated machine learning models. This capability to enrich training with detailed analytics and adaptable scenarios extends the applicability of VR technologies to various industries, including electronics, automotive, and aerospace, setting a robust foundation for future research into its expansive potential in professional training environments.

4.6. Limitations of the System

While our system demonstrates promising results in enhancing training efficiency, several limitations should be acknowledged. First, the current implementation lacks multi-user scenarios, which Shamsuzzoha et al. [59] identified as crucial for comprehensive industrial training. Future work will explore collaborative VR environments where multiple users can interact simultaneously with robotic systems, enabling team-based training scenarios that better reflect real-world industrial settings.

Second, potential tracking inaccuracies may affect motion prediction accuracy. The Meta Quest Pro's Inside-Out Body-Tracking technology faces inherent limitations, particularly for lower body tracking [46]. While "Generative Legs" provides AI-based predictions for lower body movements, it cannot track specific foot positions or individual leg movements [47]. Similarly, hand tracking encounters challenges with occlusion when hands overlap or move outside the camera's field of view [48], potentially affecting precision during complex manipulation tasks. Chen et al. [60] demonstrated how these tracking limitations can impact disassembly task evaluation, particularly when precise movements are required. Further, the inclusion of an immersive sensory experience may also affect the operator's fatigue levels [60]. Future implementations will incorporate multi-sensor fusion approaches to improve tracking robustness and accuracy.

Third, our current LSTM model, while effective for motion prediction, could benefit from comparison with alternative architectures. Transformer-based models have shown promising results in sequential prediction tasks, as demonstrated by Daneshmand et al. [34], who surveyed various deep learning approaches for robotic assembly and disassembly in Industry 4.0 contexts. Hjorth and Chrysostomou [30] highlighted the potential of hybrid approaches combining deep learning with physical models for improved prediction accuracy in human–robot collaborative disassembly. Jacob et al. [14] further emphasized the importance of exploring various prediction architectures to address the unique challenges of human–robot cooperation in disassembly tasks. Future work will explore these alternatives to potentially improve prediction accuracy and computational efficiency while maintaining the real-time performance required for immersive VR environments.

Fourth, further validation in real-world industrial settings is needed to fully assess the system's effectiveness, following evaluation frameworks proposed by Schwarz et al. [8]. This validation will include comparative studies between VR-trained and traditionally trained operators performing actual disassembly tasks, measuring the transfer of learning and retention over time. Yang et al. [52] demonstrated the importance of validating virtual haptic disassembly platforms against real-world performance metrics, particularly when considering the disassembly process and physical attributes of components.

5. Conclusions and Future Work

5.1. Conclusions

This study demonstrates the effectiveness of a VR-based training system for technical skills development in hard disk drive disassembly tasks. By integrating advanced body- and hand-tracking technologies with LSTM-based motion prediction, the system creates an immersive environment that enhances both training outcomes and human–robot collaboration.

Our implementation shows several key advantages over traditional training methods:

- **Enhanced engagement:** the interactive VR environment increases user engagement through realistic simulation of industrial tasks.
- **Accelerated learning:** immediate feedback and risk-free repetition enable faster skill acquisition without material costs.
- **Improved precision:** advanced tracking technology ensures proper technique execution, particularly for fine motor skills.
- **Data-driven insights:** the system collects comprehensive performance metrics that inform both training optimization and motion prediction models.

The LSTM-based motion prediction framework demonstrates promising results in forecasting human movements during disassembly tasks, enabling more intuitive human–robot collaboration. The system's ability to generate synthetic training data further enhances the robustness of these predictive models.

Our approach aligns with recent work by Qu et al. [61], who demonstrated the effectiveness of reinforcement learning for robotic disassembly task training and skill transfer. However, our system extends beyond purely robotic learning to incorporate human–robot collaboration in an immersive VR environment, addressing the need for integrated training solutions identified by Jacob et al. [14].

Future work will explore the integration of ergonomic design principles as suggested by Bortolini et al. [11] to further enhance the training experience and reduce physical strain during collaborative tasks. Additionally, we plan to investigate multi-user training scenarios [59] and the integration of more advanced robotic systems [35] to expand the applicability of our framework.

Furthermore, we will focus on improving the UI/UX experience through iterative design processes that incorporate user-centered design methodologies. This will include enhancing the visual feedback mechanisms, implementing more intuitive interaction patterns, and developing adaptive interfaces that respond to individual user proficiency levels. The current metrics dashboard will be expanded to provide more granular and personalized feedback, helping users identify specific areas for improvement while maintaining an uncluttered visual experience.

To validate these enhancements, we will conduct comprehensive user studies with participants of varying expertise levels, from novices to experienced technicians. These studies will employ mixed-methods approaches combining quantitative performance metrics (task completion time, error rates, and learning curves) with qualitative assessments (cognitive load measurements, user satisfaction surveys, and semi-structured interviews). Special attention will be paid to evaluating the system's learnability, efficiency, memorability, error prevention, and overall user satisfaction following established usability heuristics. The findings from these studies will not only validate the effectiveness of our UI/UX improvements but also provide valuable insights for further refinements to maximize training outcomes in human–robot collaborative disassembly tasks.

As Industry 4.0 continues to transform manufacturing processes, our VR training system provides a valuable tool for preparing workers to collaborate effectively with robotic systems in disassembly tasks. By combining immersive VR, advanced motion prediction, and synthetic data generation, we contribute to the development of more efficient, safe, and adaptable human–robot collaborative workflows.

5.2. Future Work

Building on these findings, we identify several directions for future research. We plan to implement additional industrial scenarios beyond hard drive disassembly to validate the system's flexibility across different applications. To enhance prediction capabilities, we will explore alternative deep learning architectures, such as Transformer models, that may improve motion prediction accuracy. The integration of eye-tracking technology will provide deeper insights into user attention patterns during complex tasks, potentially revealing optimization opportunities for task sequencing and interface design. We also aim to develop multi-user collaborative scenarios where multiple operators can interact simultaneously with robotic systems, better reflecting real industrial team environments. Finally, longitudinal validation studies will be conducted to measure skill transfer and retention from VR to real-world environments, establishing empirical evidence for the long-term effectiveness of our approach. These enhancements will further strengthen the connection between immersive training and predictive modeling for human–robot collaboration in industrial settings.

These enhancements will further strengthen the connection between immersive training and predictive modeling for human–robot collaboration in industrial settings.

5.3. Implications for Industry

Developing sophisticated VR training systems has significant implications for the industry, particularly in fields where precision and accuracy are paramount. The utilization of synthetic data generation within these systems is a key innovation that enhances their capability further, allowing for the training of machine learning models that can adapt to user behaviors. This makes the training process more robust and highly adaptable to the nuances of individual performance, paving the way for highly personalized training experiences.

As technology progresses, it is anticipated that VR training systems will become a staple in technical education and professional development across various industries, revolutionizing the way practical skills are taught and refined.

Author Contributions: Conceptualization, B.L.; funding acquisition, X.L., M.Z., and B.L.; methodology, Y.M. and B.L.; project administration, X.L., M.Z., and B.L.; software, Y.M. and S.T.; supervision, B.L.; visualization, Y.M.; writing—original draft, Y.M.; writing—review and editing, S.T., X.L., M.Z., and B.L. All authors have read and agreed to the published version of the manuscript.

Funding: National Science Foundation (NSF) Directorate for Engineering (grant no. 2132773/2506209 and 2132923/2422640).

Data Availability Statement: The data presented in this study are available upon request from the corresponding author.

Acknowledgments: The authors acknowledge the funding support from NSF. However, the views expressed here are those of the authors and are not necessarily those of the NSF.

Conflicts of Interest: The authors declare no conflicts of interest.

Abbreviations

The following abbreviations are used in this manuscript:

AR	Augmented reality
VR	Virtual reality
MR	Mixed reality
HRC	Human–robot collaboration
ROS	Robot Operating System
URDF	Unified Robot Description Format
LSTM	Long Short-Term Memory
RNN	Recurrent Neural Network
TCT	Task completion time
ONNX	Open Neural Network Exchange

References

1. Mourtzis, D.; Zogopoulos, V.; Vlachou, E. Augmented Reality Application to Support Remote Maintenance as a Service in the Robotics Industry. *Procedia CIRP* **2017**, *63*, 46–51. [CrossRef]
2. Webel, S.; Bockholt, U.; Engelke, T.; Gavish, N.; Olbrich, M.; Preusche, C. An augmented reality training platform for assembly and maintenance skills. *Robot. Auton. Syst.* **2013**, *61*, 398–403. [CrossRef]
3. Gavish, N.; Gutierrez, T.; Webel, S.; Rodriguez, J.; Tecchia, F. Design Guidelines for the Development of Virtual Reality and Augmented Reality Training Systems for Maintenance and Assembly Tasks. *Bio Web Conf.* **2011**, *1*, 00029. [CrossRef]
4. Li, J.R.; Khoo, L.P.; Tor, S.B. Desktop virtual reality for maintenance training: An object oriented prototype system (V-REALISM). *Comput. Ind.* **2003**, *52*, 109–125. [CrossRef]
5. Tao, F.; Zhang, H.; Liu, A.; Nee, A.Y.C. Digital Twin in Industry: State-of-the-Art. *IEEE Trans. Ind. Inform.* **2019**, *15*, 2405–2415. [CrossRef]
6. Werrlich, S.; Nitsche, K.; Notni, G. Demand Analysis for an Augmented Reality based Assembly Training. In Proceedings of the PETRA '17: 10th International Conference on PErvasive Technologies Related to Assistive Environments, Island of Rhodes, Greece, 21–23 June 2017; pp. 416–422.
7. Hořejší, P. Augmented Reality System for Virtual Training of Parts Assembly. *Procedia Eng.* **2015**, *100*, 699–706. [CrossRef]
8. Schwarz, S.; Regal, G.; Kempf, M.; Schatz, R. Learning Success in Immersive Virtual Reality Training Environments: Practical Evidence from Automotive Assembly. In Proceedings of the NordiCHI '20: 11th Nordic Conference on Human-Computer Interaction, Tallinn, Estonia, 25–29 October 2020; pp. 1–11. [CrossRef]
9. Lerner, M.A.; Ayalew, M.; Peine, W.J.; Sundaram, C.P. Does Training on a Virtual Reality Robotic Simulator Improve Performance on the da Vinci® Surgical System? *J. Endourol.* **2010**, *24*, 467–472. [CrossRef]
10. Dwivedi, P.; Cline, D.; Jose, C.; Etemadpour, R. Manual Assembly Training in Virtual Environments. In Proceedings of the 2018 IEEE 18th International Conference on Advanced Learning Technologies, Mumbai, India, 9–13 July 2018; pp. 395–401.
11. Bortolini, M.; Botti, L.; Galizia, F.G.; Mora, C. Ergonomic Design of an Adaptive Automation Assembly System. *Machines* **2023**, *11*, 898. [CrossRef]
12. Carlson, R.; Gonzalez, R.; Geary, D. Enhancing Training Effectiveness with Synthetic Data and Feedback in VR Environments. *J. Appl. Ergon.* **2015**, *46*, 56–65.
13. Liu, W.; Liang, X.; Zheng, M. Task-Constrained Motion Planning Considering Uncertainty-Informed Human Motion Prediction for Human–Robot Collaborative Disassembly. *IEEE/ASME Trans. Mechatron.* **2023**, *28*, 2056–2071. [CrossRef]
14. Jacob, S.; Klement, N.; Bearee, R.; Pacaux-Lemoine, M.P. Human-Robot Cooperation in Disassembly: A Rapid Review. In Proceedings of the 21st International Conference on Informatics in Control, Automation and Robotics, Porto, Portugal, 18–20 November 2024; SCITEPRESS-Science and Technology Publications: Setubal, Portugal, 2024; Volume 2, pp. 212–219.

15. Lee, Y.S.; Rashidi, A.; Talei, A.; Beh, H.J.; Rashidi, S. A Comparison Study on the Learning Effectiveness of Construction Training Scenarios in a Virtual Reality Environment. *Virtual Worlds* **2023**, *2*, 36–52. [CrossRef]
16. Schwarz, M.; Weser, J.; Martinetz, T.; Pawelzik, K. Immersive Virtual Reality for Cognitive Training: A Pilot Study on Spatial Navigation in the Cave Automatic Virtual Environment. *Front. Psychol.* **2020**, *11*, 579993.
17. Rocca, R.; Rosa, P.; Fumagalli, L.; Terzi, S. Integrating Virtual Reality and Digital Twin in Circular Economy Practices: A Laboratory Application Case. *Sustainability* **2020**, *12*, 2286. [CrossRef]
18. Boud, A.C.; Baber, C.; Steiner, S.J. Virtual Reality: A Tool for Assembly? *Presence* **2000**, *9*, 486–496. . [CrossRef]
19. Al-Ahmari, A.M.; Abidi, M.H.; Ahmad, A.; Darmoul, S. Development of a virtual manufacturing assembly simulation system. *Adv. Mech. Eng.* **2016**, *8*, 1687814016639824. [CrossRef]
20. Numfu, M.; Riel, A.; Noel, F. Virtual Reality Technology for Maintenance Training. *Appl. Sci. Eng. Prog.* **2020**, *13*, 274–282. [CrossRef]
21. Gutiérrez, T.; Rodríguez, J.; Vélaz, Y.; Casado, S.; Suescun, A.; Sánchez, E.J. IMA-VR: A multimodal virtual training system for skills transfer in Industrial Maintenance and Assembly tasks. In Proceedings of the 19th International Symposium in Robot and Human Interactive Communication, Viareggio, Italy, 13–15 September 2010; pp. 428–433, ISSN 1944-9437. [CrossRef]
22. Li, Z.; Wang, J.; Yan, Z.; Wang, X.; Anwar, M.S. An Interactive Virtual Training System for Assembly and Disassembly Based on Precedence Constraints. In *Advances in Computer Graphics, Proceedings of the 36th Computer Graphics International Conference, CGI 2019, Calgary, AB, Canada, 17–20 June 2019*; Gavrilova, M., Chang, J., Thalmann, N.M., Hitzer, E., Ishikawa, H., Eds.; Springer: Cham, Switzerland, 2019; pp. 81–93. [CrossRef]
23. Westerfield, G.; Mitrovic, A.; Billinghurst, M. Intelligent Augmented Reality Training for Industrial Operations. *IEEE Trans. Learn. Technol.* **2014**, *7*, 331–344.
24. Pan, X.; Cui, X.; Huo, H.; Jiang, Y.; Zhao, H.; Li, D. Virtual Assembly of Educational Robot Parts Based on VR Technology. In Proceedings of the 2019 IEEE 11th International Conference on Engineering Education (ICEED), Kanazawa, Japan, 6–7 November 2019; pp. 1–5. [CrossRef]
25. Seth, A.; Vance, J.M.; Oliver, J.H. Virtual reality for assembly methods prototyping: A review. *Virtual Real.* **2011**, *15*, 5–20. [CrossRef]
26. Devagiri, J.S.; Paheding, S.; Niyaz, Q.; Yang, X.; Smith, S. Augmented Reality and Artificial Intelligence in industry: Trends, tools, and future challenges. *Expert Syst. Appl.* **2022**, *207*, 118002. [CrossRef]
27. Zhao, X. Extended Reality for Safe and Effective Construction Management: State-of-the-Art, Challenges, and Future Directions. *Buildings* **2023**, *13*, 155. [CrossRef]
28. Lee, M.L.; Liu, W.; Behdad, S.; Liang, X.; Zheng, M. Robot-Assisted Disassembly Sequence Planning with Real-Time Human Motion Prediction. *IEEE Trans. Syst. Man Cybern. Syst.* **2022**, *53*, 438–450. [CrossRef]
29. Vongbunyong, S.; Vongseela, P.; Sreerattana-aporn, J. A Process Demonstration Platform for Product Disassembly Skills Transfer. *Procedia CIRP* **2017**, *61*, 281–286. [CrossRef]
30. Hjorth, S.; Chrysostomou, D. Human–robot collaboration in industrial environments: A literature review on non-destructive disassembly. *Robot. Comput. Integr. Manuf.* **2022**, *73*, 102208. [CrossRef]
31. Ottogalli, K.; Rosquete, D.; Rojo, J.; Amundarain, A.; María Rodríguez, J.; Borro, D. Virtual reality simulation of human-robot coexistence for an aircraft final assembly line: Process evaluation and ergonomics assessment. *Int. J. Comput. Integr. Manuf.* **2021**, *34*, 975–995. [CrossRef]
32. Kothari, A. Real-Time Motion Prediction for Efficient Human-Robot Collaboration. Master's Thesis, Massachusetts Institute of Technology, Cambridge, MA, USA, 2023. Available online: https://dspace.mit.edu/handle/1721.1/152639 (accessed on 16 March 2025).
33. Liu, W.; Liang, X.; Zheng, M. Dynamic Model Informed Human Motion Prediction Based on Unscented Kalman Filter. *IEEE/ASME Trans. Mechatron.* **2022**, *27*, 5287–5295. [CrossRef]
34. Daneshmand, M.; Noroozi, F.; Corneanu, C.; Mafakheri, F.; Fiorini, P. Industry 4.0 and prospects of circular economy: A survey of robotic assembly and disassembly. *Int. J. Adv. Manuf. Technol.* **2023**, *124*, 2973–3000. [CrossRef]
35. Kuts, V.; Cherezova, N.; Sarkans, M.; Otto, T. Digital Twin: Industrial robot kinematic model integration to the virtual reality environment. *J. Mach. Eng.* **2020**, *20*, 53–64. [CrossRef]
36. Xu, X.; Guo, P.; Zhai, J.; Zeng, X. Robotic kinematics teaching system with virtual reality, remote control and an on–site laboratory. *Int. J. Mech. Eng. Educ.* **2020**, *48*, 197–220. [CrossRef]
37. Pérez, L.; Rodríguez-Jiménez, S.; Rodríguez, N.; Usamentiaga, R.; García, D.F. Digital Twin and Virtual Reality Based Methodology for Multi-Robot Manufacturing Cell Commissioning. *Appl. Sci.* **2020**, *10*, 3633. [CrossRef]
38. Sassanelli, C.; Rosa, P.; Terzi, S. Supporting disassembly processes through simulation tools: A systematic literature review with a focus on printed circuit boards. *J. Manuf. Syst.* **2021**, *60*, 429–448. [CrossRef]
39. Zhao, F.; Deng, W.; Pham, D.T. A Robotic Teleoperation System with Integrated Augmented Reality and Digital Twin Technologies for Disassembling End-of-Life Batteries. *Batteries* **2024**, *10*, 382. [CrossRef]

40. Zhao, G.; Wang, Y. Development of Machine Tool Disassembly and Assembly Training and Digital Twin Model Building System based on Virtual Simulation. *Int. J. Mech. Electr. Eng.* **2024**, *2*, 50–59. [CrossRef]
41. Pérez, L.; Diez, E.; Usamentiaga, R.; García, D.F. Industrial robot control and operator training using virtual reality interfaces. *Comput. Ind.* **2019**, *109*, 114–120. [CrossRef]
42. Zhao, F.; Pham, D.T. Integration of Augmented Reality and Digital Twins in a Teleoperated Disassembly System. In *Advances in Remanufacturing, Proceedings of the VII International Workshop on Autonomous Remanufacturing, IWAR 2023, Detroit, MI, USA, 16–18 October 2023*; Fera, M., Caterino, M., Macchiaroli, R., Pham, D.T., Eds.; Springer Nature Switzerland: Cham, Switzerland, 2024; pp. 93–105. [CrossRef]
43. Lewczuk, K. Virtual Reality Application for the Safety Improvement of Intralogistics Systems. *Sustainability* **2024**, *16*, 6024. [CrossRef]
44. Siemens AG. ROS#: Integration of Unity3D with ROS 2019. Available online: https://github.com/siemens/ros-sharp (accessed on 16 March 2025).
45. Unity-Technologies. Unity Barracuda Documentation 2024. Available online: https://docs.unity3d.com/Packages/com.unity.barracuda%401.0/manual/index.html (accessed on 16 March 2025).
46. Meta Platforms Inc. *Inside-Out Body Tracking and Generative Legs*; Meta Platforms Inc.: Menlo Park, CA, USA, 2023. Available online: https://developers.meta.com/horizon/blog/inside-out-body-tracking-and-generative-legs/ (accessed on 16 March 2025).
47. Meta Platforms Inc. *Movement Overview*; Meta Platforms Inc.: Menlo Park, CA, USA, 2023. Available online: https://developers.meta.com/horizon/documentation/unity/move-overview (accessed on 16 March 2025).
48. Meta Platforms Inc. *Learn About Hand and Body Tracking on Meta Quest*; Meta Platforms Inc.: Menlo Park, CA, USA, 2024. Available online: https://www.meta.com/help/quest/290147772643252/ (accessed on 16 March 2025).
49. Behdad, S.; Berg, L.P.; Thurston, D.; Vance, J. Leveraging Virtual Reality Experiences with Mixed-Integer Nonlinear Programming Visualization of Disassembly Sequence Planning Under Uncertainty. *J. Mech. Des.* **2014**, *136*, 041005. [CrossRef]
50. Denavit, J.; Hartenberg, R.S. A kinematic notation for lower-pair mechanisms based on matrices. *J. Appl. Mech.* **1955**, *22*, 215–221. [CrossRef]
51. Technologies, U. Articulation Body Component Reference. 2021. Available online: https://docs.unity3d.com/2021.3/Documentation/Manual/class-ArticulationBody.html (accessed on 13 March 2025).
52. Yang, Y.; Yang, P.; Li, J.; Zeng, F.; Yang, M.; Wang, R.; Bai, Q. Research on virtual haptic disassembly platform considering disassembly process. *Neurocomputing* **2019**, *348*, 74–81. [CrossRef]
53. Tahriri, F.; Mousavi, M.; Yap, I.D.H.J. Optimizing the Robot Arm Movement Time Using Virtual Reality Robotic Teaching System. *Int. J. Simul. Model.* **2015**, *14*, 28–38. [CrossRef]
54. Quigley, M.; Conley, K.; Gerkey, B.; Faust, J.; Foote, T.; Leibs, J.; Wheeler, R.; Ng, A.Y. ROS: An open-source Robot Operating System. In Proceedings of the ICRA Workshop on Open Source Software, Kobe, Japan, 12–17 May 2009; Volume 3, p. 5.
55. Corporation, M. ONNX Runtime: A High-Performance Inference Engine for Machine Learning Models 2024. Available online: https://onnxruntime.ai (accessed on 13 March 2025).
56. Liu, Q.; Liu, Z.; Xu, W.; Tang, Q.; Zhou, Z.; Pham, D.T. Human-robot collaboration in disassembly for sustainable manufacturing. *Int. J. Prod. Res.* **2019**, *57*, 4027–4044. [CrossRef]
57. Paszke, A., Gross, S.; Massa, F.; Lerer, A.; Bradbury, J.; Chanan, G.; Killeen, T.; Lin, Z.; Gimelshein, N.; Antiga, L.; et al. PyTorch: An imperative style, high-performance deep learning library. *Adv. Neural Inf. Process. Syst.* **2019**, *32*, 10761–10771.
58. Merkel, D. Docker: Lightweight Linux containers for consistent development and deployment. *Linux J.* **2014**, *239*, 2.
59. Shamsuzzoha, A.; Toshev, R.; Vu Tuan, V.; Kankaanpaa, T.; Helo, P. Digital factory—Virtual reality environments for industrial training and maintenance. *Interact. Learn. Environ.* **2021**, *29*, 1339–1362. [CrossRef]
60. Chen, J.; Mitrouchev, P.; Coquillart, S.; Quaine, F. Disassembly task evaluation by muscle fatigue estimation in a virtual reality environment. *Int. J. Adv. Manuf. Technol.* **2017**, *88*, 1523–1533. [CrossRef]
61. Qu, M.; Wang, Y.; Pham, D.T. Robotic Disassembly Task Training and Skill Transfer Using Reinforcement Learning. *IEEE Trans. Ind. Inform.* **2023**, *19*, 10934–10943. [CrossRef]

machines

Article

Virtual and Real Occlusion Processing Method of Monocular Visual Assembly Scene Based on ORB-SLAM3

Hanzhong Xu [1], Chunping Chen [1], Qingqing Yin [1], Chao Ma [1] and Feiyan Guo [2,*]

[1] Shanghai Institute of Aerospace Technical Foundation, Shanghai 201109, China
[2] School of Mechanical Engineering, University of Science and Technology Beijing, Beijing 100083, China
* Correspondence: guofy@ustb.edu.cn

Abstract: Addressing the challenge of acquiring depth information in aero-engine assembly scenes using monocular vision, which complicates mixed reality (MR) virtual and real occlusion processing, we propose an ORB-SLAM3-based monocular vision assembly scene virtual and real occlusion processing method. The method proposes optimizing ORB-SLAM3 for matching and depth point reconstruction using the MNSTF algorithm. MNSTF can solve the problems of feature point extraction and matching in weakly textured and texture-less scenes by expressing the structure and texture information of the local images. It is then proposed to densify the sparse depth map using the double-three interpolation method, and the complete depth map of the real scene is created by combining the 3D model depth information in the process model. Finally, by comparing the depth values of each pixel point in the real and virtual scene depth maps, the virtual occlusion relationship of the assembly scene is correctly displayed. Experimental validation was performed with an aero-engine piping connector assembly scenario and by comparing it with Holynski's and Kinect's methods. The results showed that in terms of virtual and real occlusion accuracy, the average improvement was 2.2 and 3.4 pixel points, respectively. In terms of real-time performance, the real-time frame rate of this paper's method can reach 42.4 FPS, an improvement of 77.4% and 87.6%, respectively. This shows that the method in this paper has good performance in terms of the accuracy and timeliness of virtual and real occlusion. This study further demonstrates that the proposed method can effectively address the challenges of virtual and real occlusion processing in monocular vision within the context of mixed reality-assisted assembly processes.

Keywords: occlusion handling; mixed reality; ORB-SLAM3; assisted assembly

1. Introduction

The background of an aero-engine assembly site is highly complex, featuring numerous overlapping components and intricate spatial relationships. This complexity renders it challenging to accurately capture and process depth information. Existing methods often fail to handle these complexities due to their reliance on sparse depth maps or limited computational capabilities, resulting in inaccurate occlusion relationships and poor virtual–real fusion effects [1,2]. Mixed reality (MR)-assisted assembly guidance methods are widely used in the assisted assembly process of major equipment scenarios such as aero-engines [3–5]. After the virtual and real objects have finished tracking and registering, the virtual model or other guidance information can be correctly superimposed on the real scene position in MR glasses, but with the movement of the operator, the viewing angle of the MR glasses and the positional relationship between the virtual and real models

 https://doi.org/10.3390/machines13030212

will change, and the occlusion relationship between them will also change. Assembly sites for other complex manufacturing tasks such as aero-engines, satellite manufacturing, automobile manufacturing, etc. have complex backgrounds and complex occlusion relationships between assembly objects [6–9]. The virtual–real occlusion processing method is crucial for the virtual–real fusion display of the assembly scene, and the correct virtual–real occlusion relationship can help the operator understand the guided information more easily when using the MR operation guidance assistance system [10]. Therefore, in the process of virtual–reality fusion for MR-assisted assembly of an aero-engine, the correct occlusion processing between virtual objects and objects in the real environment is a problem that needs to be solved urgently.

At present, there are three commonly used methods for real-image occlusion processing: real-image occlusion processing based on 3D model reconstruction [11], image depth information computation [12], and image features [13]. However, the virtual occlusion processing method based on 3D model reconstruction requires a large amount of preliminary 3D modeling of the assembly scene, which is a large amount of work and requires too much accuracy in model construction for such major and complex equipment as aero-engines. The advantage of the image feature-based virtual masking processing method is that it requires less equipment, but the disadvantage is also obvious: it needs the real assembly scene to have enough features to identify and detect, but the external accessories of an aero-engine, such major equipment, has the characteristics of a metal surface lacking textural features. Therefore, the method based on image depth information computation for virtual masking processing is more suitable for complex assembly scenes such as aero-engines. However, the traditional virtual masking processing method has poor robustness, low accuracy, and poor timeliness for this kind of assembly scenario.

The aero-engine assembly scene is characterized by a complex background, intricate relationships between assembly models, and an inability to use binocular vision cameras. To address the virtual and real occlusion problem in the MR-assisted assembly process of aero-engine external accessories, we propose a monocular vision-based virtual and real occlusion processing method using an improved ORB-SLAM3 framework. The proposed method reconstructs the depth points of the assembly scene using the enhanced ORB-SLAM3 algorithm, densifies the sparse depth map through bicubic interpolation, and integrates the depth information from the 3D model in the digitized process model to generate a complete depth map of the real scene. By comparing the depth values of each pixel in the real and virtual scene depth maps, our method accurately determines the spatial relationship between virtual and real models, correctly handles virtual–real occlusion, and optimizes the virtual–real fusion display effect for MR-assisted assembly of aero-engine external attachments.

- Propose a novel method based on ORB-SLAM3 for handling virtual–real occlusion in MR environments, specifically tailored for the complex assembly scenes of aero-engines.
- Propose the use of the MNSTF algorithm for matching optimization of ORB-SLAM3 and depth point reconstruction of assembly scenes, which is suitable for feature point extraction and matching in weakly textured or untextured regions.
- Propose a bicubic interpolation-based method to densify sparse depth maps and integrate them with the depth information from the 3D model in the digitized process model, generating a complete and accurate depth map of the real scene.

The rest of the paper is organized as follows. Section 2 presents an overview of the current state of domestic and international research on virtual occlusion processing. Section 3 describes the monocular vision virtual occlusion processing flow based on ORB-SLAM3. Section 4 describes the virtual occlusion rendering flow based on depth images. Section 5 uses an assembly scene of space engine piping connectors for experimental validation and analyzes the results.

2. Related Work

The virtual and real occlusion method based on depth image information calculation first acquires the depth information of the real object, and through the to-be-superimposed region of the virtual object, the depth information of the virtual object and the real object are compared and judged to obtain the correct occlusion relationship [14]. The current way of depth image information acquisition is mainly accomplished by binocular vision and monocular vision [15,16].

In terms of binocular vision to obtain depth image information, Kim [17] used the depth information of a real assembly scene to obtain the occlusion relationship between real and imaginary objects, searched for smooth and accurate parallax vector fields with clear object boundaries in binocular image pairs, and reconstructed the three-dimensional surface of the real scene using the vector fields to obtain the occlusion relationship between the real and imaginary models. Zhen [18] proposed a new algorithm based on binocular stereo vision (BSV) for recognition and depth estimation of inland vessels and proposed a sub-pixel-level feature point detection and matching algorithm based on the ORB algorithm, which further improved the density of image feature points and detection accuracy and was more conducive to the computation of the parallax value of the image. Yang [19] proposed a three-dimensional reconstruction system combining binoculars and depth cameras that effectively improved the accuracy of the three-dimensional reconstruction and accurately recognized the distance from the camera. Luo [20] proposed a method to improve the accurate acquisition of depth information by combining the multi-channel information of RGB-D, which can determine the depth relationship between the virtual object and the real object pixel point by pixel point. Zhang [21] realized the virtual masking processing of virtual and real scenes through the layering of the depth information and the optimization of the mapping.

The above research was based on a binocular vision sensor to obtain depth image information, which has certain performance requirements. With the current focus on lightweight devices and convenience [22], many scholars have begun to study monocular vision to obtain scene depth information of [23–25]. Simon [26] presented MonoNav, a fast 3D reconstruction and navigation stack that leverages recent advances in deep predictive neural networks to enable accurate 3D scene reconstruction from monocular images and pose streams. MonoNav uses off-the-shelf pretrained monocular depth estimation and fusion techniques to build maps. Chang [27] proposed a combination of monocular visual depth estimation and multiview depth estimation to complement the strengths of the two methods. Multiview depth is very accurate, but only for highly textured regions and high parallax values. Single-view depth captures the local structure of mesoscopic regions, including untextured regions, but the estimated depth lacks global consistency. Luo [28] proposed a deep neural network that can simultaneously estimate camera pose and reconstruct full-resolution depth information about the environment using only monocular continuous images. Fink [29] proposed taking a monocular depth map as input, scaling that depth map to absolute distances based on structure from motion, and converting the depth to a triangular surface mesh. This depth mesh is then refined in a local optimization that enforces photometric and geometric consistency. Zhang [30] presented a new unsupervised learning framework for estimating scene depth and camera pose from video sequences that can be used as a basis for 3D reconstruction and augmented reality (AR). Li [31] used monocular SLAM to reconstruct an assembly scene, converted the reconstructed sparse 3D points into depth points in the depth map, and thickened the sparse map processing. Finally, the depth relationships between the real assembly scene, assembly objects, and the virtual model were compared. Holynski [32] proposed a new monocular depth estimation algorithm that propagates the sparse depth to each pixel, obtains the sparse depth of the key points using the SLAM method, and optimizes the propagation of the sparse depth with the image edges. The method can compute the depth map in near-real time.

In summary, when addressing the MR virtual–real occlusion problem in complex assembly scenarios, such as those of aero-engines, both binocular and monocular vision methods exhibit certain limitations in computational efficiency, occlusion accuracy, and adaptability to complex scenes. To reduce overreliance on hardware performance, a monocular vision approach is selected herein to acquire depth image information of an assembly site during MR virtual–real occlusion processing for a complex aero-engine assembly site.

3. Monocular Vision-Based Occlusion Handling Method

3.1. Virtual and Real Occlusion Handling Framework

Owing to the suboptimal performance of contemporary MR glasses, with many lacking binocular vision capabilities, the application of binocular vision algorithms is constrained. Consequently, in this study, we adopt a ORB-SLAM3 monocular vision-based virtual–reality occlusion processing method. In the domain of MR-assisted assembly, monocular vision technology offers notable advantages over depth-camera technology. Monocular vision technology obviates the need for specialized depth cameras, thus alleviating the weight burden of the overall hardware configuration. In this paper, considering the characteristics of MR-assisted assembly scenarios for aero-engine external attachments, a real-time virtual–reality occlusion processing method grounded in monocular vision is proposed. The processing flow framework is depicted in Figure 1.

The framework mainly describes how to fuse the virtual scene and the real scene using the improved ORB-SLAM3 monocular vision technique to realize the processing of virtual and real occlusion. First, the virtual camera data and virtual scene model are constructed from 3D digitized process model information, and then the virtual scene depth map is generated from these data. In the real scene, the real scene model and depth map are obtained by the improved ORB-SLAM3 sparse depth point reconstruction technique, and the depth map is densified. Next, the virtual scene depth map is matched with the real scene depth map to determine the occluded objects in the virtual scene. According to the depth map information of the virtual and real scenes, the depth value images of the virtual and real scenes are obtained by RGB pixel value filling, and then the values of the scenes are calculated and judged. The judgment rule is that if the depth value of the real scene is greater than the depth value of the virtual scene, then render the depth value of the virtual scene, and vice versa render the depth value of the real scene. Thus, the corresponding depth images are generated to realize the accurate rendering of dynamic assembly objects in the MR environment. Finally, in the MR-assisted assembly system for aero-engine external accessories, the depth information relationship between the real scene and the virtual scene is judged, and the final virtual–reality fusion scene is rendered and processed to obtain the final accurate virtual–reality fusion assembly scene.

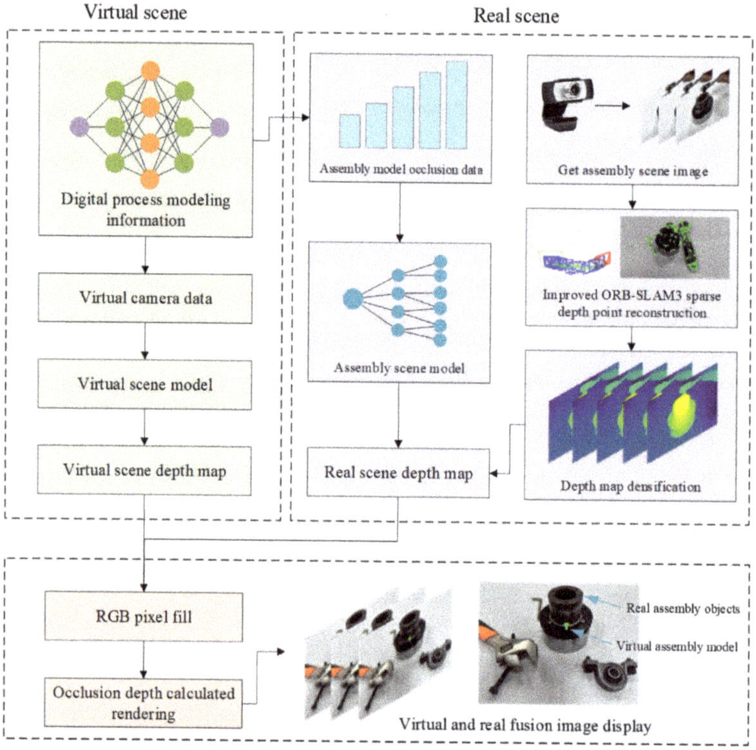

Figure 1. Monocular vision-based occlusion processing flow framework.

3.2. The Relationship Between Virtual and Real Object Occlusion

In the MR-assisted assembly system for aero-engine external accessories, the occlusion relationship between virtual and real objects represents a complex and crucial factor that directly impacts the efficiency of assisted assembly. As illustrated in Figure 2, the virtual–real object occlusion relationship is present in the aero-engine external attachment assembly scene. In the MR-assisted assembly system for aero-engine external accessories, virtual objects pertain to the visualization elements and virtual models derived from the 3D digital process model. These virtual objects are grounded in pre-established model data and projected into the real assembly environment via MR technology to furnish functions such as assembly guidance, information display, and process simulation. Virtual objects, relying on computer-generated 3D models, possess no physical entity in the real world. However, in the MR-assisted assembly scenario, they interact with real objects to form a virtual–real integrated working environment. Conversely, real objects denote the elements existing in the assembly scene, encompassing real assembly components (e.g., aero-engine parts, assemblies, and connectors) and real-world scenes (e.g., work platforms, workshop backgrounds, etc.). These real objects are physical entities with actual shapes, sizes, and positions during the assembly process, and their states change as the assembly operation progresses.

In MR-assisted assembly systems, the occlusion relationship between virtual and real objects is critical. This relationship must accurately reflect the physical occlusion in the real world to ensure that the operator receives correct visual guidance. For example, if a virtual object should be occluded by a real object in the real world, the virtual object must also appear behind the real object in the MR-assisted assembly scenario. By establishing precise occlusion relationships, operators can better understand the assembly process and avoid errors caused by visual misdirection.

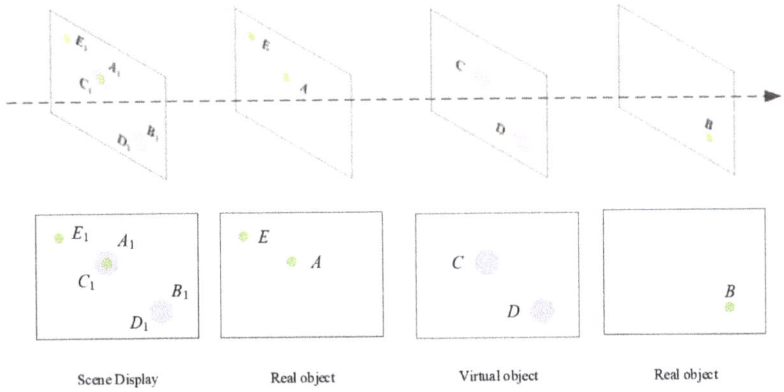

Figure 2. Schematic diagram of occlusion relationships.

4. Depth Image-Based Voxel Occlusion Rendering

4.1. Assembly Scene Sparse Depth Point Reconstruction

In the visual guidance process of aero-engine external accessory assembly, images captured by monocular vision systems lack depth information, which restricts their application in virtual occlusion processing. Nevertheless, the acquisition of depth information from monocular vision images can be achieved through the utilization of SLAM technology. Initially, the monocular camera mounted on the MR glasses must be calibrated to determine the camera's internal and external reference matrices K and M. This calibration involves the determination of parameters such as focal length, principal point position, and aberration coefficient. Subsequently, a sequence of images of the assembly scene is captured using the calibrated camera. These images will be employed for subsequent depth point reconstruction and voxel-masking processing.

The acquired images undergo feature extraction using the fast and rotated oriented brief—simultaneous localization and mapping (ORB-SLAM3) algorithm. ORB is an algorithm for image feature detection and description capable of rapidly and accurately extracting key points in an image. Feature points of neighboring images are matched, and feature trajectories are constructed from the matching results. This process facilitates the establishment of spatial relationships between consecutive image frames. Based on the matched feature points and feature trajectories, sparse depth points of the assembly scene are reconstructed using the triangulation principle. These depth points offer information regarding the approximate position and shape of objects within the scene. ORB-SLAM3 is a monocular vision-based SLAM algorithm that can estimate the camera pose and reconstruct the 3D structure of the environment from the video information captured by the monocular camera. As the monocular camera cannot directly measure depth, ORB-SLAM3 employs the structure from motion (SFM) method to estimate the depth information of points in the scene.

In SFM, when the same assembly scene is captured by the camera at different locations, the camera motion as well as the 3D position of the point P in the scene can be estimated by tracking the movement of the feature point between consecutive frames. The polar geometry constraint is the correspondence between two rays formed by the assembly scene space point P and the camera optical centers O_1 and O_2 in both views. Point P in the assembly scene space has projection points p_1 and p_2 in the camera views at two different locations, and then the lines connecting these two points to the respective camera optical centers O_1 and O_2 intersect at one point in space. This is shown in Figure 3.

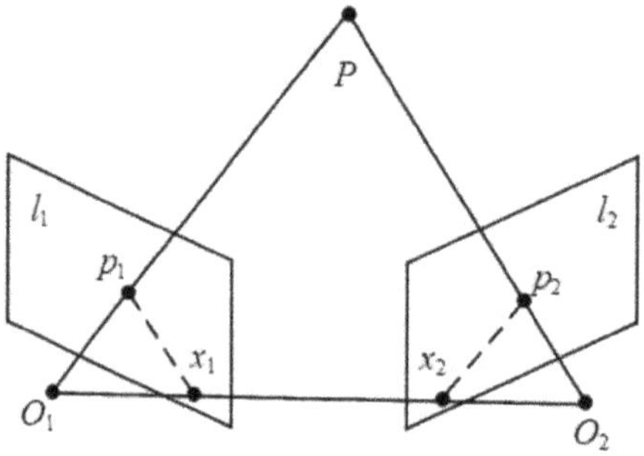

Figure 3. Feature point pair-pole geometry constraints.

The positions of p_1 and p_2 are calculated based on the camera's small-hole imaging model:

$$p_1 = K \cdot [R|t] \cdot P \tag{1}$$

$$p_2 = K \cdot [R'|t'] \cdot P \tag{2}$$

where K is the internal reference matrix of the camera containing the focal length and the position of the optical center. r, t, and R', t' denote the forward and backward rotation and translation matrices from the first camera coordinate system to the second camera coordinate system in the world coordinate system, respectively. The 3D coordinates of the spatial point P in the world coordinate system are the corresponding 2D-pixel coordinates.

For each pair of matching points, the following equations can be established based on their homogeneous coordinates and the projection relationship of the camera:

$$p_2^T F p_1 = 0 \tag{3}$$

where p_1 and p_2 are homogeneous coordinate vectors of the matching points in the first and second views, respectively, and F is the basis matrix.

The basis matrix F and the essence matrix E are of the form:

$$F = K^{-T} E K^{-1} \tag{4}$$

$$E = t \times R \tag{5}$$

The internal reference K is obtained by camera calibration, and the rotation matrix R and translation matrix t of the camera in the SLAM coordinate system can be obtained according to the above equations.

Considering only the translation along the baseline direction and ignoring the motion perpendicular to the baseline direction, the problem can be reduced to a two-dimensional problem. Therefore, the above system of equations can be simplified as:

$$x_1 = f\frac{X}{Z} \tag{6}$$

$$x_2 = f\frac{X + \Delta X}{Z + \Delta Z} \tag{7}$$

where x_1 and x_2 are the transverse coordinates of the projections of the point in the two views, ΔX is the amount of translation of the camera between the two positions along the direction of the baseline, ΔZ is the amount of depth change of point P between the two views, and f is the focal length of the camera.

With the above system of equations, it is possible to solve for the depth value Z. First, the difference between the transverse coordinates of the two projected points is utilized to express ΔX:

$$\Delta x = x_2 - x_1 = f\left(\frac{X + \Delta X}{Z + \Delta Z} - \frac{X}{Z}\right) \tag{8}$$

This can be transformed into a linear equation about ΔZ:

$$f\frac{\Delta X}{Z + \Delta Z} - f\frac{\Delta X}{\Delta Z} = \Delta x - f\frac{X}{Z} + f\frac{X}{Z + \Delta Z} \tag{9}$$

Further simplification leads to:

$$\Delta Z = \frac{fZ^2}{f\Delta X - Z\Delta x} \tag{10}$$

where ΔX is the actual travel distance between the two camera positions, provided by the inertial measurement unit (IMU). Above, the depth value Z can be solved by solving for Δx according to Equation (10) and the focal length f obtained from the camera calibration.

4.2. Improved ORB-SLAM3 Feature Point Matching

The aero-engine and external accessories themselves have smooth surfaces, similar colors, and a lack of texture. When using the traditional ORB feature point matching, it shows some limitations: the correct point pairs to be matched become fewer, and it faces the problem of fewer matching points and increased false-matching rate. This will lead to incorrect or non-generation of depth maps. Therefore, to obtain enough matched point pairs in the aero-engine external accessory assembly scenario, in this paper, the ORB matching link in the ORB-SLAM3 algorithm is improved. Using the multi-neighborhood structure tensor feature (MNSTF) algorithm, which is an optimization method for feature point matching [33], the matching accuracy is improved by considering the local structure information around the feature points, and it is suitable for feature point extraction in weakly textured or untextured regions and matching.

The core of the MNSTF algorithm is to extract structural tensor features in multiple neighborhoods of the feature points and combine these features for matching. The following are the steps in the computational process of the MNSTF algorithm.

For each feature point pi after each matching, define K neighborhoods of different scales, each containing a certain number of neighboring points. These neighborhoods can be circular, square, or other shapes. For each neighborhood, the structure tensor T_j is computed, which is a tensor reflecting the distribution of points within the neighborhood of second or higher order. The structure tensor can be computed from the positional information of the points within the neighborhood.

$$T_j(k) = \sum\nolimits_{p \in N_j} w(p, p_i) \cdot (p - p_i) \otimes (p - p_i) \tag{11}$$

Here, N_j is the set of points in the j neighborhood, p is a point in the neighborhood, $w\ (p, p_i)$ is a weight function to adjust the contribution of each point to the structure tensor, and $\otimes$ denotes the tensor product.

The structural tensor features of each neighborhood are combined to construct a comprehensive feature descriptor D_i for representing the local structural information of the feature point p_i:

$$D_i = \{T_1(p_i), T_2(p_i), \ldots, T_K(p_i)\} \tag{12}$$

For all feature point pairs (p_i, p_j) in both images, their descriptors D_i and D_j are used to compute the similarity score S_{ij}:

$$S_{ij} = \text{Similarity}(D_i, D_j) \tag{13}$$

where similarity is the similarity measure based on the similarity score S_{ij}. The best-matching pairs are selected and the matching results are further optimized by the random sample consensus (RANSAC) robust method to eliminate outliers and improve the accuracy of matching.

The modified ORB-SLAM3 algorithm is used to construct a sparse depth point cloud in the aero-engine external accessory MR-assisted assembly system for virtual occlusion processing. The specific process is as follows. First, images of the assembly scene from multiple angles are captured by a moving camera. Next, feature points in neighboring images are matched to determine their correspondence in space. The result of the matching forms a feature trajectory, which describes the motion trajectory of the feature points in consecutive image frames. Using the principle of triangulation, the 3D positions of the objects in the scene can be calculated from the feature points and the feature trajectory. Finally, all the calculated 3D position points are combined to form a sparse depth point cloud, and the corresponding sparse depth image is generated. The detected feature points and camera positions of the assembled scene are shown in Figure 4, where the data of the 10th, 20th, 30th, 50th, 70th, 100th, 125th, and 150th frames are shown.

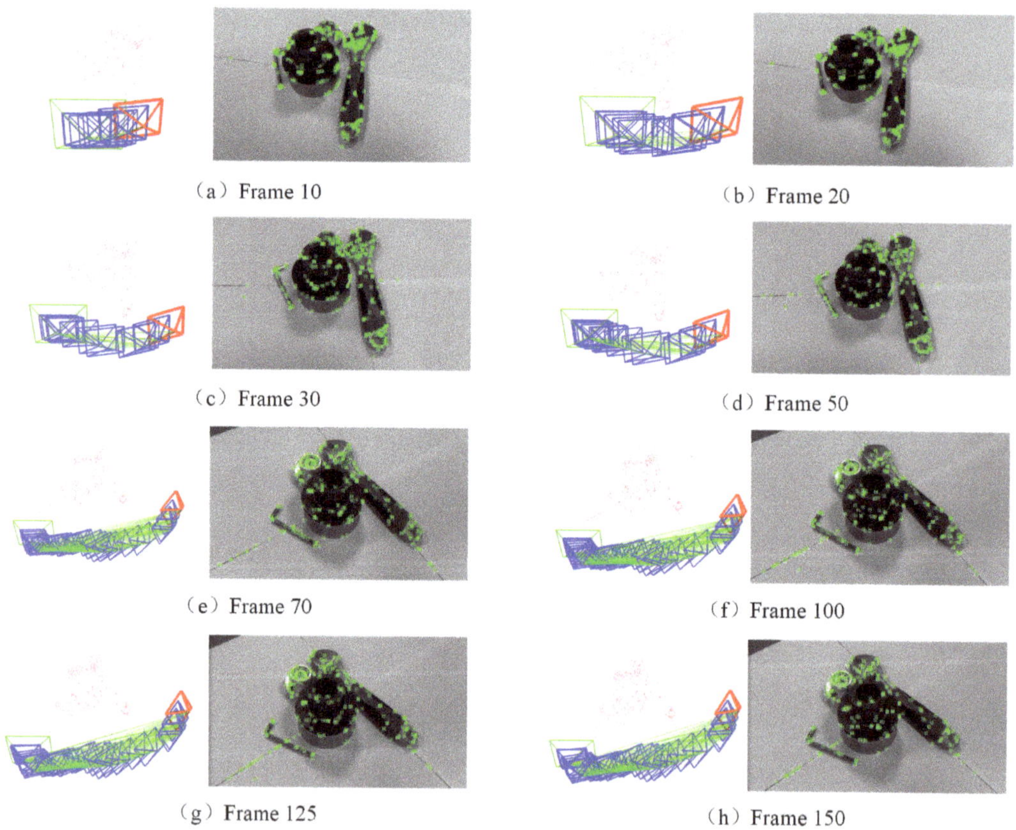

Figure 4. Key points and camera positions.

Figure 5 shows the comparison of the localization effect of different algorithms for camera tracking, and the 3D trajectories are extracted by quantitative analysis of the camera position data. This is compared and analyzed with the 3D trajectories of other algorithms. As can be seen in Figure 5a, b, in the comparison of the trajectory translations and rotations, the highest error of the improved ORB-SLAM3 algorithm proposed in this paper is 3.2 mm and the smallest is only 0.42 mm, with an average accuracy error of 1.6 mm, while the average errors of the other two algorithms, ORB-SLAM and ORB-SLAM2, are 2.3 mm and 2.1 mm. It is proved that the proposed algorithm in this paper has higher localization accuracy compared with other algorithms. This is also mainly due to the improvement in the matching algorithm in this paper, which makes the number of matched pairs of points increase and the correct rate of matching increase such that it obtains higher accuracy.

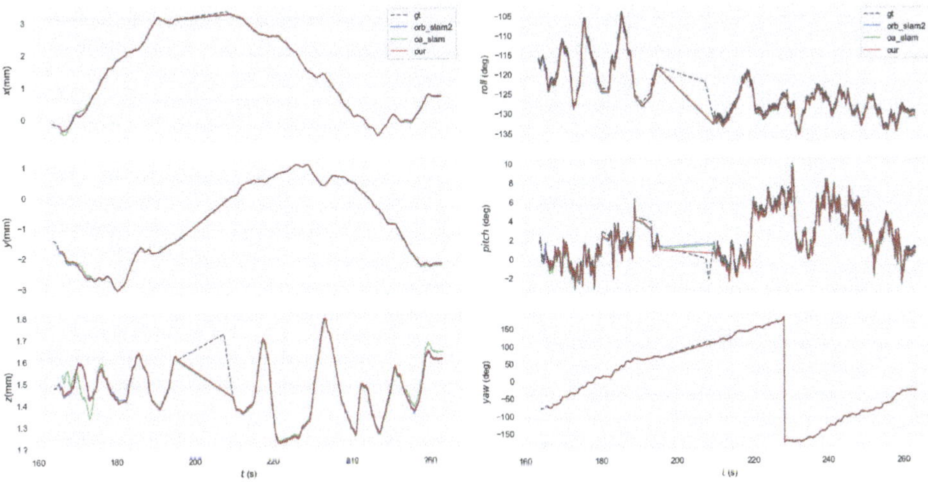

Figure 5. Comparison of tracking and positioning effects of different algorithms. (**a**) Comparison of trajectory translations; (**b**) Comparison of trajectory rotations.

4.3. Depth Map Densification

The matching method of ORB-SLAM3 is improved by the MNSTF algorithm, which makes the matched correct point pairs more uniform and accurate. Next, the obtained sparse depth map needs to be densified. The densification process used in this paper utilizes the bilinear interpolation (BI) method. Using the weighted average of the 16 surrounding pixel points to calculate the value of the new pixel provides smoother and more accurate results when the image is zoomed in or out.

It is necessary to compute the pixel value d of the point (x, y) in the new image I_{new}, which has the floating-point coordinates (x', y') in the original image and can be found surrounded by 16-pixel points with coordinates (x_i, y_j), where i and j range from 0 to 3, respectively, and each of the pixel points has a corresponding depth value $d_{i,j}$.

Next, the weighted value of each pixel point in this 4 × 4 grid is calculated:

$$d(x', y') = (1-u)(1-v)d_{0,0} + u(1-v)d_{1,0} + (1-u)vd_{0,1} + uvd_{1,1} + \ldots + (1-u)u)(1-v)d_{3,3} \quad (14)$$

where $u = x' - \lfloor x' \rfloor$ and $v = y' - \lfloor y' \rfloor$ are the positions of the new pixel points (x, y) concerning the nearest integer pixel point. $f(x, y)$ is the pixel value of point (x, y) in the original image.

Two interpolations were performed on the formula. Interpolating each y-value in the x-direction yields a one-dimensional interpolation result:

$$g(y') = \sum_{3}^{i=0} f(i, y0)B_{i,1}(u) + \sum_{3}^{i=0} f(i, y1)B_{i,2}(u) + \sum_{3}^{i=0} f(i, y2)B_{i,3}(u) + \sum_{3}^{i=0} f(i, y3)B_{i,4}(u) \quad (15)$$

Next, the one-dimensional interpolation results obtained above are interpolated in the y-direction to obtain the final pixel values:

$$h(x') = \sum_{3}^{j=0} g(y0)B_{j,1}(v) + \sum_{3}^{j=0} g(y1)B_{j,2}(v) + \sum_{3}^{j=0} g(y2)B_{j,3}(v) + \sum_{3}^{j=0} g(y3)B_{j,4}(v) \quad (16)$$

where $B_{i,k}(u)$ and $B_{j,k}(v)$ are one-dimensional bicubic interpolation basis functions corresponding to the u and v directions, respectively. Through the above process, the pixel value $h(x')$ of the new pixel point (x, y) can be obtained. This process needs to be repeated for each new pixel point until the entire image is interpolated. Bicubic interpolation [34] provides smoother and more natural results than bilinear interpolation [35] and trilinear interpolation [36].

The densified depth image can be obtained by using the method based on double-trilinear interpolation, as shown in Figure 6. Figure 6a shows the original RGB images of different scenes and Figure 6b shows the obtained densified depth image. The depth image provides the depth information of each pixel point in the scene, which reflects the distance and positional relationship between the objects in the scene and expresses the process of depth from near too far in yellow and blue. Figure 6c–f shows the results of the densified depth maps for different assembly scenes, and the use of the method based on bi-trilinear interpolation has the characteristics of high resolution, smooth transition, local detail preservation, adaptability, computational efficiency, and accuracy in generating the densified depth images. These features enable the method to provide accurate depth information in MR-assisted assembly systems for aero-engine external accessories.

4.4. Assembly Scene Occlusion Rendering

To realize accurate virtual and real occlusion effects in MR-assisted assembly of external attachments of an aero-engine, real scene depth maps and virtual scene depth maps are obtained according to the above method. These maps are used to represent the depth information at different locations in the scene. A uniform coordinate system is used to process these two scene depth maps. This helps to convert the different depth maps to the same reference frame for subsequent processing, as shown in Figure 7. During the virtual occlusion rendering process, the depth maps of the real scene are computed using an improved algorithm based on ORB-SLAM3. It is ensured that these depth maps are consistent with the depth map coordinate system of the virtual scene model in the same viewpoint. Next, for each pixel point, the depth values of the corresponding positions in the two depth maps are compared to determine the front–back relationship between the real and virtual objects. If the depth value of the real pixel point is greater than that of the virtual point, it indicates that the virtual object is closer to the camera, and therefore the pixel value of the virtual pixel point should be rendered and displayed. Conversely, if the depth value of the real pixel point is less than or equal to that of the virtual point, the pixel value of the real pixel point should be rendered. Finally, the final rendered image is presented by traversing all the pixel points on the image and merging them for display.

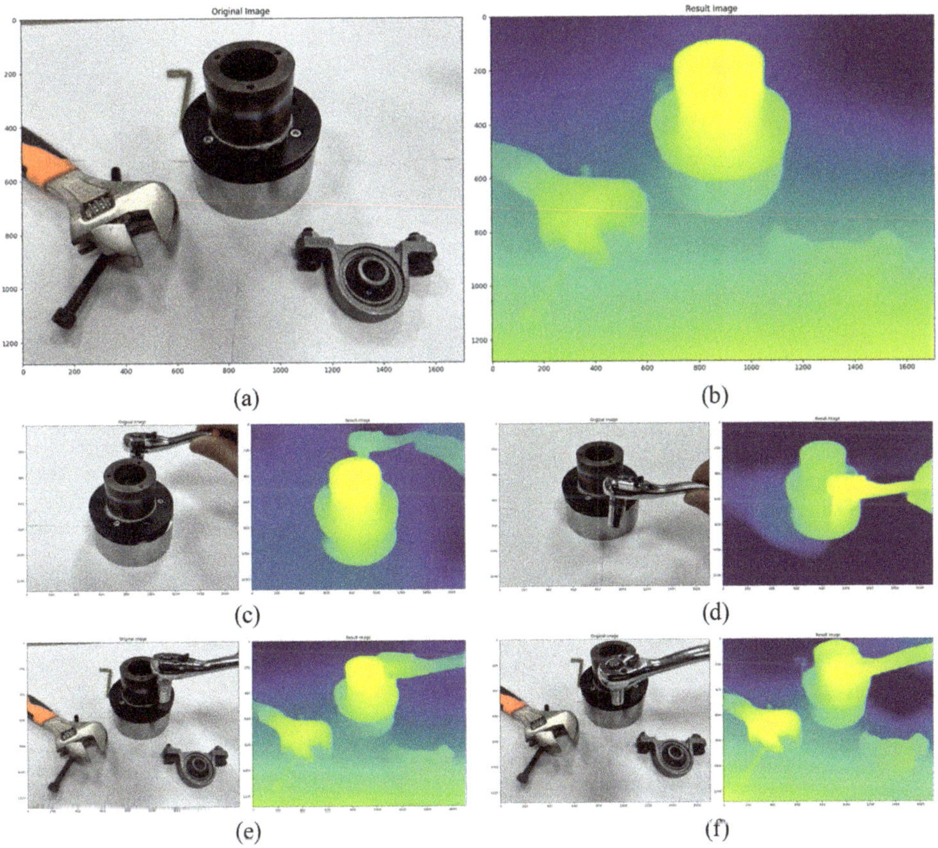

Figure 6. RGB image and depth diagram of the assembly scene. (**a**) Original RGB images; (**b**) Densified depth image; (**c**–**f**) Results of the densified depth maps for different assembly scenes.

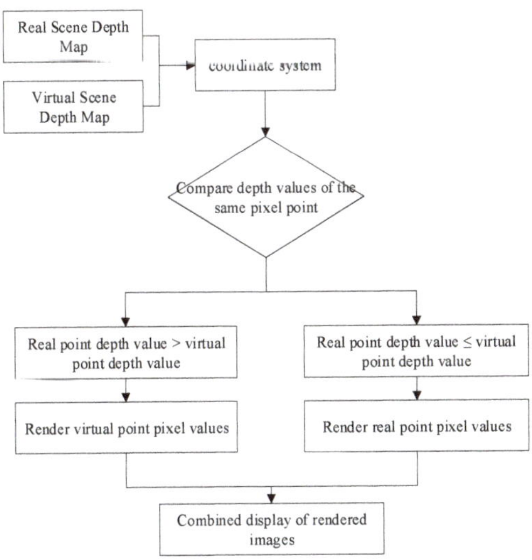

Figure 7. Rendering process for virtual–real occlusion of assembly objects.

5. Case Study

To test the effectiveness of the false-real occlusion processing method based on improved ORB-SLAM3 monocular vision proposed in this paper, this section compares and analyzes it with a variety of algorithms in terms of false and real occlusion effect, accuracy, and timeliness. The computer configuration for this experiment was a Windows 11 operating system, i5-13500H, 32 GB RAM, Nvidia RTX4050 (8 GB graphics memory), and the image transmission used a USB mobile camera with the resolution set to 680 × 480. Taking the assembly of the external accessory piping connectors of the aero-engine as an example, the MR-assisted assembly process of the external accessory piping using mutual virtual and real occlusion is studied. The pipeline attachment assembly mainly includes the pipeline base body, assembly wrench, small bolt, small nut, and other parts. According to the test requirements, a 3D digitized model is designed to obtain the depth map information of the external accessory piping model. The test process involves the digitized process manual of MR-assisted assembly of the external accessory pipeline and associates the involved model information with the corresponding nodes to obtain the virtual and real mutual occlusion data about the pipeline-assisted assembly information.

5.1. Virtual and Real Occlusion Effects

In the MR-assisted assembly process of the external accessories of the aero-engine, the effect of virtual–reality fusion for the static virtual model and the real scene is shown in Figure 8. In the case of no virtual–real masking, the masking relationship between the assembly object and the virtual model is not obvious. This may lead to misunderstandings or errors in the assembly process. After the virtual–real occlusion process, the occlusion relationship between the assembly objects and the virtual model is more obvious. This means that in MR-assisted assembly systems, when a real object is in the way of a virtual object, the virtual object will appear behind the real object and vice versa. This processing accurately reflects the physical occlusion situation in the real world, provides accurate visual guidance to the operator, and helps to avoid assembly errors.

Figure 8. Aero-engine external accessories exhaust pipe bolt virtual model and scene fusion.

Figure 8 shows the guided assembly process of the pilot tube bolts, and Figure 8a and 8b show the guided assembly process of pilot tube bolts 1 and 2, respectively. These two figures are in the assembly process of the typical occlusion relationship intercepted for the virtual model and the real scene superimposed analysis. When the bolts are overlaid with the virtual model and the real scene, it can be seen in Figure 8b that the bolts are located at the very front of the scene. The normal situation should be the location of the bolt hole, but when the assembly tool wrench is positioned over it, the bolt is still located in the front of the scene, which is not processed by the virtual reality masking. As shown in Figure 8c, when the bolt is superimposed on the virtual model and the real scene, firstly the wrench can be seen blocking the virtual model, and secondly the real scene is also blocking part of the virtual model of the bolt. Figure 8e,f presents the relationships before and after the virtual and real occlusion treatments during the assembly of pilot tube bolt 2. From the results, these occlusion relationships are consistent with the real occlusion relationships, which proves the effectiveness of the virtual model and real scene-based virtual occlusion processing algorithm proposed in this paper, and further verifies the reliability of the method in this paper.

5.2. Accuracy Analysis of Virtual and Real Occlusion

To analyze the accuracy of the proposed algorithms in aero-engine external attachment assembly scenarios in terms of virtual and real occlusion, the depth image contours computed by different algorithms in aero-engine external attachments were analyzed by comparing them with the real contour values and comparing with Holynski's [32] and Kinect's [37] depth camera methods. Both methods could generate depth maps of assembly scenes in previous studies and have achieved good results. To compare with the depth images in this paper, the contour line of the connecting tube of the external attachment assembly of the aero-engine in Figure 9a is taken as an example. Each time, the contour line is sampled by uniformly selecting 30 sampling points. Each of the three methods is tested five times, and the average value is taken as the value of the sampling points for comparative analysis.

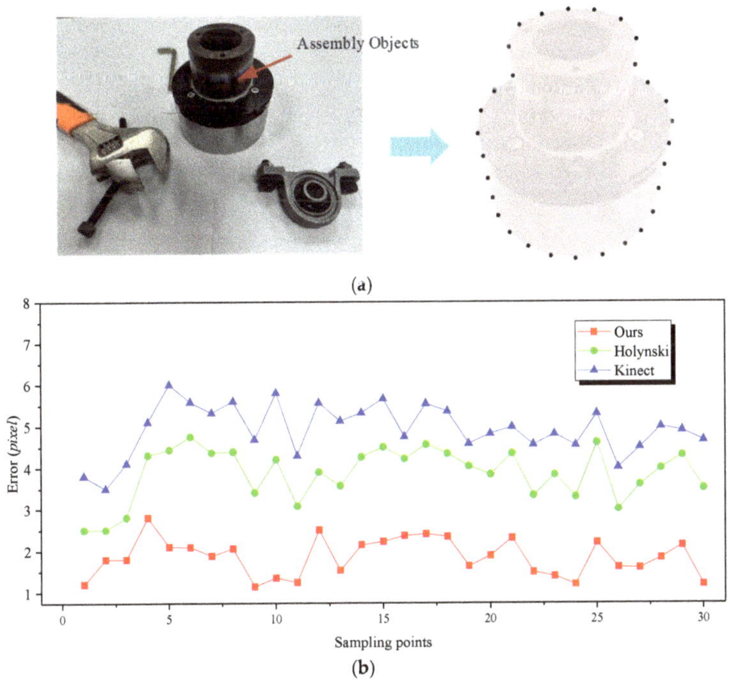

Figure 9. Comparison of edge errors in depth images. (**a**) Schematic diagram of the accessory connecting tube and its contour extraction. (**b**) Depth map sampling point error.

Figure 9b shows the error curves of different algorithms for the depth image sampling points, from which the error values of Holynski [32] and Kinect [37] methods are relatively high, with an average error of 3.9 pixels and 5.1 pixels, respectively. In contrast, the algorithm based on the improved ORB-SLAM3 method proposed in this paper has an average error of 1.7 pixels. Mainly, the matching algorithm with improved ORB is used in the depth image processing process, which can identify and extract the contour feature points of the connecting tube more accurately. By optimizing the descriptors and matching strategies of the feature points, the accuracy of feature point matching is improved and false matching is reduced. Moreover, it has better robustness in dealing with environmental factors such as light change and perspective change and can work stably under different conditions to obtain lower error values.

5.3. Timeliness Analysis of Occlusion

Timeliness of occlusion refers to the speed and efficiency of the algorithm to process the real-void occlusion depth image, which is crucial for the MR-assisted assembly guidance system of aero-engine external accessories because it is directly related to the timeliness of the whole MR guidance system in rendering the before-and-after relationship of the real-void model, which affects the real-time performance and experience of the operator on the MR guidance system. To verify the real-time performance of the virtual–reality occlusion processing algorithm proposed in this paper, the virtual–reality occlusion processing algorithms of Li [31], Holynski [32], and Kinect [37] are compared and analyzed in terms of the processing time and running real-time frame rate. We performed 30 repetitions of the three methods in our experiments and calculated the mean, standard deviation, and 95% confidence intervals for processing time and frame rate.

Table 1 shows the processing time and real-time frame rate of different methods. From the table, the processing time of the virtual and real masking algorithms proposed by Holynski [32] and Kinect [37] are 41.8 ms $\pm$ 1.2 ms and 44.2 $\pm$ 1.5 ms, respectively, and the frame rates are 23.9 $\pm$ FPS and 22.6 $\pm$ FPS. These two methods need to perform preprocessing of the acquired image, which affects the efficiency of the subsequent processing, thus increasing the computation time. Li [31] proposed a method that improves ORB-SLAM2 to achieve better real-time results, with a processing time of 28.9 $\pm$ 0.9 ms and a real-time frame rate of 34.6 $\pm$ 1.1 FPS, while the method proposed in this chapter has an operation time of 23.6 $\pm$ 0.7 ms and a real-time frame rate of greater than 42 $\pm$ 1.3 FPS, which improves the real-time frame rate over the methods of Holynski and Kinect by 77.4% and 87.6%, respectively. The algorithm can effectively improve the efficiency of depth image processing, mainly since the method in this paper utilizes the MNSTF algorithm for optimization in feature point extraction matching. This method can quickly eliminate the wrong matching points, reduce the computation amount for feature point matching, and improve the processing efficiency of the algorithm for the assembly scene image. Thus, it can provide a higher real-time frame rate for the occlusion rendering of the depth map of the assembly scene and meet the real-time requirements of the MR-assisted assembly guidance system for the external accessories of the aero-engine.

Table 1. Processing time and real-time frame rate of different methods.

Methods	Holynski [32]	Azure Kinect [37]	Li [31]	Ours
Time (ms)	41.8 $\pm$ 1.2	44.2 $\pm$ 1.5	28.9 $\pm$ 0.9	23.6 $\pm$ 0.7
Frame (FPS)	23.9 $\pm$ 0.8	22.6 $\pm$ 0.7	34.6 $\pm$ 1.1	42.4 $\pm$ 1.3

6. Conclusions

To address the challenge of obtaining depth information for the assembly scene of aero-engine external accessories via monocular vision, which in turn hinders virtual–real occlusion processing, this paper presents an MR-assisted assembly virtual–real occlusion processing method grounded in ORB-SLAM3 monocular vision technology. This method aims to resolve the virtual–real occlusion issue in the MR-assisted assembly of aero-engine external accessories. The main conclusions of this paper are as follows.

- By incorporating the MNSTF algorithm, we enhanced the feature matching and optimization capabilities of ORB-SLAM3, enabling more accurate reconstruction of the assembly scene with sparse depth points. This improvement significantly reduces computational overhead while maintaining high precision in depth estimation.
- Our method compares the depth values of each pixel in the real and virtual scene depth maps to determine the spatial relationship between virtual and real objects. This ensures accurate occlusion handling and optimizes the visual fusion effect in MR-assisted assembly scenarios.
- The MR-assisted assembly guidance process of aero-engine piping connectors was used as experimental validation and compared with the methods of Holynski and Kinect, and the results show that the method in this paper can effectively solve the problem of dealing with the real–virtual occlusion in the MR-assisted assembly process. The proposed method of false-reality occlusion processing based on improved ORB-SLAM3 monocular vision performs well in terms of false-reality occlusion effect, accuracy, and timeliness.

The work in this paper still has some limitations, such as inaccuracies in capturing depth information for parts smaller than 20 mm. With the rapid development of deep learning models, future work will consider leveraging deep learning techniques to recognize and supplement depth information for small target parts. Deep learning is particularly well suited for this task due to its ability to learn complex patterns and features from large datasets, which can enhance the accuracy of depth estimation for small and intricate structures. Additionally, deep learning can be integrated with the existing ORB-SLAM3 framework to refine sparse depth maps and improve the overall robustness of the system. This hybrid approach has the potential to significantly enhance the accuracy and reliability of depth information capture for small parts, addressing the current limitations of the proposed method.

Author Contributions: Conceptualization, H.X. and F.G.; methodology, Q.Y.; software, C.M.; validation, C.C. and H.X.; writing—original draft preparation, H.X.; writing—review and editing, H.X. and F.G.; funding acquisition, F.G. and C.M All authors have read and agreed to the published version of the manuscript.

Funding: This research was funded by Shanghai Specialized Program for Promoting High-Quality Development (2023-GZL-RGZN-01024), National Defense Industrial Technology Development Program of China (JCKY2023205B006), and National Natural Science Foundation of China (52175450).

Data Availability Statement: The data are contained within the article.

Acknowledgments: The authors gratefully acknowledge the support of the Intelligent Interactive Innovation Studio of Shanghai Aerospace Technology Foundation for the experimental conditions.

Conflicts of Interest: The authors declare that they have no known competing financial interests or personal relationships that could have appeared to influence the work reported in this paper.

References

1. Mal, F.; Karaman, S. Sparse-to-Dense: Depth Prediction from Sparse Depth Samples and a Single Image. In Proceedings of the 2018 IEEE International Conference on Robotics and Automation (ICRA), Brisbane, QLD, Australia, 21–25 May 2018.
2. Fu, H.; Gong, M.; Wang, C.; Batmanghelich, K.; Tao, D. Deep Ordinal Regression Network for Monocular Depth Estimation. In Proceedings of the 2018 IEEE/CVF Conference on Computer Vision and Pattern Recognition, Las Vegas, NV, USA, 18–23 June 2018.
3. Raj, S.; Murthy, L.R.; Shanmugam, T.A.; Kumar, G.; Chakrabarti, A.; Biswas, P. Augmented reality and deep learning based system for assisting assembly process. *J. Multimodal User Interfaces* **2024**, *18*, 119–133. [CrossRef]
4. Yuan, M.L.; Ong, S.K.; Nee, A.Y.C. Augmented reality for assembly guidance using a virtual interactive tool. *Int. J. Prod. Res.* **2008**, *46*, 1745–1767. [CrossRef]
5. Subramanian, K.; Thomas, L.; Sahin, M.; Sahin, F. Supporting Human-Robot Interaction in Manufacturing with Augmented Reality and Effective Human–Computer Interaction: A Review and Framework. *Machines* **2024**, *12*, 706. [CrossRef]
6. Mu, X.; Wang, Y.; Yuan, B.; Sun, W.; Liu, C.; Sun, Q. A New assembly precision prediction method of aeroengine high-pressure rotor system considering manufacturing error and deformation of parts. *J. Manuf. Syst.* **2021**, *61*, 112–124. [CrossRef]
7. Li, J.; Wang, S.; Wang, G.; Zhang, J.; Feng, S.; Xiao, Y.; Wu, S. The effects of complex assembly task type and assembly experience on users' demands for augmented reality instructions. *Int. J. Adv. Manuf. Technol.* **2024**, *131*, 1479–1496. [CrossRef]
8. Patricio, A.; Valente, J.; Dehban, A.; Cadilha, I.; Reis, D.; Ventura, R. AI-Powered Augmented Reality for Satellite Assembly, Integration and Test. *arXiv* **2024**, arXiv:2409.18101.
9. Wolfartsberger, J.; Hallewell Haslwanter, J.D.; Lindorfer, R. Perspectives on Assistive Systems for Manual Assembly Tasks in Industry. *Technologies* **2019**, *7*, 12. [CrossRef]
10. Li, W.; Wang, J.; Liu, M.; Zhao, S.; Ding, X. Integrated registration and occlusion handling based on deep learning for augmented-reality-assisted assembly instruction. *IEEE Trans. Ind. Inform.* **2022**, *19*, 6825–6835. [CrossRef]
11. Tian, Y.; Long, Y.; Xia, D.; Yao, H.; Zhang, J. Handling occlusions in augmented reality based on 3D reconstruction method. *Neurocomputing* **2015**, *156*, 96–104. [CrossRef]
12. Zhu, J.; Pan, Z.; Sun, C.; Chen, W. Handling occlusions in video-based augmented reality using depth information. Computer Animation and Virtual Worlds. *Comput. Animat. Virtual Worlds* **2010**, *21*, 509–521. [CrossRef]
13. Hayashi, K.; Hirokazu, K.; Shogo, N. Occlusion Detection of Real Objects Using Contour Based Stereo Matching. In Proceedings of the ICAT05: The International Conference on Augmented Tele-Existence, Christchurch, New Zealand, 5–8 December 2005; pp. 180–186.
14. Walton, D.R.; Steed, A. Accurate Real-Time Occlusion for Mixed Reality. In Proceedings of the VRST' 17: 23rd ACM Symposium on Virtual Reality Software and Technology, Gothenburg, Sweden, 8–10 November 2017; pp. 1–10.
15. Lee, J.H.; Kim, C.S. Monocular depth estimation using relative depth maps. In Proceedings of the 2019 IEEE/CVF Conference on Computer Vision and Pattern Recognition (CVPR), Long Beach, CA, USA, 15–20 June 2019; pp. 9729–9738.
16. Liu, F.; Zhou, S.; Wang, Y.; Hou, G.; Sun, Z.; Tan, T. Binocular light-field: Imaging theory and occlusion-robust depth perception application. *IEEE Trans. Image Process.* **2019**, *29*, 1628–1640. [CrossRef] [PubMed]
17. Kim, H.; Yang, S.J.; Sohn, K. 3D reconstruction of stereo images for interaction between real and virtual worlds. In Proceedings of the Second IEEE and ACM International Symposium on Mixed and Augmented Reality, Tokyo, Japan, 10 October 2003.
18. Zheng, Y.; Liu, P.; Qian, L.; Qin, S.; Liu, X.; Ma, Y.; Cheng, G. Recognition and depth estimation of ships based on binocular stereo vision. *J. Mar. Sci. Eng.* **2022**, *10*, 1153. [CrossRef]
19. Yang, Y.; Meng, X.; Gao, M. Vision system of mobile robot combining binocular and depth cameras. *J. Sens.* **2017**, *2017*, 4562934. [CrossRef]
20. Luo, T.; Liu, Z.; Pan, Z.; Zhang, M. A virtual-real occlusion method based on GPU acceleration for MR. In Proceedings of the 2019 IEEE Conference on Virtual Reality and 3D User Interfaces (VR), Osaka, Japan, 23–27 March 2019; pp. 1068–1069.
21. Zhang, C.; Xu, R.C.; Han, C.; Zhai, H.Y. An Occlusion Consistency Processing Method Based on Virtual-Real Fusion. In *Frontier Research and Innovation in Optoelectronics Technology and Industry*; CRC Press: Parkway, NW, USA, 2018; pp. 47–56.
22. Ibrahim, M.M.; Liu, Q.; Khan, R.; Yang, J.; Adeli, E.; Yang, Y. Depth map artefacts reduction: A review. *IET Image Process.* **2020**, *14*, 2630–2644. [CrossRef]
23. Simon, N.; Majumdar, A. Mononav: Mav navigation via monocular depth estimation and reconstruction. In Proceedings of the 18th International Symposium on Experimental Robotics (ISER 2023), Chiang Mai, Thailand, 26–30 November 2023; Springer Nature: Cham, Switzerland, 2023; pp. 415–426.
24. Chaplot, D.S.; Gandhi, D.; Gupta, S.; Gupta, A.; Salakhutdinov, R. Learning to explore using active neural slam. *arXiv* **2022**, arXiv:2004.05155.
25. Chaplot, D.S.; Salakhutdinov, R.; Gupta, A.; Gupta, S. Neural topological slam for visual navigation. In Proceedings of the 2020 IEEE/CVF Conference on Computer Vision and Pattern Recognition (CVPR), Seattle, WA, USA, 13–19 June 2020; pp. 12875–12884.

26. Muravyev, K.; Bokovoy, A.; Yakovlev, K. tx2_fcnn_node: An open-source ROS compatible tool for monocular depth reconstruction. *SoftwareX* **2022**, *17*, 100956. [CrossRef]
27. Chang, M.F. Monocular Depth Reconstruction Using Geometry and Deep Convolutional Networks. Master's Thesis, Carnegie Mellon University, Pittsburgh, PA, USA, May 2018.
28. Luo, Y.; Liu, G.; Liu, H.; Liu, T.; Tian, G.; Ji, Z. Simultaneous Monocular Visual Odometry and Depth Reconstruction with Scale Recover. In Proceedings of the 2019 IEEE International Conference on Robotics and Biomimetics (ROBIO), Dali, China, 6–8 December 2019; pp. 682–687.
29. Fink, L.; Franke, L.; Keinert, J.; Stamminger, M. Refinement of Monocular Depth Maps via Multi-View Differentiable Rendering. *arXiv* **2024**, arXiv:2410.03861.
30. Zhang, X.; Zhao, B.; Yao, J.; Wu, G. Unsupervised monocular depth and camera pose estimation with multiple masks and geometric consistency constraints. *Sensors* **2023**, *23*, 5329. [CrossRef]
31. Li, W.; Wang, J.; Liu, M.; Zhao, S. Real-time occlusion handling for augmented reality assistance assembly systems with monocular images. *J. Manuf. Syst.* **2022**, *62*, 561–574. [CrossRef]
32. Holynski, A.; Kopf, J. Fast depth densification for occlusion-aware augmented reality. *Trans. Graph.* **2018**, *37*, 1–11. [CrossRef]
33. Han, X.; Chen, X.; Deng, H.; Wan, P.; Li, J. Point Cloud Deep Learning Network Based on Local Domain Multi-Level Feature. *Appl. Sci.* **2023**, *13*, 10804. [CrossRef]
34. Dengwen, Z. An Edge-Directed Bicubic Interpolation Algorithm. In Proceedings of the 2010 3rd International Congress on Image and Signal Processing, Yantai, China, 16–18 October 2010; Volume 3, pp. 1186–1189.
35. Kirkland, E.J.; Kirkland, E.J. Bilinear Interpolation. In *Advanced Computing in Electron Microscopy*; Springer: Boston, MA, USA, 2010; pp. 261–263.
36. Bai, Y.; Wang, D. On the comparison of trilinear, cubic spline, and fuzzy interpolation methods in the high-accuracy measurements. *IEEE Trans. Fuzzy Syst.* **2010**, *18*, 1016–1022.
37. Azure Kinect DK. Available online: https://learn.microsoft.com/zh-cn/azure/kinect-dk/depth-camera (accessed on 15 May 2024).

machines

MDPI

Article

Modular Design and Layout Planning of Tooling Structures for Aircraft Assembly

Zhanghu Shi [1,2], Chengyu Li [1], Junshan Hu [1,*], Xingtao Su [1], Hancheng Wang [1] and Wei Tian [1]

1 College of Mechanical and Electrical Engineering, Nanjing University of Aeronautics and Astronautics, Nanjing 210016, China
2 Aviation Industry Chengdu Aircraft Industry (Group) Co., Ltd., Chengdu 610092, China
* Correspondence: hujunshan@nuaa.edu.cn; Tel.: +86-159-5168-3586

Abstract: Aircraft structures consist of numerous complex components that require a high level of precision to assemble. Tooling plays a crucial role in the assembly of aircraft components, providing the functions of positioning, shape maintenance, and support to guarantee the accuracy of the product. Aiming to obtain reusable assembly tooling that can be rapidly reconfigured, this study focuses on the modular design and layout of tooling structures. The concept of functional elements for the characterization of tooling parts is proposed, and the relationship between each pair of elements is established to clarify the similarities and dependencies among various tooling structures. Based on the analysis of functional elements and their relationships, the tooling structures are divided and recombined into several modules. The detailed module designs are demonstrated by using typical structures such as platforms, columns, and locators as examples. A parametric representation of the multi-source information of tooling modules is proposed, and optimization methods for the layout and configuration of locators and platforms are developed using their parametric information. A reconfigurable tooling process integrated with a monitoring system is designed, realized, and successfully applied to the assembly of a practical type of fuselage. The results from verifying these methods' efficiencies show that the modular design and reconfiguration planning of tooling only takes about 10 min and a few seconds, respectively, which is far less than the time consumed during traditional tooling design (from several days to weeks). The work in this study provides an engineering paradigm for the serialization and reconfiguration of assembly tooling in aviation manufacturing.

Keywords: aircraft assembly; reconfigurable tooling; modular design; layout planning

1. Introduction

There are numerous structural components to an aircraft that are diverse and have high requirements for precision assembly [1]. In the assembly process, tooling provides functions including positioning, shape maintenance, and fixed support for aircraft components [1,2]. Traditional assembly tooling is mostly made of robust steel parts that are permanently welded together to ensure the structural rigidity and stability required for accurate positioning. These positioning and supporting parts are densely arranged and are typically used for a specific aircraft product. This tooling design scheme results in poor openness of the assembly space and means that the tooling cannot be reused when the product is changed. Thus, it is difficult to adjust to a manufacturing mode that can realize the small-batch production of multiple varieties in the future. While ensuring structural stability is important, investigating the modularization and reconfigurability of the tooling process is also beneficial for enabling a rapid response to product change and assessing the efficiency of tooling usage.

Focusing on the reconfigurability and adaptability of assembly tooling, aviation enterprises around the world have conducted considerable research on digital measurement and control technologies. For instance, various types of reconfigurable tooling have been developed; these are also known as flexible tooling and can be configured according to different assembly tasks. The currently used flexible tooling can be classified into three types: high-speed column tooling [1,3], multi-point forming tooling [1,4], and positioning tooling based on parallel mechanisms [1,5]. High-speed column tooling is mainly used to position and support slender parts, such as wing beams and stringers; multi-point forming tooling is suitable for the shaping and single-coordinate positioning of thin-walled parts, and flexible tooling based on parallel mechanisms is used for the posture alignment of large components. Bojite developed two types of tooling, the Eco Positioner and Flexible Section Assembly Cell, which have been applied to fuselage positioning in Boeing and Airbus aircraft assemblies [1,6]. Shanghai Aircraft Manufacturing applied the Eco Positioner to the automated fuselage assembly of the C919 [1,7]. Ke et al. [8] have developed an attitude adjustment system for fuselage, improving the quality and efficiency of aircraft assembly through numerical control locators. Huang et al. [9] used a parallel mechanism based on numerical control locators for the assembly of fuselage and wings. Li et al. [1,10] have developed a monitoring system for detecting the positioning error and clamping force of a type of assembly tooling that can be reconfigured for different fuselages. Muller et al. [1,11] have developed a reconfigurable tooling based on parallel mechanisms for the assembly of fuselage walls. Thomas et al. [1,12] realized the modularization of parallel mechanisms for adaptation to various types of assembly parts. The above types of tooling have some adaptability to the assembly of different products. However, their current design still relies on human experience, and there is no unified standard or routinized process for their design. The development process of tooling is time-consuming, and its adaptation to various products is limited.

Aiming to achieve the goal of a standardized design, some institutions have integrated modular structures with reconfigurable technologies and proposed the concept of low-cost reconfigurable tooling based on box connections [1,13]. This type of tooling incorporates all of the modular parts, including the main frame, the mechanisms for positioning, the connection structures, and measurement devices. The overall structure of this tooling is adjustable and detachable, which can significantly improve the utilization rate of assembly tooling resources [1,14,15]. Löchte [1,16] proposed a six-SPS parallel mechanism to achieve adjustment through the reconfigurable connection of parallel mechanism branches. Prause [1,17] developed a three-PUU parallel mechanism for the positioning of fuselage panels during assembly and obtained an optimized bearing capability through the reconfiguration of parallel mechanism branches. Saab and Linkoping University have proposed the concept of Affordable Reconfigurable Tooling (ART) [1,13,18], which uses a series of modular mechanisms and effectively reduces the precision requirements for installation, thereby saving costs. It was successfully applied to the assembly of the "Neuron" unmanned aerial vehicle [1,19]. In 2009, Saab and Airbus established the first box-joint tooling based on the ART concept and successfully applied it to the assembly of wing boxes [1,20]. Considering that ART adopts an offline reconfiguration strategy that cannot realize the posture adjustment of aircraft components, some reconfigurable tooling systems have been developed using a combination of servo control and modular structures to improve their adaptability. Xiong [1,21] proposed a tooling system consisting of a mobile base, parallel mechanisms, and clamping mechanisms for the capability of online reconfiguration. Mannheim [1,22] developed a flexible gripper that can automatically reconfigure itself by adjusting the configuration of parallel mechanisms in real time. Molfino [1,23] developed an assembly platform called SwarmItFIX that can automatically adjust its positioning units online, and it was successfully applied in clamping and holding processes. The Airbus company proposed a flexible assembly system called FAHF [1,24] that can respond to customer needs in a timely manner and has a lower cost. Francesco et al. [1,14] developed an automated synthesis of multiple robotic cells on reconfigurable tooling to realize the cooperation thereof for aircraft assembly, and they validated this in the European LOCOMACHS (Low-Cost Manufacturing and Assembly of Composite and Hybrid Structures) project. Based on this project, Ilker et al. [1,25] also presented a framework for automated flexible tooling development and presented their results on a hexapod fixture as a case study.

In summary, focusing on the reconfigurability, adaptability, and low-cost design of aircraft assembly tooling, considerable research has been conducted on digital measurement, mechanism design, and servo control in academia and industry, and various types of flexible tooling have been developed. However, the modular design of reconfigurable tooling lacks a general guiding framework, and the degrees of modularization and serialization of tooling structures are still insufficient. Moreover, in order to improve the adaptability and reconfigurability of tooling, a large number of studies focus on online measurement and servo control while neglecting the importance of rigidity and stability. The application of a multiple-degree-of-freedom mechanism driven by a servo motor makes it impossible for reconfigurable tooling to have adequate rigidity and repeatable accuracy of positioning. An automated servo system is indeed useless for the assembly of small parts, since the parts can be placed and fixed on tooling rapidly and manually by workers. So, there is a demand for attention to the development of assembly tooling that is reconfigured using some releasable structures, as this would have more application prospects for the assembly of small parts due to both its stiffness and its adaptability.

Therefore, this study focuses on a general framework of structural design and rapid reconfiguration for tooling that is configured using releasable connections and is used in the assembly of small parts. A modular design theory based on the concept of functional elements and a planning method for reconfiguration based on optimization algorithms are proposed. Through the analysis of so-called functional elements, the differences and similarities between various structures of the tooling are clarified, and the structures are reorganized and divided into several modules. A parametric representation and a storage strategy for the design information of modules are proposed. Depending on the parametric information, the layouts of two typical structures, namely, a locator and platform, are optimized to realize the rapid reconfiguration of tooling. It is hoped that these works can provide a paradigm for the serialization and modularization of assembly tooling. The main contributions of this study are as follows:

(1) The first contribution is the modularization and routinized design of assembly tooling structures that have both adaptability and rigidity. Unlike reconfigurable tooling based on a servo mechanism, the type of tooling developed in this study is reconfigured with releasable connections and is more applicable due to its structural robustness and the stability of its accuracy. Benefiting from reasonable classification and reorganization of mechanical structures, the modularization and rapid design of assembly tooling, which are neglected by academics but need more attention in engineering, are realized in this study.
(2) The second contribution is the planning of the layout of assembly tooling structures. Based on the robust and adaptable tooling developed, the authors propose two intelligent planning algorithms for the layouts of platform units and locating points to enhance the efficiency and automation of configuration. The efficient plan-

ning of reconfiguration further enhances the adaptability and usage of this tooling in engineering.

2. Modular Design Method for Tooling Based on Functional Elements

2.1. Problem Configuration

Traditional assembly tooling mainly consists of a main frame and functional parts, including positioning and clamping components, as shown in Figure 1a. The main frame is highly rigid, supporting the entire tooling structure and ensuring the positional stability of numerous locators and clamps, which are in direct contact with the aircraft product. In order to guarantee the positioning accuracy and support for the aircraft product, the locators and clamps are densely distributed and permanently fixed to the main frame, resulting in poor openness of the assembly space and weak adjustability. Therefore, this study aims to develop a type of tooling that uses reconfigurable locators and supports that are lightweight and low in density, as shown in Figure 1b. The reconfigurable tooling mainly consists of a platform, columns, cantilever locators, and auxiliary clamps. These structures are all connected through detachable bolts or movable slides, allowing the tooling's layout to be changed according to the product. Compared with the types based on servo mechanisms, this reconfigurable tooling is adaptable while retaining adequate rigidity and stability; thus, it has more applications in practice.

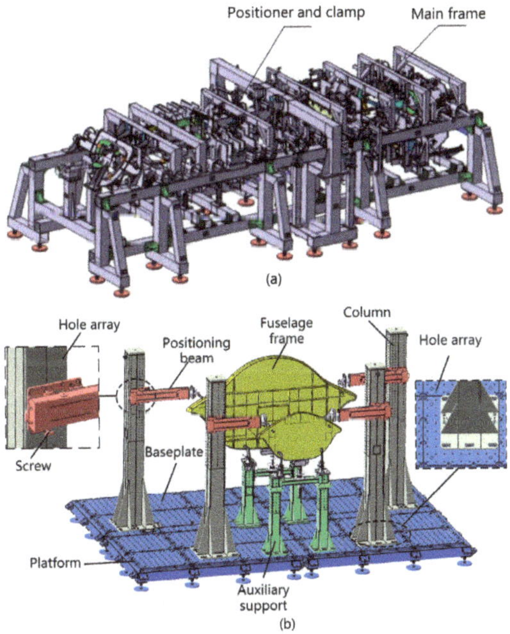

Figure 1. (**a**) Traditional rigid tooling in cage form; (**b**) modular and reconfigurable tooling developed in this study.

Although the new generation of tooling improves the openness of the assembly space and the reusability of the structure, there is still a lack of a general framework for the routinized design of tooling. Designs have relied on human experience so far and are time-consuming. Moreover, when a product is modified, the reconfiguration scheme lacks a fast and efficient optimization strategy, thus limiting the speed of response to the assembly tasks. This study draws attention to the general design method and rapid layout technology for reconfigurable tooling. Based on the analysis of similarities between the functions of parts, tooling structures are divided and reorganized into several modules, forming a routinized and general design paradigm. For the purpose of rapid reconfiguration, a parametric representation of structural information and optimization algorithms for layout planning are proposed.

2.2. Functional Element Analysis and Modular Division of Tooling

For the convenience of functional analysis, the i th function of tooling is defined as FR_i, and a tree diagram is used to clarify the relationships between multiple functions provided by the assembly tooling from top to bottom, as shown in Figure 2. A function may be composed of multiple subfunctions that can be separately realized by different parts of the tooling, and the common function from which these subfunctions are derived is called their parent function. The functions are decomposed stage by stage until each subfunction can be implemented by an independent structure of the tooling. The subfunction located at the end of each branch in the diagram is called as a functional element and is marked in red in Figure 2. This figure gives an example of such a decomposition, where the functional elements are FR_4, FR_6, FR_8, FR_{10}, and FR_{11}~FR_{17}, and marked in red.

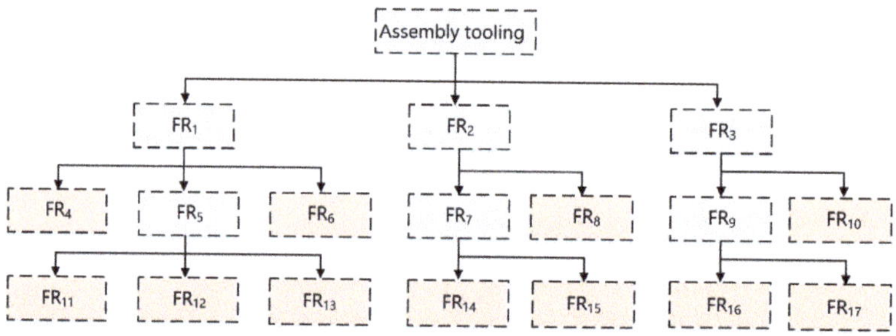

Figure 2. An example of the relationships between functional elements of assembly tooling.

We denote the distances from the functional elements FR_i and FR_j to their nearest common parent function as l_i and l_j, respectively; then, the relevance between FR_i and FR_j is defined as the following:

$$r(FR_i, FR_j) = r(FR_j, FR_i) = \begin{cases} x & x \geq 0.25 \\ 0 & x < 0.25 \end{cases}, \quad x = \frac{1}{2^{l_i+l_j-2}} \tag{1}$$

where $r(FR_i, FR_j)$ is the relevance between FR_i and FR_j, and l_i and l_j are the stages of distances between FR_i, FR_j, and their nearest common parent function. Based on the combination of the relatedness of each pair of functional elements, the following relevance matrix is defined as the following:

$$\boldsymbol{T} = \{r_{ij}\}_{n \times n} = \begin{bmatrix} r_{11} & r_{12} & \cdots & r_{1n} \\ r_{21} & r_{22} & \cdots & r_{2n} \\ \vdots & \vdots & \ddots & \vdots \\ r_{n1} & r_{n2} & \cdots & r_{nn} \end{bmatrix} \tag{2}$$

where $r_{ij} = r(FR_i, FR_j)$ is the relevance between FR_i and FR_j, and n is the total number of functional elements. Taking FR_6 and FR_{12} in Figure 2 as an example for calculation, their nearest common parent function is FR_1; then, $l_6 = 1$, $l_{12} = 2$, and, thus, $r_{6,12} = r(FR_6, FR_{12}) = 0.5$.

Based on the relevance matrix of functional elements, a binary matrix can be obtained by setting a threshold for the relevancies. This matrix represents a binary judgment of the similarity between each pair of functional elements. For example, if the threshold is taken as λ, then the corresponding binary matrix $^*T$ is shown in Equation (3), which can be expressed as the following:

$$^*T = \{t_{ij}\}_{n\times n},\ t_{ij} = \begin{cases} 1\ r_{ij} \geq \lambda \\ 0\ r_{ij} < \lambda \end{cases} \tag{3}$$

Based on the binary matrix $^*T$ of functional element relevance, multiple similar elements are integrated into an entirety accordingly. In the matrix $^*T$, if $t_{ij} = t_{i+1,\,j} = t_{i+2,\,j} = \ldots = t_{i+k,\,j} = 1$, then it is considered that the functions of FR_i, FR_{i+1}, … FR_{i+k}, FR_j are approximately equivalent. These functional elements are integrated into a new single element denoted as M_i, and their corresponding tooling structures are combined into a single module. Taking the original diagram of the functional elements in Figure 2 as an example, if the relevance threshold is $\lambda = 0.5$, the result of module division is shown in Figure 3, where the functional elements are reorganized into three modules denoted as M_1, M_2, and M_3.

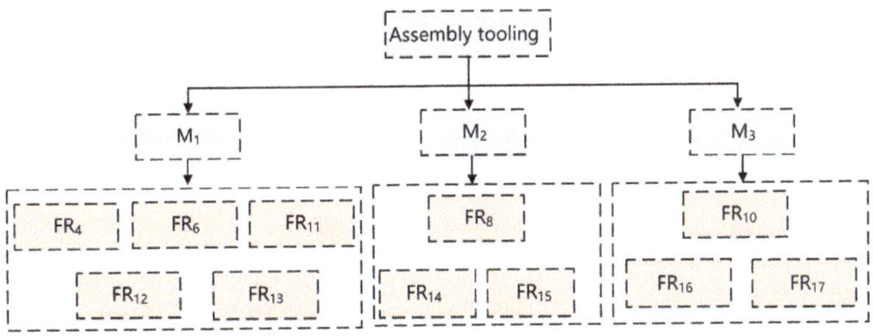

Figure 3. The division and integration of functional elements into modules.

2.3. Design Methods of Typical Modular Structures

As can be seen from the example of the reconfigurable assembly tooling shown in Figure 1, the typical modular structures of the tooling mainly include support platforms, columns, and locators. In this section, detailed structural design methods for these three typical types of modules are proposed.

2.3.1. Design of the Support Platform

The platform is used for the support of the whole structure of the assembly tooling and serves as a reference for installation. Unlike traditional tooling, which is fixed on the ground, the platform of the reconfigurable tooling provides adjustment capabilities for the structures installed on it. The platform has an array of bolt holes for the connections between itself and the tooling structures, allowing them to be moved conveniently. The platform is a combination of multiple units, and its total size depends on the maximum size of the working space envelope, which further depends on the maximum size of the product envelope according to the reference posture and top view, as well as the requirement for openness. The boundary of the platform should be greater than or equal to the boundary of the working space from the top view during assembly. Considering that various products have different requirements for the platform size, this study aims to meet these requirements to cover all types of products while using as few platform units as possible, thus saving time and supplies. Therefore, the platform is divided into units with a size of $l \times w$, such that a single-unit module can cover the envelope of the smallest products, and multiple units can be combined to meet the assembly requirements for the largest ones, as shown in Figure 4a.

According to the geometric relationship shown in Figure 4a, the boundary size condition of the platform can be expressed as the following:

$$\begin{cases} n_1 l \leq L + a \\ n_2 w \leq W + a \end{cases} \tag{4}$$

where n_1 is the number of units divided along the length direction, n_2 is the number of units divided along the width direction, and a is an experience value, which is usually taken as 700 mm.

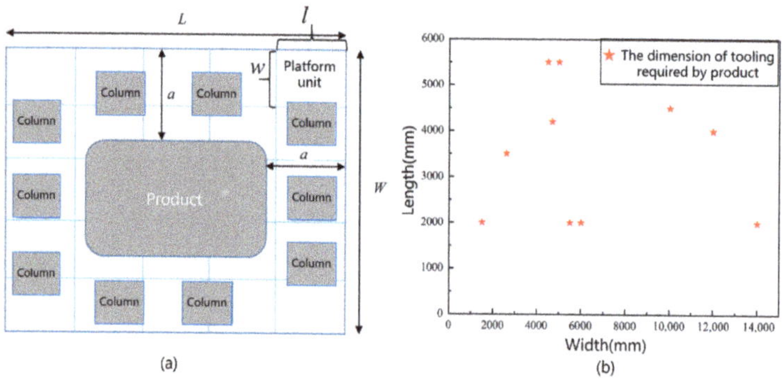

Figure 4. (**a**) Division and layout of the platform units; (**b**) statistics of the boundary sizes in product requirements.

By statistically analyzing the requirements for assembly space from 10 types of aircraft products, as shown in Figure 4b, the width direction is greater than or equal to 2000 mm. The greatest common divisor of all types of products from the top view is calculated, and the dimensions of the units are determined to be 3 m × 2 m after rounding, i.e., l = 3000 mm, w = 2000 mm.

In most cases, the platform is composed of two or more units, and the platform influences the horizontal accuracy of the structures installed on it. So, it must be possible for the posture of the platform to be adjusted. Regarding this requirement, this study further carries out modular and series designs of a matching surface, installation ports, and the integration of a leveling function, as shown in Figure 5. A single 3 m × 2 m platform unit consists of a mounting plate, support frame, leveling bolts, and other splicing parts, which are integrated and packaged. The mounting plate is drilled with a series of threaded holes in intervals of 100 mm, with a specification of M16 bolts. The parts of a single unit are fixed by splicing parts and finished bolts, and multiple units are attached by connecting plates and sockets to form an entire platform. In order to ensure the universality and interchangeability of the above parts, all dimensions and installation interfaces are designed with uniform specifications.

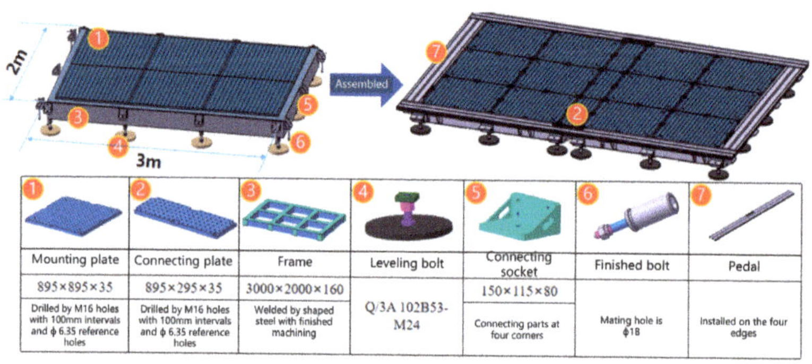

Figure 5. Design and combination of the platform unit.

2.3.2. Design of the Column Structure

The columns of the tooling are used for the installation of locators and clamps, and they act as the skeleton of the assembly tooling. Unlike the skeleton structure of traditional tooling, which is in a closed and welded form, the columns are distributed dispersedly in an open form in the design in this study. According to the requirements of the assembly process, the columns are located on both sides and the bottom of the product, and their sizes are determined by the height of the product in the assembly posture, as shown in Figure 6. The maximum envelope principle is used to determine the column specifications, and the columns are designed in series, producing six types of modules with heights of 5 m, 3 m, 1.1 m, 0.6 m, 0.4 m, and 0.2 m.

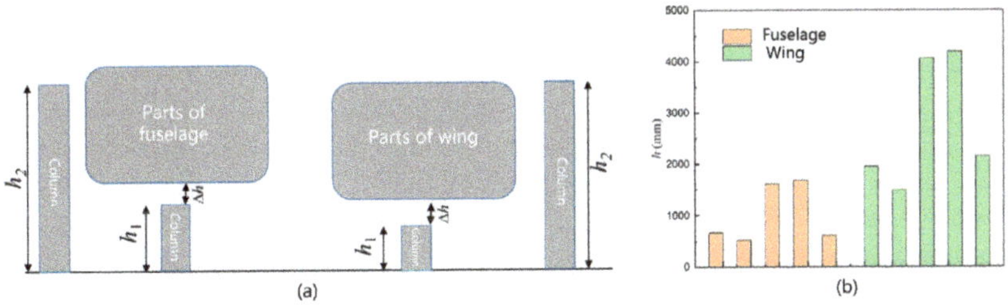

Figure 6. (**a**) Relationship between the column height and product size; (**b**) statistics of heights of multiple products.

Figure 7 shows the designed series of column modules. The bottom of a column is an installation interface that is in contact with the platform and is arranged with bolt holes. Because of the interchangeable design, the bolt holes have a diameter of 18 mm, and the distance between them is in multiples of 50 mm. Considering the rigidity of the column, a reinforcing rib is designed between the bottom and the column body. To provide a universal design, the column body is drilled with an array of screw holes for the installation of locators in different positions. In order to ensure the positioning accuracy of the locators, the flatness tolerance of the column plane does not exceed 0.1 mm, and its roughness is less than or equal to 1.6 μm. Considering the requirement for digital measurement during installation, placement holes for measurement targets for laser trackers are drilled on the top of the columns.

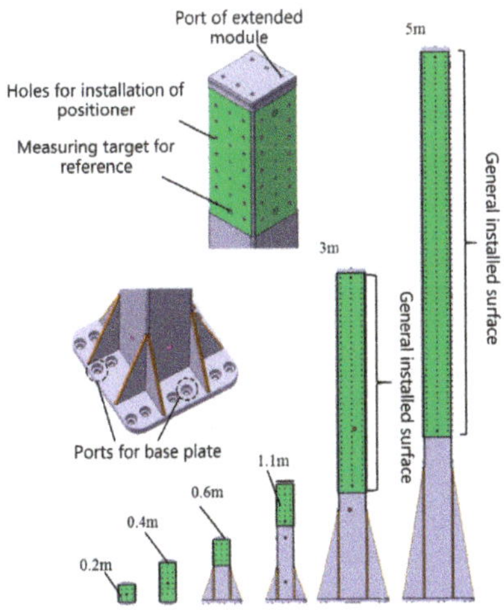

Figure 7. Design of the column structure in the tooling.

2.3.3. Locator Design

Locators are used for the precise positioning of products and provide some support functions for them. The modular structure of a locator mainly consists of three parts, namely, a beam, socket, and end effector. Each part has multiple types of modules, as shown in Figure 8a. Since the socket and end effector of a locator are in direct contact with the product and are designed according to the product, this study only carries out the modular design of the beam to achieve the maximum reuse rate of the structure. According to the structural form and usage requirements, the beams can be classified into two separate types based on sliding rails and guide posts. Guide post beams are often used when they are in contact with the product on the top of the post and can be further classified into two types based on whether they have a single guide post or double guide post. A single guide post is suitable for narrow spaces and is used as a support rather than a positioner due to its low rigidity. A double guide post is suited for the requirements of rigidity and positioning accuracy, such as when a product hole is being positioned after precision machining. Sliding rail beams are attached to the columns and can be classified into two types: single rails and double rails. The single rail type is only used in cases where the assembly space is narrow. In order to ensure guidance accuracy and rigidity, the scheme of double rails is generally adopted. According to actual requirements, a locator does not need to contain all three parts and can be classified into three types according to the combination of the parts. The first type consists of a beam, socket, and end effector; the second type consists of a beam and end effector; and the third type consists of a socket and end effector, as shown in Figure 8b. Considering that the double-sliding-rail beam is the most commonly used type of module in locators, this study only elaborates on the series design of this type of module.

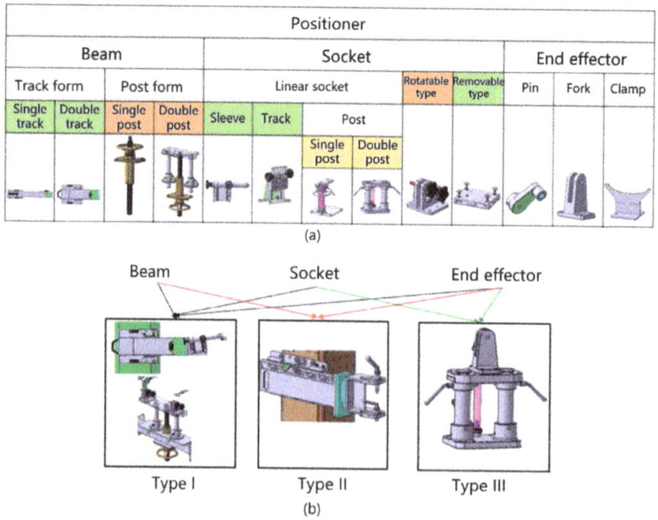

Figure 8. (**a**) Classification of the locator parts; (**b**) three types of modular locators.

Figure 9 shows the series design of a double-sliding-rail beam. According to the statistical data of the beam length required for usage, the beam can be classified into six specifications according to the different cantilever lengths, namely, 450 mm, 800 mm, 1000 mm, 1200 mm, 1400 mm, and 1600 mm, corresponding to different suspension demands. Each specification adopts the same structural form as that of the installation plate that is connected to the columns, is drilled with Φ18 mm through-holes, and is placed on a sliding block in contact with the rail. The sliding rail provides mobility for the beam to adapt to different products and assembly spaces. The beam can be in two states: the working state and non-working state. When the product is placed on or off the tooling, the beam slides into the non-working state to make room for openness, and the beam enters the working state to meet positioning requirements during assembly. A locking pin is also integrated into the beam; this is used to ensure the stability of the tooling in different states. In addition, it also involves some holes for the placement of measurement targets used during digital installation and accuracy evaluation.

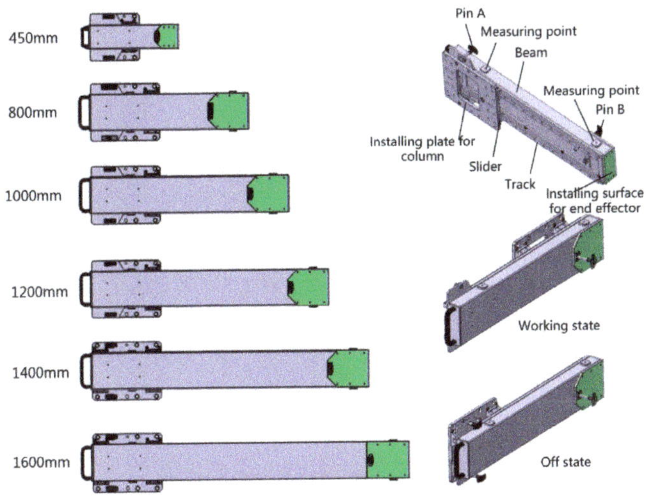

Figure 9. Series design of the locator arms based on a double sliding rail.

3. Optimization of the Layout of Reconfigurable Tooling Modules

On the basis of the modular design of the tooling structure, quickly configuring each functional module and guaranteeing the positioning and support while achieving the simplification of the tooling structure are the keys to the rapid response to product modifications and efficient completion of assembly tasks. The efficient layout of positioning points and the rapid paving of platform units are the two main tasks in the reconfiguration process of modular tooling. This section focuses on researching intelligent planning technology to achieve rapid decision making and the optimization of tooling reconfiguration schemes.

3.1. Parametric Representation and Storage of Modular Structural Information

The modular structures of tooling and assembly products have contact constraints, geometric shapes, and other multi-source features. Achieving a parametric representation and the efficient storage of their information is the basis for intelligent reconfiguration planning. Therefore, this section first defines a modular unit symbol and product assembly symbol for the digital models of tooling and products. The modular unit symbol (*MUS*) is a set that parametrically represents the geometric characteristics, processing requirements, and executable functions of tooling modules and is banded with digital models, as shown in Equation (5), which can be expressed as the following:

$$MUS = \left\{ A_{Typ}^{Fea}, S_c, T_c, T_f \right\} \quad (5)$$

where A_{Typ}^{Fea} is the identifier of functions, including positioning, clamping, and support. Since a single module may have multiple functions, this identifier is defined as a 0–1 binary vector containing three elements, each of which separately represents whether the module has one of the above three functions. S_c represents the roughness requirement of the module surface, including a rough surface, semi-finished surface, and finished surface. In general, the roughness requirements of various surfaces of a single module are consistent, so this identifier is defined as an integer variable with a range of $-1, 0, 1$. T_c is the type of contact between the module and the product, and it includes no contact, point contact, line contact, and surface contact. Considering that a single module generally has only one type of contact with the product, this identifier is also defined as an integer variable from -2 to 2. T_f is the type of assembly scene applicable to the module, such as fuselage skeleton assembly, automatic drilling, and the riveting of panels. It is defined as a 0–1 binary vector where each element represents the applicability of the module to a certain scene, where 0 indicates inapplicability and 1 indicates applicability.

The product assembly symbol (*PAS*) is defined for the digital model of the product, and it maps the function identifiers of all tooling modules to the corresponding sum of constraint points, the constraint surface normal, and the positioning, clamping, and support functions. It is defined using Equation (6). Equation (6) can be expressed as the following:

$$PAS = \left\{ F_{Poi}, D_{Fea}, \sum_{i=1}^{n} \left(A_{Typ}^{Fea}\right)_i \right\} \tag{6}$$

where $\boldsymbol{F}_{Poi}$ is a $3 \times n$ matrix representing the coordinate set of multiple constraint points between the tooling and the product. $\boldsymbol{D}_{Fea}$ is also a $3 \times n$ matrix representing the unit normal set of multiple constraint surfaces between the tooling and the product. $\sum_{i=1}^{n} \left(A_{Typ}^{Fea}\right)_i$ represents the sum of the binary vector of functions corresponding to each module. The modular unit identifier MUS has some limiting conditions that are determined by the designer according to experience and product models; it usually requires that the constraint points cannot be on the same line, and the number of modules with positioning functions cannot exceed three.

The composition of reconfigurable tooling is complex, as there are many types of structures. Therefore, based on the modular unit symbol of the tooling, a knowledge base for reconfigurable tooling design is established to manage the modules and facilitate the update and iteration of subsequent versions of structures. The process of designing the knowledge base is shown in Figure 10a. Firstly, the assembly tooling is divided into functional modules according to the method proposed in Section 2.2. Then, the design information of each module in the tooling and the product information are, respectively, converted into the MUS and PAS, as shown in Equations (5) and (6), and a database containing these symbols is established. Finally, by combining the database with prior knowledge of the assembly process, a mapping from the parameters of the PAS to the parameters of the MUS is established to build a knowledge base that associates the product information with the tooling structures. The knowledge base allows the sharing of various standard resources needed for rapid design and lays the foundation for the efficient planning of reconfiguration. The composition and architecture of the knowledge base are shown in Figure 10b, including the libraries of instances, rules, and parametric information. The instance library is composed of digital models of tooling modules that cover their geometric information, the rule library contains information on the constraints of connections and layouts between structures, and the parametric information library mainly consists of the MUSs corresponding to modules.

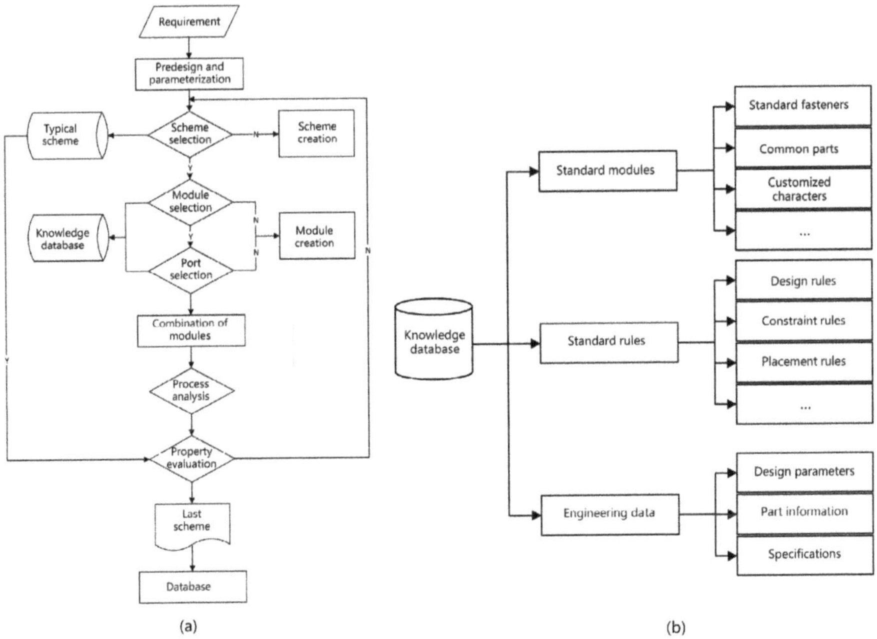

Figure 10. (**a**) Design process of the knowledge base for modular structures; (**b**) architecture of the knowledge base.

3.2. Layout Planning for Locators

In order to improve the degree of automation in the planning of the locator layout according to the product, this study proposes a locator planning method based on the DTW (Dynamic Time Warping) trajectory comparison algorithm using the parametric representation of the tooling structure. Different types of aircraft product components are classified, and their outer contours are simplified. The DTW algorithm can align two sequences of different lengths in time and can compare the similarities between them. The planning process for the locator layout based on the DTW algorithm consists of the following seven steps:

(1) Obtaining curve C_1: The outer contour curve C_1 of the aircraft component projected on the support platform of the tooling is extracted, and the smallest rectangle with size a that can envelop the curve C_1 on the platform plane is obtained. Then, a local coordinate system xoy is created, where the original point o is located at the midpoint of the rectangle, and the x and y axes are in the plane, as shown in Figure 11. Furthermore, the posture matrix T of the local coordinate system xoy relative to the aircraft coordinate system is obtained.

(2) Obtaining curve C_2: The typical positioning layouts suitable for the product in the knowledge base are projected onto the platform plane with the matrix T, and the projected contour curve and the points of positioning location are extracted. The smallest rectangle enveloping the projected contour is created, and its length a′ is obtained in the same way as in step (1); then, a local coordinate system x′o′y′ is created. In order to facilitate subsequent trajectory comparisons, the projections of the contour curve and the location points are scaled with reference to the origin point o′, separately outputting curve C_2 and point set Z. The scaling ratio is a′/a. Then, the nearest point on C_2 corresponding to each of the points in Z is obtained, and these are

combined into a point set Y while recording the translation vectors from the points in Y to the corresponding points in Z.

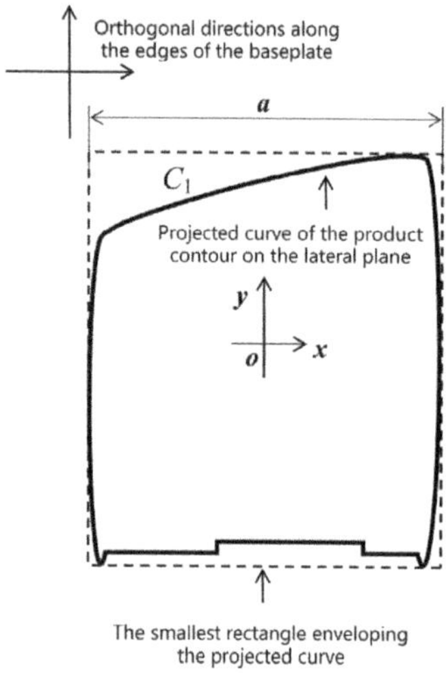

Figure 11. The projection of aircraft components on the supporting platform.

(3) Obtaining point sequences Q_1 and Q_2: The curve C_1 is discretized to obtain its point sequence Q_1, and the curve C_2 is discretized into points that are combined with the points in set Y to create sequence Q_2. The lengths of sequences Q_1 and Q_2 are denoted as n and m, respectively, as shown in Equation (7). Equation (7) can be expressed as the following:

$$Q_1 = \{p_1, p_2, \ldots p_n\}\ Q_2 = \{q_1, q_2, \ldots q_m\} \tag{7}$$

(4) Creating distance matrix **D**: An $(n + 1) \times (m + 1)$ matrix **D** is created, where the element D_{ij} is the distance $dis(p_i, q_j)$ between the points p_i and q_j, which can be expressed as the following:

$$D_{ij} = dis(p_i, p_j) = \sqrt{(x_{pi} - x_{qj})^2 + (y_{pi} - y_{qj})^2} \tag{8}$$

(5) Calculating loss matrix **M**: Based on the distance matrix **D**, another $(n + 1) \times (m + 1)$ matrix **M** is created. This can be expressed as the following:

$$M_{ij} = \begin{cases} D_{ij} & i = 0, j = 0 \\ D_{ij} + \min\{D_{i-1,j-1}, D_{i,j-1}, D_{i-1,j}\} & 1 \le i \le n, 1 \le j \le m \end{cases} \tag{9}$$

(6) Calculating the distance between sequences Q_1 and Q_2: The distance between sequences Q_1 and Q_2 is the 2-norm of the loss matrix **M**. The smaller the value, the higher the similarity between curves C_1 and C_2.

(7) Solving for the initial positioning location: Using the above method, the curve C_2 that is in highest similarity with the product contour C_1 is extracted. Then, according to the shortest distance in the loss matrix **M**, the points in sequence Q_1 corresponding to those points in set Y are obtained. Finally, by moving these obtained points from Q_1 via the translation vectors recorded in step (2), the positioning point set X is obtained.

In summary, the locator planning method based on the DTW trajectory comparison algorithm is used to compare the projected contour curve C_1 of the product with the projected curve C_2 of the typical positioning schemes in the knowledge base and extract the optimized version of C_2 that has the most similarity with C_1.

3.3. Layout Planning for the Platform Unit

The paving of the platform unit refers to the combination of platform units according to the length and width (L, W) of a unit and the gap D between connections, as shown in Figure 12. The automatic and rapid completion of this process requires the automatic calculation of the direction and quantity of the platform units according to the determined platform size with a proper planning algorithm. This study designs a paving method for platform units based on dynamic programming.

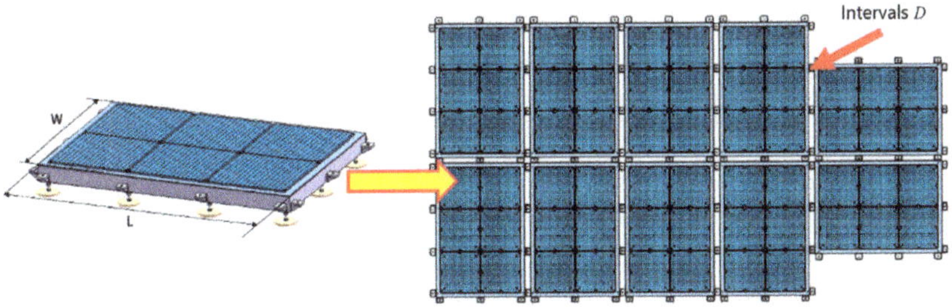

Figure 12. Paving of the platform unit.

Dynamic programming is a decision-making process for optimization by decomposing a complex problem into relatively simple sub-problems, and it is often used for problems with overlapping sub-problems and optimal substructure properties. This study combines this algorithm with a layered paving method to achieve maximum efficiency. The layered paving method refers to several platform units along the same paving direction (longitudinal or transverse) forming a layer, and multiple layers that have different paving directions construct an entire platform. According to the size of the entire platform determined by the product envelope, the initial outer contour of the platform is created, as illustrated in Section 2.3.1. The V and H axes of the outer contour are separately selected as the paving directions of the platform units in layers, and the paving direction of each layer is optimized according to the simulation result. Taking the V-axis as an example, the paving method for one layer based on dynamic programming is firstly introduced as follows.

Firstly, the bottom line of the first layer is obtained, as shown in Figure 13a. Aiming at ensuring that the platform is as regular as possible, the paving direction of all units in each layer is consistent. The platform units are paved along the longitudinal or transverse direction on the bottom line. When the longitudinal direction is selected, the top line of the first layer is the line segment L_{l1} obtained by moving the bottom line upward by the sum of L and D within the initial outer contour, as shown in Figure 13a. Similarly, when the transverse direction is selected, the top line of the first layer is the line segment L_{w1} obtained by moving the bottom line of the first layer upward by the sum of W and D within the initial outer contour.

In order to save costs and ensure the complete coverage of the initial outer contour, the paved units of the first layer should be the fewest and should cover the top line. If the units are paved along the longitudinal direction, the number n_{l1} of platform units is the result of the division of L_{l1} by $L + D$ and is rounded up. This can be expressed by the following:

$$n_{l1} = [L_{l1}/(L+D)] + 1 \quad (10)$$

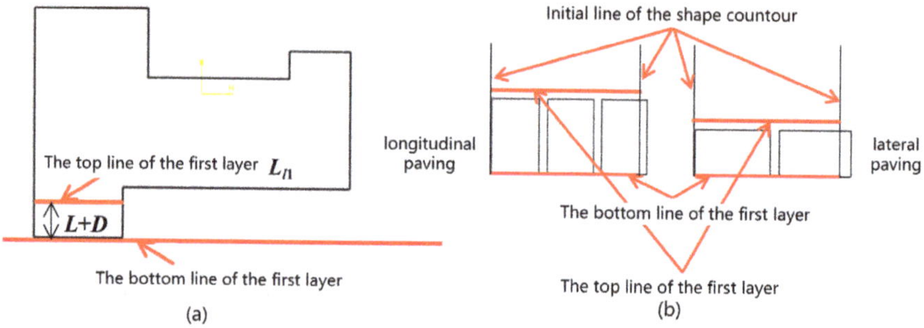

Figure 13. (**a**) Schematic diagram of the bottom and top lines of the first layer; (**b**) schematic diagram of the first layer laid along two directions.

If the platform units are paved along the transverse direction, the number n_{w1} of platform units is the result of the division of L_{w1} by $W + D$ and is rounded up. This can be expressed by the following:

$$n_{w1} = [L_{w1}/(W + D)] + 1 \tag{11}$$

The paving of the first layer of the platform along two directions is shown in Figure 13b.

To ensure the standardization of the design, for adjacent layers, if the top line of the former layer coincides with that of the latter layer after being offset by a certain distance, then the paving methods and quantities of the platform units of these adjacent layers are consistent. As shown in Figure 14, if the n th layer is paved horizontally and its top line coincides with that of the $n + 1$ th layer after being offset upward by $W + D$, then the paving of the $n + 1$ th layer is also along the horizontal direction, and the quantity of units is the same as that in the n th layer. On the contrary, if the top line of the n th layer does not coincide with that of the $n + 1$ th layer after being offset by an arbitrary distance, then there are two choices of the paving direction for the $n + 1$ th layer, and the quantity of units needs to be recalculated according to the new top line of the $n + 1$ layer.

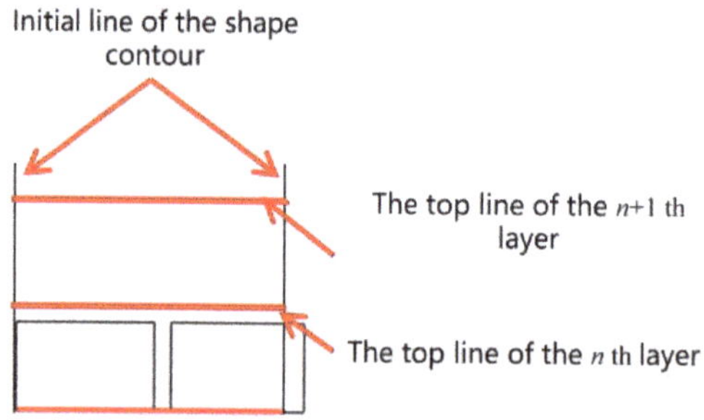

Figure 14. Schematic diagram of the adjacent paving layers.

On the basis of the above method for platform paving in layers, a dynamic programming algorithm is used to solve for the optimized scheme that uses the fewest units to construct the entire platform. The total number of units used is denoted as *Sum*. The whole process of optimization is divided into the following three steps:

(1) The process is divided into different stages.

Boundaries are set to divide the process into different stages when the top line of the previous paving layer does not coincide with that of the next paving layer after being offset.

This means that if the paving direction and unit quantity of a layer need to be redetermined, then the paving of the layer is in a new stage after the previous one.

(2) The paving scheme of each stage is determined.

In general, each stage has two schemes for choices, which are classified according to the paving direction. During the i th stage, the numbers of platform units separately calculated from the paving directions along the longitude and transverse are represented as N_{ij} and $N_{i,j+1}$. Taking the first stage shown in Figure 15 as an example, when the longitudinal direction is chosen, the value for the paved layer during this stage is only one, and the quantity of units in N_{11} is calculated as 3. When the transverse direction is chosen, the value for the paved layer during the first stage is two, and the quantity of units in N_{12} is calculated as 4.

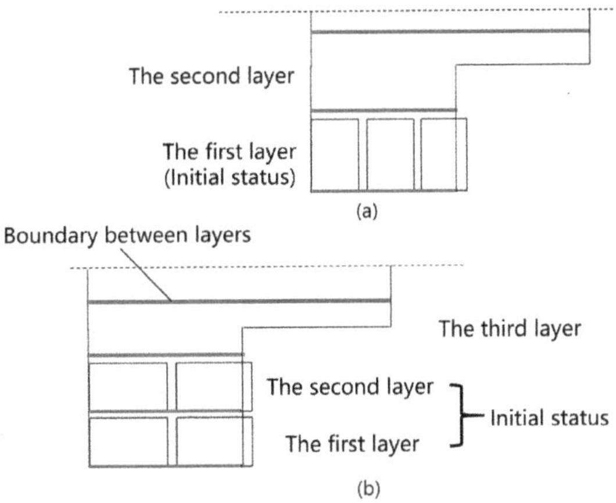

Figure 15. An example of the first paving stage: (**a**) longitudinal paving; (**b**) horizontal paving.

(3) The state transition equation is determined.

Based on the determination of each stage in two paving directions, a paving strategy diagram can be drawn, and one of its examples is shown in Figure 16, where $N[i][j]$ represents the total number of platform units needed for the chosen direction during the i th stage, and j is the ordinal number of the term.

As described previously, the optimal paving strategy involves letting the total number Sum of units reach its minimum value. Let $MINSum_{ij}$ represent the minimum number of platform units needed from node $N[i][j]$ in the i th stage to one of the nodes in the last stage. The corresponding path of $MINSum_{ij}$ from $N[i][j]$ to the last stage is part of the optimal strategic path that will be determined. Then, if the entire paving process contains a total of m stages, the state transition equation can be derived as follows:

$$MINSum_{ij} = \begin{cases} N[i][j] & i = m \\ \min\{N[i+1][2j-1], N[i+1][2j] + N[i][j]\ \} & i \neq m \end{cases} \quad (12)$$

According to the state transition equation, the optimal strategic path is solved by starting from the node with the smallest value in the first stage and reaching the last stage. For the above example, since $N_{11} < N_{12}$, the optimal strategy is the calculation process of solving $MINSum_{11}$, and this can be easily obtained by traversing all of the quantities of units represented by the nodes in the first stage.

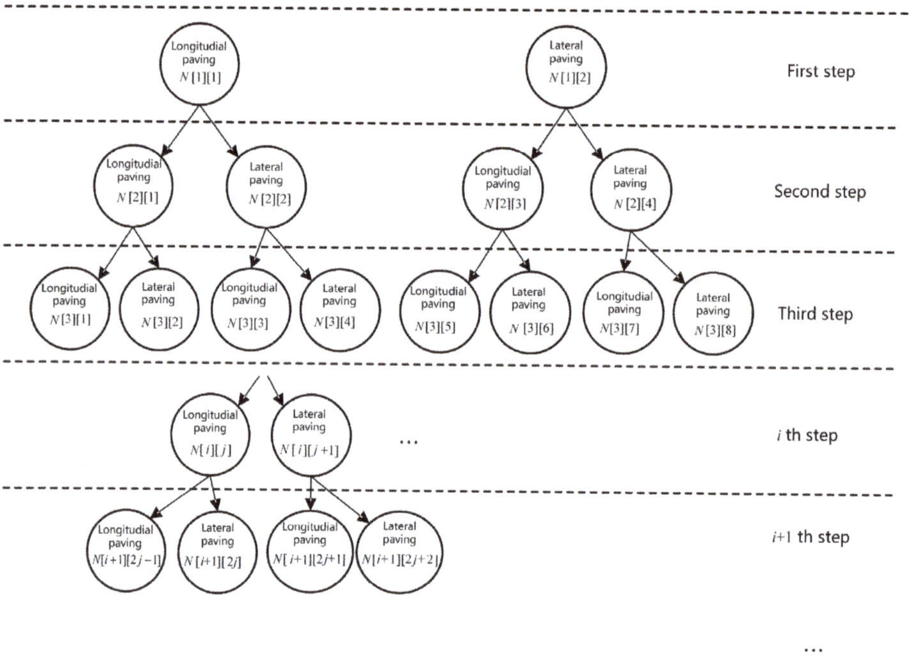

Figure 16. Diagram of the paving strategy.

4. Verification Example

4.1. Planning of the Positioning Layout

The assembly of two types of products was taken as an example for verification. The design of the reconfigurable tooling was completed using software that was developed based on CATIA and contained a function for reconfiguration planning.

Figure 17 shows the geometric features and dimensions of the two products, which are named A and B. Considering their sizes and structural characteristics, product A is assembled in a vertical posture, while product B is assembled in a horizontal posture. For the determinations of the positioning and support during assembly, the outer contours of products A and B are extracted from the top view.

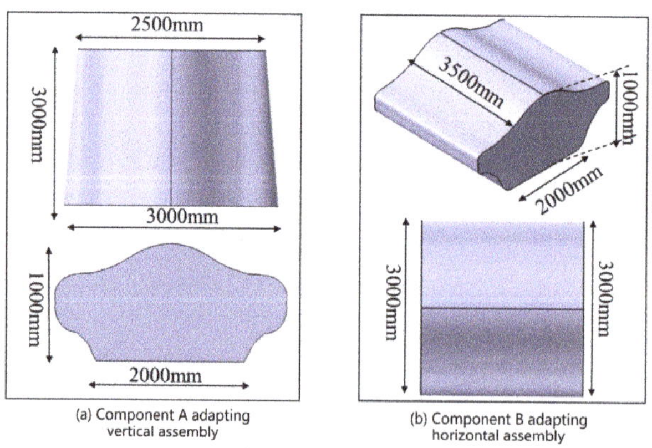

Figure 17. Assembly scheme and process features of components A and B.

Based on the DTW trajectory comparison algorithm, the initial points for the bottom support and positioning during assembly are calculated. The number of initial positioning points is set to 11, and then a comparative calculation is carried out. According to the projection curve C of component A's outer contour, the smallest rectangle that can envelop curve C on the platform surface (with the platform surface approximated as a horizontal plane) is obtained, and then the midpoint of the rectangle is used as the origin point o to create a local coordinate system: xoy. The initial positioning points calculated with the DTW algorithm are projected onto the platform surface to obtain their coordinates within the coordinate system xoy. The calculation results are shown in Table 1.

Table 1. Calculation results for the initial positioning points for product A.

Positioning Point	P1	P2	P3	P4	P5	P6	P7	P8	P9	P10	P11
x coordinate	−1325	−535	−297	291	536	1390	1335	1201	60	−1180	−1344
y coordinate	189	492	563	575	512	188	−296	−444	−688	−455	−305

Similarly, based on the DTW trajectory comparison algorithm, the initial positioning points for the horizontal assembly of product B are calculated. The number of positioning points is set to 10, and then the calculation is carried out. The coordinates of these points within the local system defined on the platform plane are shown in Table 2.

Table 2. Calculation results for the initial positioning points for product B.

Positioning Point	P1	P2	P3	P4	P5	P6	P7	P8	P9	P10
x coordinate	8255	8233	9745	10,958	10,857	10,844	10,962	9771	8211	8269
y coordinate	−1019	−1798	−1775	−1764	−622	645	1799	1755	1760	1000

The determined positioning points for products A and B from the top view are shown in Figure 18a, and the time consumed by the planning process for the positioning points based on the DTW trajectory comparison algorithm is shown in Figure 18b. The time consumed is only a few seconds, thus validating the efficiency of the planning of the positioning layout.

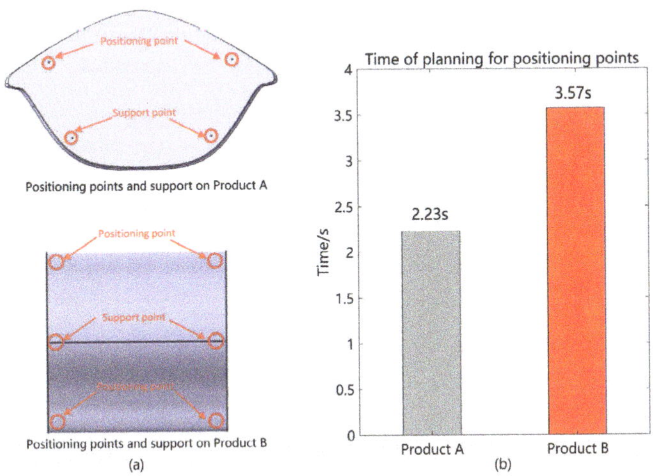

Figure 18. (**a**) The planning results for the positioning and support points for products A and B; (**b**) the time consumed by the planning process.

The stress distributions of products A and B under gravity and the layouts of the support planned by the DTW algorithm and personnel are shown in Figure 19. As demonstrated by the comparisons between the stress distributions resulting from the support layouts planned by the algorithm and personnel, the stress of products A and B under manually planned support is greater and more concentrated, since the distributions of support and positioning points planned by the algorithm are more dispersed and more appropriate than those planned manually using human experience. Benefiting from the integration and analysis of multiple layout schemes in the database, the DTW algorithm can output an optimal positioning scheme and support layout that is best suited for the current product.

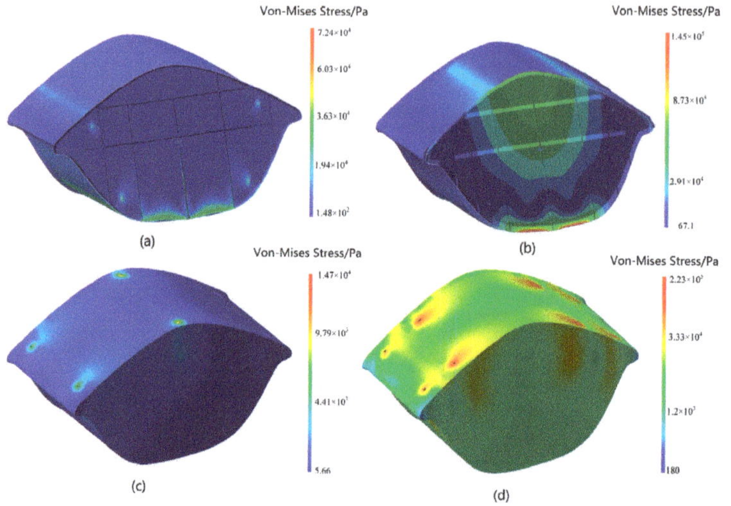

Figure 19. The stress distributions of products A and B under gravity and different layouts of supports: (**a**) product A under the support layout planned using the DTW algorithm; (**b**) product A under the manually planned support layout; (**c**) product B under the support layout planned using the DTW algorithm; (**d**) product B under the manually planned support layout.

4.2. Rapid Design of Reconfigurable Tooling

Firstly, the projection of product A is extracted onto the platform plane, as shown in Figure 20a. The points on the projection curve are used as inputs, the planning algorithm is run based on dynamic programming, and the optimized scheme of a 3×3 layout for the paving units of the platform is provided as an output. Then, the platform planning module of the design software is run to carry out the rapid assembly of digital models. The user specifies the *X/Y* direction for the platform's installation in the digital environment and inputs the number of units in the two directions. The software automatically inputs models of platform units with assembly constraints. When the platform units are assembled, the software automatically starts the rapid assembly of the remaining parts, such as pedals and connecting pieces, to complete the entire process, as shown in Figure 20b.

Then, the initial positioning points calculated using the planning algorithm are input into the current digital model. The user can design the installation positions of columns according to these points and then initiate the corresponding module in the software, where they can select the types of columns and their installation positions by clicking the points on the platform model as shown by the red circles in Figure 21a. The column models can be rapidly input, as shown in Figure 21b. Finally, the user can enable the installation module of the locator to achieve the rapid assembly of the beam and sliding rail units on the column through a similar operation. The final design model in digital space is shown in Figure 21c. The time taken by a group of processing personnel to complete the whole design process is shown in Figure 21d. As shown in Figure 21d, the design time was only around 10 min at most, and it was far less than that of the traditional design process, which usually requires several days.

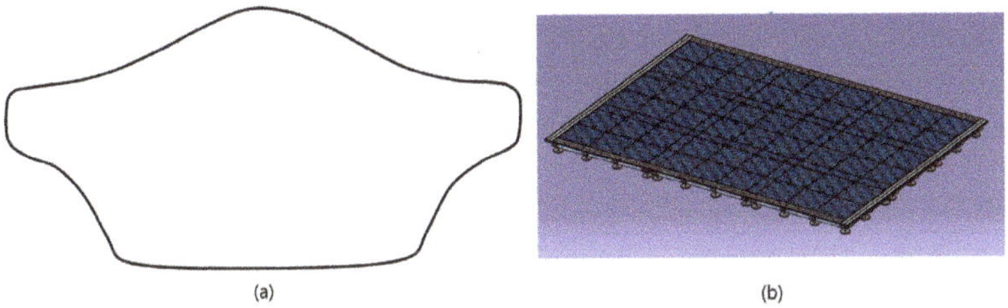

Figure 20. (**a**) Horizontal projection of product A; (**b**) paving results of platform units.

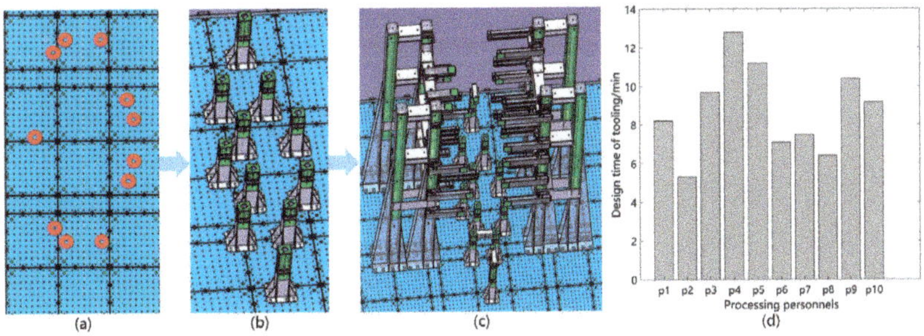

Figure 21. Design process of the assembly tooling for product A in a vertical posture: (**a**) selection of column installation positions; (**b**) importing a column model; (**c**) importing beam and sliding rail units; (**d**) time taken for the design process to be completed by 10 processing personnel.

Similarly, the projection of product B in a horizontal posture is extracted on the platform plane. Then, the planning algorithm runs and outputs the scheme of platform paving, for which a 3 × 3 layout is also adopted, and the unit models are automatically input to form the platform. Finally, the column and locator models are imported and rapidly assembled according to the positioning points planned for product B. The process and result of the tooling design for the horizontal assembly of product B are shown in Figure 22a–c. The time taken by 10 personnel to complete the design process is shown in Figure 22d. As in the assembly tooling of product A, the process of tooling design for product B only takes 10 min, which is notably less than the several days required for a traditional design process.

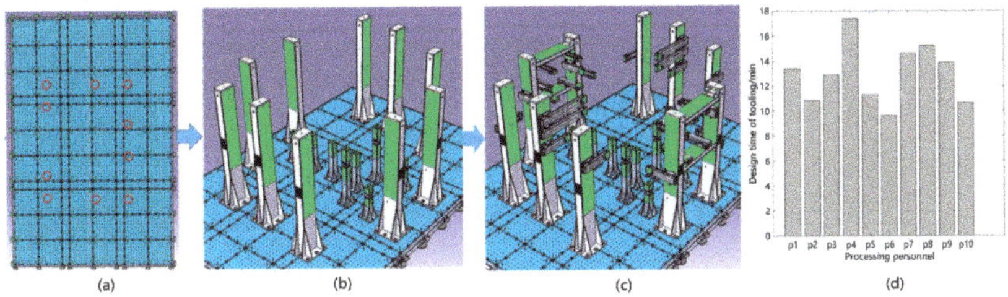

Figure 22. Design process and result of the assembly tooling for product B: (**a**) selection of column positions; (**b**) importing a column model; (**c**) importing beams and sliding rails; (**d**) time consumed by the design process when completed by 10 personnel.

4.3. Monitoring System for Assembly Tooling

On the basis of modular design and reconfiguration planning technology, tooling integrated with multi-source sensors is manufactured and applied to assemble a practical type of aircraft fuselage. As shown in Figure 23a, three-dimensional force sensors and fiber Bragg gratings are arranged on locator beams to separately sense their loads and deformation in real time. In addition, temperature and vibration sensors are installed on the columns and beams to detect the thermal deformation and mechanical vibration of the tooling structure. Details of the sensor layout are shown in Figure 23b. More detailed information on the use of fiber Bragg gratings and deformation sensors can be found in our previous work [10].

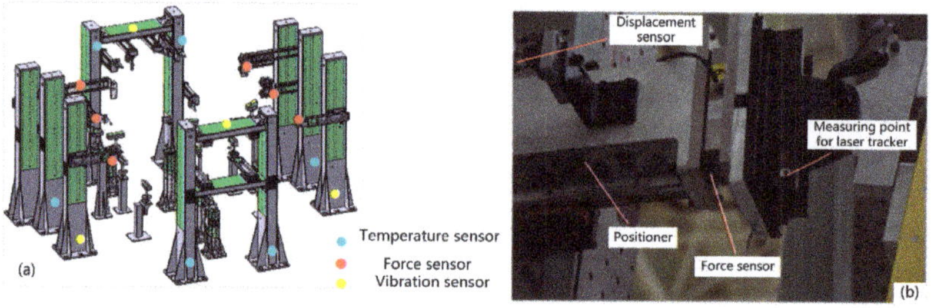

Figure 23. (**a**) Schematic diagram of the sensor layout; (**b**) details of the sensor layout.

An online monitoring software system containing information on the vibrations, clamping force, and environmental temperature of the tooling structure was developed to provide warnings about stress and positioning errors during the assembly process. The monitoring software interface developed in this study is shown in Figure 24. In this interface, the deformation and temperature distributions of the tooling structure are intuitively presented through a three-dimensional model with a nephogram, and the current clamping force, temperature, and vibration are presented using waveforms on the right side of the interface, providing a good visualization of the online service state of the tooling. When the measures exceed their thresholds, the system issues a warning and makes recordings. This provides an example for the realization of intelligent assembly based on reconfigurable tooling.

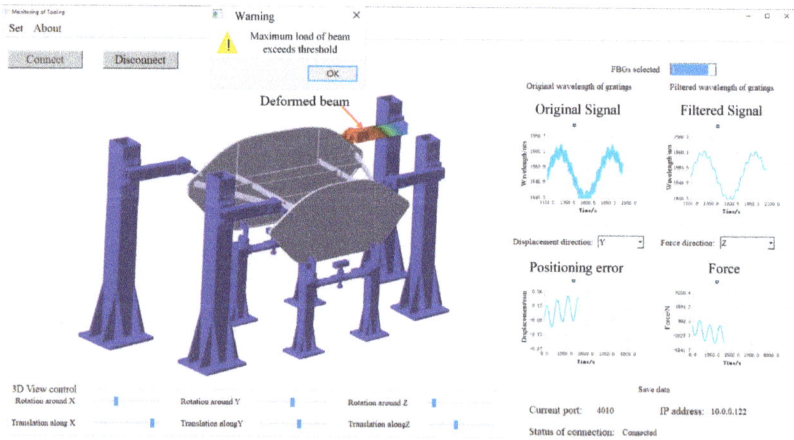

Figure 24. Interface of the monitoring system for reconfigurable tooling.

5. Conclusions

This study investigated the modular design of aircraft assembly tooling, as well as how to optimize its configuration, to enhance the generality of designs and their ability to rapidly respond to product changes. The main conclusions are as follows:

(1) In terms of the modular structure design, the concept of functional elements is proposed, and a diagram revealing the relationships between functional elements of all tooling parts is established. Based on this diagram, the similarity between structures is defined and calculated. The tooling structure is divided and reorganized into modules according to similarity, and a detailed design of the modules is presented using three typical structures, namely, platforms, columns, and locators. Modularization and serialization make the structures lightweight and simplified, enhancing the design efficiency and openness of the assembly space while guaranteeing structural rigidity and stability.

(2) In terms of reconfiguration technology, a parametric representation of modular structures is proposed, and a knowledge base is constructed to achieve efficient transmission and storage of structural information. By using the parametric information of structures, the locator layout and platform configuration are separately optimized based on the DTW trajectory comparison and dynamic programming algorithms, thus allowing for the rapid reconfiguration of tooling. The usage of releasable screws for the adjustment of structures guarantees both adaptability and stability.

Reconfigurable tooling is built and successfully applied to the assembly of a practical type of aircraft fuselage. A system for monitoring multi-source information, including structural vibrations, environmental temperature, and clamping force, is developed, thus enabling early warnings about assembly stress and positioning errors. This engineering verification provides a paradigm for the serialization, modularization, and intellectualization of aircraft assembly tooling.

Author Contributions: Conceptualization, Z.S. and J.H.; methodology, Z.S.; software, C.L.; validation, Z.S.; formal analysis, C.L.; resources, Z.S.; writing—original draft preparation, C.L.; writing—review and editing, J.H., X.S. and H.W.; supervision, W.T.; project administration, W.T. All authors have read and agreed to the published version of the manuscript.

Funding: The work reported herein was financially supported by the National Key Research and Development Program of China (Grant No. 2019YFB1707403), Natural Science Foundation of Jiangsu Province (BK20202007), and Fundamental Research Funds for the Central Universities (Grant No. NT2024013).

Data Availability Statement: Data will be made available upon request from the corresponding author.

Conflicts of Interest: Author Zhanghu Shi was employed by the company Aviation Industry Chengdu Aircraft Industry (Group) Co., Ltd. The remaining authors declare that the research was conducted in the absence of any commercial or financial relationships that could be construed as a potential conflict of interest. The funders had no role in the design of the study, the collection, analyses, or interpretation of data; the writing of the manuscript; or the decision to publish the results.

References

1. Grieves, M. *Virtually Perfect: Driving Innovative and Lean Products Through Product Lifecycle Management*; USA Space Coast Press: Merritt Island, FL, USA, 2011.
2. Hu, F.; Li, D. Process Planning and Simulation Strategies for Perimeter Milling of Thin-walled Flexible Parts Held by Reconfigurable Fixturing System. In Proceedings of the Third International Conference on Measuring Technology and Mechatronics Automation, Shangshai, China, 6–7 January 2011; pp. 922–926.
3. Xie, F.; Liu, X.J.; You, Z.; Wang, J. Type synthesis of 2T1R-type parallel kinematic mechanisms and the application in manufacturing. *Robot. Comput. Integr. Manuf.* **2014**, *30*, 1–10. [CrossRef]

4. Bi, Z.M.; Lang, S.Y.T.; Verner, M.; Orban, P. Development of reconfigurable machines. *Int. J. Adv. Manuf. Technol.* **2008**, *39*, 1227–1251. [CrossRef]
5. Naing, S. Feature based Design for Jig-Less Assembly. Ph.D. Thesis, Crandfield University, Bedfordshire, UK, 2004.
6. Schwake, K.; Wulfsberg, J. Robot-based system for handling of aircraft shell parts. *Procedia Cirp* **2014**, *23*, 104–109. [CrossRef]
7. Zhu, Y.G.; Huang, X.; Li, S.G. A novel six degrees-of-freedom parallel manipulator for aircraft fuselage assemble and its trajectory planning. *J. Chin. Inst. Eng.* **2015**, *38*, 928–937. [CrossRef]
8. Esser, M.; Vette, M. Reconfigurable handling systems as an enabler for large components in mass customized production. *J. Intell. Manuf.* **2013**, *24*, 977–990.
9. Tao, F.; Cheng, Y.; Zhang, L.; Nee, A.Y.C. Advanced manufacturing systems: Socialization characteristics and trends. *J. Intell. Manuf.* **2015**, *7*, 1079–1094. [CrossRef]
10. Li, C.; Hu, J.; Kang, R.; Yang, Y.; Jin, J.; Tian, W. Structural deformation and clamping force monitoring of reconfigurable tooling motivated by strain data in aircraft assembly. *Smart Mater. Struct.* **2024**, *33*, 055046. [CrossRef]
11. Muller, R.; Esser, M.; Mbarek, T.; Biyiklioglu, N.; Quinders, S. *Development of Modular Low-Cost Automation Systems for Aircraft Assembly*; SAE Technical Papers; SAE: Warrendale, PA, USA, 2011.
12. Schneider, T. Innovative approach for modular and flexible positioning systems for large aircraft assembly. In Proceedings of the SAE 2015 AeroTech Congress & Exhibition, Washington, DC, USA, 22–24 September 2015.
13. Kihlman, H. Affordable Automation for Airframe Assembly Development of Key Enabling Technologies. Ph.D. Thesis, Linkoping University, Linkoping, Sweden, 2005.
14. Basile, F.; Caccavale, F.; Chiacchio, P.; Coppola, J.; Marino, A.; Gerbasio, D. Automated synthesis of hybrid Petri net models for robotic cells in the aircraft industry. *Control. Eng. Pract.* **2014**, *31*, 35–49. [CrossRef]
15. Jonsson, M.; Ossbahr, G. Aspects of reconfigurable and flexible fixtures. *Prod. Eng.* **2010**, *4*, 333–339. [CrossRef]
16. Löchte, C.; Dietrich, F.; Raatz, A. *A Parallel Kinematic Concept Targeting at More Accurate Assembly of Aircraft Sections*; Springer: Berlin/Heidelberg, Germany, 2011; pp. 142–151.
17. Prause, I.; Mbarek, T.; Corves, B. Increasing the stiffness of a 3-PUU parallel kinematic positioning device for high payloads by modifying the leg configuration. In Proceedings of the IFToMM Workshop on Fundamental Issues and Future Research Directions for Parallel Mechanisms and Manipulators, Tianjin, China, 7–8 July 2014.
18. Jefferson, T.; Crossley, R.; Smith, T. *Review of Reconfigurable Assembly Systems Technologies for Cost Effective Wing Structure Assembly*; SAE Technical Papers; SAE: Warrendale, PA, USA, 2013.
19. Jonsson, M.; Kihlman, H.; Ossbahr, G. Coordinate controlled fixturing for affordable reconfigurable tooling. In Proceedings of the 2nd Conference on Assembly Technologies and Systems, Tianjin, China, 10–12 May 2008; pp. 1–11.
20. Kihlman, H. *Flexapods-Flexible Tooling at SAAB for Building the NEURON Aircraft*; SAE Technical Papers; SAE: Warrendale, PA, USA, 2010.
21. Xiong, L.; Molfino, R.; Zoppi, M. Fixture layout optimization for flexible aerospace parts based on self-reconfigurable swarm intelligent fixture system. *Int. J. Adv. Manuf. Technol.* **2013**, *66*, 1305–1313. [CrossRef]
22. Mannheim, T.; Riedel, M.; Hüsing, M.; Corves, B. A new way of grasping: PARAGRIP—The fusion of gripper and robot. *Grasping Robot.* **2013**, 433–464.
23. Molfino, R.; Zoppi, M.; Zlatanov, D. Reconfigurable swarm fixtures. In Proceedings of the Asme/Iftomm International Conference on Reconfigurable Mechanisms and Robots, London, UK, 22–24 June 2009; IEEE: Piscataway, NJ, USA, 2009; pp. 730–735.
24. Ramirez, J.; Wollnack, J. Flexible automated assembly systems for large CFRP-structures. *Procedia Technol.* **2014**, *15*, 447–455. [CrossRef]
25. Erdema, I.; Helgosson, P.; Kihlman, H. Development of Automated Flexible Tooling as Enabler in Wing Box Assembly. *Procedia CIRP* **2016**, *44*, 233–238. [CrossRef]

machines

MDPI

Article

A Fitness Landscape-Based Method for Extreme Point Analysis of Part Surface Morphology

Jinshan Sun and Wenbin Tang *

School of Mechanical and Electrical Engineering, Xi'an Polytechnic University, Xi'an 710600, China; 42202310108@stu.xpu.edu.cn
* Correspondence: tangwb@xpu.edu.cn

Abstract: Advancements in Industry 4.0 and smart manufacturing have increased the demand for precise and intricate part surface geometries, making the analysis of surface morphology essential for ensuring assembly precision and product quality. This study presents an innovative fitness landscape-based methodology for extreme point analysis of part surface morphology, effectively addressing the limitations of existing techniques in accurately identifying and analyzing extremum points. The proposed approach integrates adaptive Fitness-Distance Correlation (FDC) with a roughness index to dynamically determine the number and spatial distribution of initial points within the pattern search algorithm, based on variations in surface roughness. By partitioning the feasible domain into subregions according to FDC values, the algorithm significantly reduces optimization complexity. Regions with high ruggedness are further subdivided, facilitating the parallel implementation of the pattern search algorithm within each subregion. This adaptive strategy ensures that areas with intricate surface features are allocated a greater number of initial points, thereby enhancing the probability of locating both regional and global extremum points. To validate the effectiveness and robustness of the proposed method, extensive testing was conducted using five diverse test functions treated as black box functions. The results demonstrate the method's capability to accurately locate extremum points across varying surface profiles. Additionally, the proposed method was applied to flatness error evaluation. The results indicate that, compared to using only the raw measurement data, the flatness error increases by approximately 3% when extremum points are taken into account.

Keywords: surface morphology; fitness landscape; extreme point analysis; pattern search algorithm

1. Introduction

In contemporary manufacturing paradigms, the geometric characteristics of a part's surface are key to assembly precision and quality. With Industry 4.0 and smart manufacturing, surface quality standards have risen, especially for high-precision parts. Analyzing surface geometry, including extremal points such as peaks, valleys, and edges, is crucial for assessing machining inaccuracies, predicting assembly precision, and improving product performance and reliability.

The analysis of extreme points in surface morphology relies heavily on surface modeling, which has gained significant attention as products advance in precision, miniaturization, and integration [1]. Recent research in 3D tolerance analysis has introduced several methodologies, including the degree-of-freedom method, skin model shapes, fractal geometry, and data-based reverse modeling. The degree-of-freedom method treats the surface as an ideal feature within a tolerance zone, useful for representing macroscopic errors, but it cannot capture surface morphology's intricacies [2–4]. The skin model shapes discretize geometric features, generating errors through techniques like modal analysis and mesh deformation; however, challenges remain in preserving error correlations across points and keeping errors within tolerance [5–8]. The fractal method, while effective for characterizing surface roughness, provides excessive geometric detail, limiting its application to assembly accuracy studies [9–12]. Data-based reverse modeling, leveraging high-precision measurement data, enhances accuracy by creating digital representations and integrating real surface errors into propagation analysis [13–19]. These studies have evolved from "ideal geometric variations" to "microscopic geometric expressions", reflecting a transition from macroscopic to microscopic scales and providing a more accurate depiction of part surface morphology. However, extreme points, as crucial feature information of the part's surface morphology, are not explicitly represented in these models. Taking the evaluation of surface machining errors as an example, the part's surface exhibits complex morphology with alternating valleys and peaks, resulting in multiple extreme points. Current methods for machining error evaluation typically rely on limited measurement data and the least squares method, as demonstrated in references [20–23]. Due to the restricted number of measurement points, these methods may overlook critical extreme points, leading to inaccurate error estimations.

To identify extreme points on part surface morphologies, a common strategy is the Multi-start algorithm, which involves random generation of initial points followed by local searches to locate respective extrema [24]. Wang et al. [25] introduced the Memory Particle Swarm Optimization (MPSO) algorithm, combining Particle Swarm Optimization (PSO) with local search techniques to identify multiple global and local optima in Multi-Objective Optimization Problems (MMOPs). Gnandt et al. [26] developed a hybrid method for global optimization by combining random search with descent methods, using region control to refine the search process with local minima, similar to step-size control in deterministic optimization. Minaee et al. [27] improved extreme point search efficiency on digital image surfaces by integrating image segmentation and edge detection, extending its applicability to other surface types. Koessler et al. [28] advanced the global extreme point search algorithm for faster identification and established a theoretical foundation for extreme point searches on part surfaces. A review of the existing literature reveals that most studies concentrate on optimizing algorithms to locate all local extrema across the entire surface. However, few investigations address surface ruggedness and attempt to mitigate optimization complexity by partitioning the surface into regions based on ruggedness characteristics.

This paper addresses the challenge of analyzing extreme points on the surface morphology of a part by introducing the concept of the fitness surface. The proposed method decomposes the surface by calculating its roughness and employs a mode-search algorithm to identify all local extreme points. This approach accounts for the geometric characteristics of the part's surface and leverages the advantages of the Multi-start algorithm. The structure of the paper is organized as follows. Section 2 introduces the concept of the fitness surface; Section 3 details the working principles and algorithmic flow for analyzing extreme points on the part's surface morphology; Section 4 validates the proposed method through both theoretical test functions and engineering examples.

2. The Concept of Fitness Landscapes

2.1. Method Overview and Basic Concepts

The concept of a fitness landscape is extensively utilized in fields such as biology, genetics, and optimization theory. It primarily describes the relationship between the solution space of an optimization problem and its corresponding fitness (or objective function) values. This concept draws inspiration from the undulating terrains found in nature to visually represent the quality of solutions, where each point signifies a potential solution, and the elevation of the point indicates the fitness value of that solution. Fitness Landscapes are particularly significant in genetic algorithms, evolutionary computation, and other heuristic search methods, as they offer an intuitive understanding of the characteristics inherent in a problem's search space. By analyzing the Fitness Landscape, researchers and engineers can more effectively design and refine their optimization strategies to address complex optimization challenges.

The concept of the fitness landscape is also applicable to the analysis of part surface morphology. Fitness landscape analysis has become an effective method for exploring the parameter information of part surface morphology. In the context of part surface morphology, each point on the local surface is treated as a potential solution. Consequently, the part surface morphology can be viewed as a surface with complex fitness values. By systematically searching this fitness landscape, the main features of the part surface morphology can be extracted. The main features of a fitness landscape include:

Height/Depth: In a fitness landscape, the height or depth of a point represents the fitness value of the corresponding solution. In minimization problems, valleys indicate optimal solutions, whereas in maximization problems, peaks signify optimal solutions.

Local Extremes: Local peaks or valleys on the fitness landscape correspond to local optimal solutions. These points can cause the search process to become easily trapped, hindering the discovery of the global optimum.

Global Extremes: The highest or lowest points on the fitness landscape represent the global optimal solutions.

Ruggedness: The ruggedness or irregularity of the fitness landscape reflects the complexity of the optimization problem. A highly rugged fitness landscape typically contains numerous local extremes, increasing the difficulty of locating the global optimal solution.

The analysis of extreme points in part surface morphology can be formulated as an optimization problem, with the solution process resembling strategic exploration on a complex terrain. In this framework, part surface morphology is abstracted as a terrain composed of numerous peaks and valleys, where the elevation of each point represents the fitness of the corresponding solution, i.e., its degree of alignment with the optimization objective. The role of the search algorithm in this terrain is akin to that of an explorer seeking the deepest valleys, aiming to find the optimal solution for a minimization problem. In this process, the optimization algorithm functions as a strategic explorer, continuously navigating the terrain to identify the most efficient path toward its optimization goal—extreme points with the lowest fitness values. The search process is illustrated in Figure 1 below.

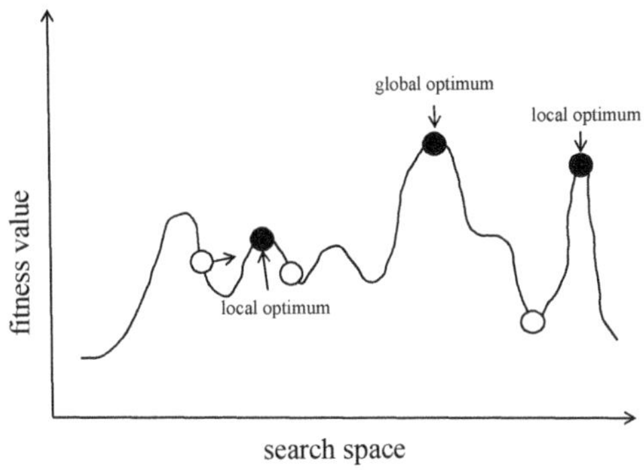

Figure 1. Schematic of extreme point search on the cross-sectional curve of part surface morphology.

2.2. *Mathematical Modeling of the Fitness Landscapes*

In the context of optimization problems, the fitness landscape offers a visual framework for understanding the complexity of the search space, while heuristic algorithms serve as methods for identifying the optimal solution within this landscape. The definition of the fitness landscape is as follows.

A fitness landscape can be formally defined as a triplet (S, f, d), where S denotes the solution space of the problem, encompassing all possible solutions. Depending on the specific problem, this space may be either continuous or discrete. The objective function $f(x)$ is a mapping defined on the search space $x \in S$, assigning a metric value to each solution to evaluate its quality. The distance metric $d(x_1, x_2)$ measures the distance between any two solutions x_1 and x_2 within the search space, serving as a function to quantify the distance or similarity between two points. This formalization provides a comprehensive framework for analyzing and navigating the complex morphology of the solution space, facilitating the design and application of effective heuristic algorithms for optimization.

Fitness landscape analysis offers an intuitive approach for optimizing low-dimensional functions. However, for optimization problems involving multi-dimensional variables, the complexity of the fitness landscape becomes challenging to represent through straightforward visualization. In such cases, Fitness-Distance Correlation (FDC) emerges as a crucial tool for quantifying and comprehending the ruggedness of the fitness landscape. FDC provides a systematic method for evaluating the relationship between the fitness of a solution and its distance to other solutions, thereby enhancing our understanding of the landscape's structure and aiding in the development of more effective optimization strategies.

Fitness-Distance Correlation (FDC) is the correlation coefficient between the fitness of a solution and its distance within the solution space. It quantifies the ruggedness of the fitness landscape by calculating the statistical correlation between fitness and distance. In relatively smooth and continuous fitness landscapes, the fitness of a solution and its distance to the optimal point typically exhibit a strong correlation. In such cases, the FDC value tends to be large, usually close to 1 or -1. This indicates that points near the optimal solution generally possess higher fitness, while those farther from the optimal solution tend to have lower fitness. Conversely, in rugged and highly fluctuating fitness landscapes, the correlation between a solution's fitness and its distance to the optimal point is weaker. In these scenarios, the FDC value is smaller, often approaching 0. This suggests that even solutions in close proximity to the optimal point may not exhibit ideal fitness levels, reflecting the presence of numerous local optima.

In practical scenarios involving complex part surface morphology models, determining the global optimum is often challenging and sometimes even infeasible. This difficulty arises because the exact location of the global optimum may be unknown or difficult to ascertain accurately due to the model's complexity and computational limitations. In such cases, the traditional Fitness-Distance Correlation (FDC) metric, which relies on knowledge of the global optimum, may no longer be applicable. To address this limitation, an improved definition of FDC has been proposed, which utilizes optimal solutions sampled from a dataset constructed from a precise part surface morphology model, thereby replacing the reliance on the traditional global optimum.

In this study, we used an improved fitness distance correlation coefficient (FDC) to quantify the characteristics of the part surface. This improved FDC measures the correlation between the fitness of sample points and their distance to the optimal solution, by considering the optimal points within the sample set, rather than the global optimum. We calculated the distance from each sample point to the optimal points within the sample set and analyzed the correlation between these distances and their corresponding fitness values, in order to assess the complexity and optimization difficulty of the part surface geometry. It is defined as follows. There is a set of selected sample points on the fitness surface $P = \{x_i | i = 1, 2, \cdots, n\}$, where each point x_i represents a possible solution in the solution space. Within this set of samples, we identify the point with the optimal fitness value $S^* = x_{\max}$ (assuming the maximum value $x_{\max}$ as the optimal point), while the point $x_{\max}$ with the minimum fitness value on the set P represents the optimal solution found within the current sample set. The FDC on the sample set P is given by:

$$fdc(s) = \frac{\sum_{i=1}^{n} (y_i - \overline{y})(d_i - \overline{d})}{\sqrt{\sum_{i=1}^{n} (y_i - \overline{y})^2 \sum_{i=1}^{n} (d_i - \overline{d})^2}} \quad (1)$$

where y_i represents the corresponding value of point x_i; the average value of all sample points y_i is represented as $\overline{y}$, that is, $\overline{y} = \sum_{i=1}^{n} y_i$; the Euclidean distance from x_i to the optimal point S^* is represented as d_i, that is, $d_i = \|x_i - x_{\max}\|$; and $\overline{d}$ represents the average value of d_i, that is, $\overline{d} = \sum_{i=1}^{n} d_i$.

FDC is a metric used to describe the relationship between the fitness value y_i of a solution and the distance d_i to the global optimum within the search space S of an optimization problem. The range of values for this metric is $FDC \in [-1, 1]$, where different values reflect different relationships between fitness and distance; when $|FDC| = 1$, it indicates a linear relationship between fitness and distance. For optimization problems, the closer the solution is to the optimal point (the smaller the d_i value), the better the fitness (the smaller the y_i value). This suggests that there is a clear path within the search space leading to the optimal solution. When the absolute value of FDC is small, it signifies the absence of a significant linear relationship between fitness and distance. In such scenarios, fitness does not consistently increase as the distance to the optimal solution decreases. Local peaks or valleys may impede the path toward the optimal solution, thereby diminishing the correlation between fitness and distance to the optimal point. Consequently, smaller absolute values of FDC typically indicate that the fitness landscape in the region is more rugged, rendering the optimization process more complex. The FDC analysis for different intervals of the same function is illustrated in Figures 2–5. The red dots on the curves in the chart represent the lowest points among all the points.

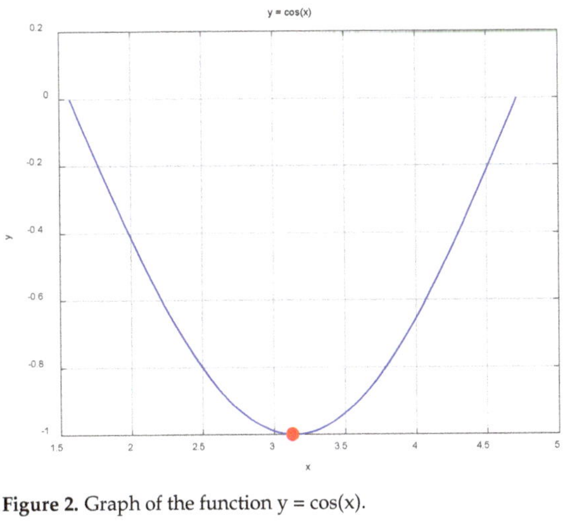

Figure 2. Graph of the function y = cos(x).

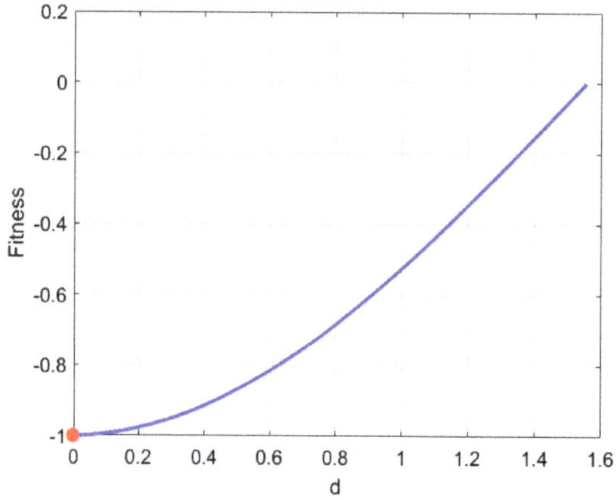

Figure 3. Relationship between fitness value and distance to the optimal point.

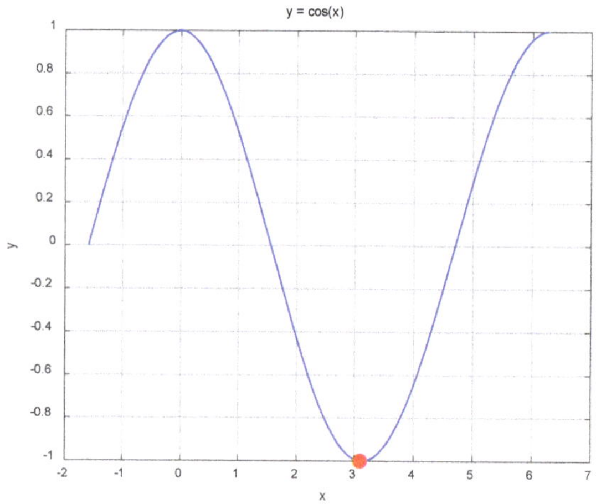

Figure 4. Curve of the function y = cos(x).

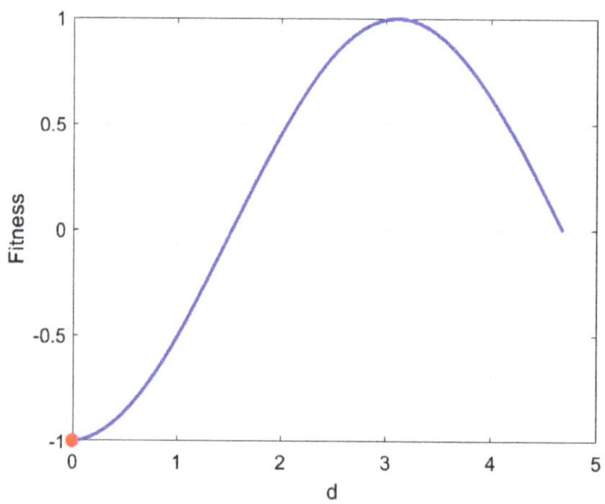

Figure 5. Relationship between fitness value and distance to the optimal point.

Figure 2 shows $y = \cos(x), x \in \left[\frac{\pi}{2}, \frac{3\pi}{2}\right]$, where 150 sample points are evenly selected over the entire feasible interval, resulting in $P = \left\{\frac{\pi}{2}, \frac{1.01\pi}{2}, \ldots, \frac{3\pi}{2}\right\}$, and the optimal point $x_{\min} = \pi$ is found. Figure 3 shows the relationship between the fitness value and the distance to the optimal point within the interval. As can be seen from the graph, the closer the distance to the optimal point, the smaller the fitness value, with $FDC = 0.9786$.

Calculations were performed on different intervals of the same function, and the results indicate that solving this function is relatively straightforward due to its low surface irregularity, facilitating the identification of extreme points. As illustrated in Figure 4, for function $y = \cos(x), x \in \left[-\frac{\pi}{2}, 2\pi\right]$, using the same analysis method, we are able to identify the optimal solution point $x_{\min} = \pi$ within the sample range.

In addition, Figure 5 illustrates that the closer a solution is to the optimal point, the lower its fitness value. However, within a certain region, a high-fitness point emerges, disrupting the previously mentioned trend—that is, when the distance from the solution to the optimal point is below a certain threshold, fitness decreases as distance decreases; conversely, when the distance exceeds this threshold, fitness decreases as distance increases. The further calculation result is $FDC = 0.8155$. Compared to Figure 3, the FDC value depicted in Figure 5 has decreased, indicating that as surface ruggedness increases, the FDC value correspondingly decreases. Nevertheless, the surface ruggedness in this range is not highly complex. Although the FDC value is lower than in Figure 3, the correlation coefficient remains relatively high, suggesting that even with slight surface irregularities, the impact on locating extreme points is still limited.

However, there are several limitations associated with using the FDC metric for evaluation. As discussed in the literature, the calculation of FDC relies on a limited sample set, which is often significantly smaller than the overall scale of the search space, potentially leading to biased results. Furthermore, according to Formula (1), the computation of FDC assumes that the position of the global optimum is known, thereby restricting its utility as a predictive tool and rendering it more suitable for post hoc verification. To address these limitations, an improved FDC metric approach is adopted in this study, which divides the search space into subintervals and analyzes the FDC values within each subinterval. Through extensive experiments, it is generally recommended that the FDC value be set between 0.4 and 0.7. Additionally, for black-box function optimization problems of part surface morphology, where gradient information of the unknown function is unavailable, determining the global optimum in advance is challenging. Therefore, this method samples surface information within each subinterval and designates the point with the highest fitness value among all sampled points as the local optimum for that region. This approach not only overcomes the limitations of the traditional FDC metric but also effectively elucidates the ruggedness of the surface.

3. Extreme Point Detection in Part Surface Morphology

3.1. Algorithm Principles

When evaluating the validity of the starting point A within the defined region P, the reference criterion can be set as follows. If point A, as the initial configuration of the search algorithm, leads the algorithm to successfully identify at least one local optimum within region P, then point A is considered a valid initial configuration point within region P. Conversely, if using point A as the starting configuration of the algorithm fails to identify a local optimum within region P, then point A is regarded as an invalid initial configuration point within region P. Figure 6 illustrates the valid initial points within the region. Point A: It is located in region P1 and serves as the starting point for the pattern search. It successfully locates the local optimal solution S1, so it is an effective initial point for region P1.

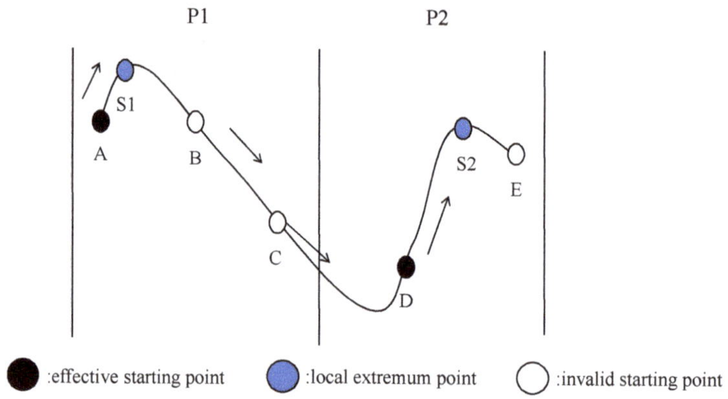

Figure 6. Region of valid initial points.

Points B and C: They are located in region P1, but as starting points, they cannot locate the local optimal solution S1. Therefore, they are invalid initial points for region P1.

Point D: It is located in region P2. As the starting point, it successfully locates the local optimal solution S2, so it is an effective initial point for region P2.

Point E: It is located in a highly rugged region, such as P3. More initial points are required to locate the extreme point. It is either an invalid initial point or a starting point from which the extreme point has not been successfully searched.

According to the definition of local extremum search, if point A, serving as the starting point of the search, successfully locates the local optimum S1 within region P1, then A is considered a valid initial point for P1. In contrast, if points B or C, as starting points, cannot identify the local optimum S1 within P1, then B and C are considered invalid initial points for P1. Similarly, if point D, as the starting point in region P2, is able to successfully locate the local optimum S2, then D is considered a valid initial point for P2.

As shown in Figure 6, the complex terrain of region P1 requires more initial points to increase the probability of finding valid starting points. On the other hand, region P2 has relatively low ruggedness, so a few initial points are sufficient to ensure the identification of extremum points.

In general, for the P1 region shown in Figure 6, which has higher ruggedness, the probability of valid initial points appearing is relatively high. High-ruggedness regions may also contain multiple peaks and valleys, as shown in Figure 7. In this case, regions such as P1 and P3, with higher ruggedness and multiple peaks and valleys, require more initial points, while the P2 region, which is relatively smooth with fewer peaks and valleys, requires fewer initial points to search. Therefore, regions with higher ruggedness require more initial points to search for local extrema.

However, reducing the number of initial points is one of the key conditions for improving the efficiency of extremum search algorithms. Therefore, regions with higher ruggedness can be further divided into smaller sub-regions, which will reduce the number of local extrema in each sub-region. As shown in Figure 7, when the P1 region is divided into two sub-regions, P1.1 and P1.2, the reduction in ruggedness of these sub-regions leads to an increase in their FDC values. Based on this, by continuously dividing the region, the ruggedness of each sub-region can be reduced, and the FDC value can be increased. When the FDC value reaches an appropriate level, the search efficiency for extremum points in each sub-region will improve, and the overall efficiency of the global extremum search algorithm will also increase.

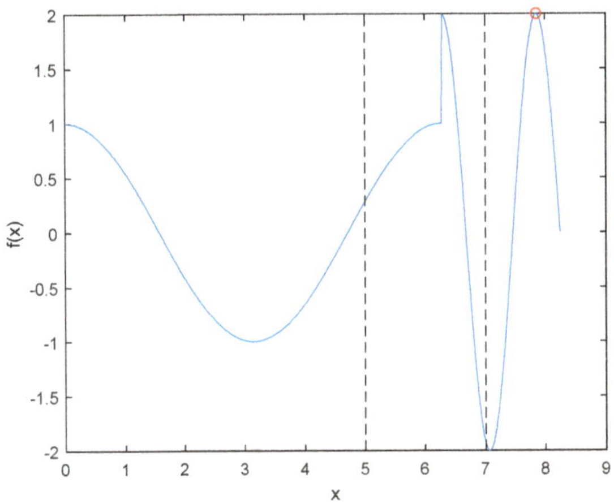

Figure 7. Piecewise function curve graph.

Next, a piecewise function will be used to illustrate the region division method based on FDC:

$$f(x) = \begin{cases} \cos(x), 0 \le x \le 2\pi \\ 2\cos(4(x - 2\pi), 2\pi \le x \le 2.625\pi \end{cases} \quad (2)$$

As shown in the figure above, the piecewise function exhibits a high degree of ruggedness, and from Figure 8, it can be seen that the pattern of adaptation values for this piecewise function is relatively irregular. Additionally, the computed values of $FDC(s) = -0.091$ for this region show that directly solving the problem is relatively difficult. Due to the presence of multiple peaks and valleys, it is challenging to directly search for the extrema points.

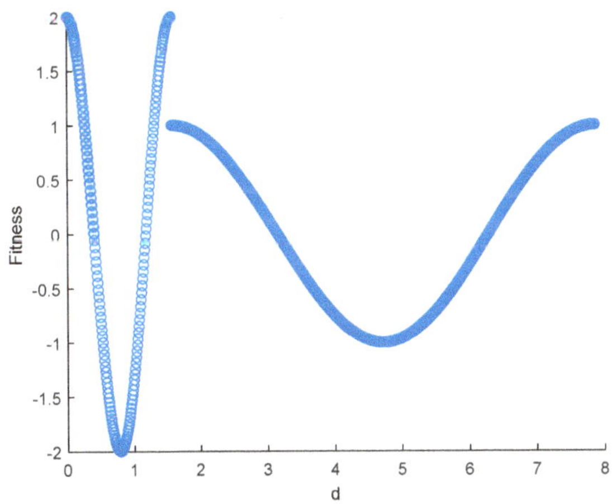

Figure 8. Relationship between fitness value and distance to the optimal point.

Therefore, the region is divided into two equal parts, and sample points are uniformly taken with an interval of 0.01 in the P1 and P2 regions. To refine the analysis of the surface's ruggedness, we split the entire interval into two sections, labeled S1 and S2. For these two intervals, uniform sampling is performed at an interval of 0.01, and the respective FDC values are calculated. The results show that $FDC(P1) = 0.83$ and $FDC(P2) = 0.49$, indicating that the ruggedness of region P2 is significantly greater than that of region P1. Assuming an acceptable FDC threshold of 0.5, the P2 region needs further subdivision.

Next, the P2 region is subdivided into two sub-regions, P2.1 and P2.2. After performing uniform sampling and analysis on these two sub-intervals, the results show that the FDC value in P2.1 is 0.31, while in P2.2, it reaches 0.95. This indicates that the terrain corresponding to S2.2 meets our expected ruggedness standard, while the S2.1 region shows a need for further refinement. Therefore, we subdivide the P2.1 region into P2.1.1 and P2.1.2. After sampling and calculation, $FDC(P2.1.1) = 0.750$ and $FDC(P2.1.2) = 0.870$, the ruggedness indices of both sub-intervals meet the standards, and no further subdivision is necessary.

As a result, the entire interval is effectively divided into three parts: P1, P2.1.1, P2.1.2, and P3. Among these, the regions with higher ruggedness, P2.1.1 and P2.1.2, are accurately identified and effectively subdivided, ensuring that the ruggedness of each sub-region meets our standards. Through this method, we can precisely identify and handle the regions with higher ruggedness, enabling effective analysis and subdivision of the entire surface.

3.2. Algorithm Flow

The pattern search algorithm is a method employed to identify the optimal solution of a function, particularly in scenarios with multiple local extrema. Similar to global extremum search algorithms, such as Multi-start, and gradient descent methods, pattern search algorithms belong to the category of direct search algorithms. A key advantage of pattern search algorithms is that they do not require gradient information of the objective function. The fundamental concept involves iteratively exploring and locating local optimal solutions to an optimization problem. The algorithm utilizes a grid-based search approach rather than relying on gradient information, rendering it suitable for complex functions that are nonlinear, non-differentiable, or noisy. By intelligently adjusting search directions and step sizes, the algorithm can effectively navigate within the feasible domain of the objective function to locate local optimal solutions.

The algorithm starts by randomly selecting a point or determining an initial point based on some heuristic strategy. It then constructs a grid around the current point. This grid is formed by adding scalar multiples of a fixed set of vectors (referred to as patterns) to the current point. These vectors determine the search directions and step sizes. A pattern is a collection of unit vectors, and the number of these vectors is determined by the number of decision variables in the problem. If the number of decision variables is m, the pattern can be represented as $\{x_1, x_2, \cdots, x_m\}$, where $x_i (i = 1, 2, \cdots, m)$ is a unit vector whose ith element is 1 and the other elements are 0, or $x_i (i = m+1, m+2, \cdots, 2m)$ can also represent a unit vector where the ith element is -1 and the rest are 0.

The pattern search algorithm operates by systematically exploring points on a grid to identify those that can enhance the objective function value. When such points are discovered on the grid, they are adopted as the new points for the subsequent iteration. If no superior points are identified within the current grid, the algorithm reduces the step size, effectively shrinking the grid scale to perform a more detailed local search. Conversely, if continuous improvements are observed, the compensation parameter may be increased to accelerate the search process. The pattern search algorithm terminates when predefined termination conditions are satisfied.

For the identification of extreme points on a part's surface morphology, the number of initial points in the pattern search algorithm directly influences the number of local extreme points detected. Typically, a single initial point in the pattern search algorithm leads to the discovery of one extreme point. Therefore, to identify all extreme points on the surface, a greater number of initial points is required. By integrating the surface roughness calculation method, the number of initial points can be determined based on the roughness of the surface area. Higher roughness necessitates a larger number of initial points. Consequently, combining the pattern search algorithm with fitness surface analysis to compute the local extreme points of the part surface morphology constitutes an effective strategy.

For regions exhibiting higher roughness, regional division is implemented, and the FDC values of the subdivided subregions are recalculated. If the FDC values of these subregions are relatively high, it indicates that fewer initial points are required for that area, thereby reducing the search difficulty within the subregion. Conversely, if the FDC values fall below a predetermined threshold, the following Formula (3) is employed for initial point allocation, adding more initial points and thereby increasing the probability of identifying global extreme points.

Initial point allocation:

$$n_i = \left\lceil \frac{n_0}{|fdc(P_i)|} \right\rceil \quad (3)$$

In the formula, n represents the number of initial points, P denotes the partitioned region, and $\lceil \bullet \rceil$ represents the ceiling function (rounding up to the nearest integer).

The steps for identifying the global extreme values of the surface morphology using the improved mode search algorithm are illustrated in Figure 9 below:

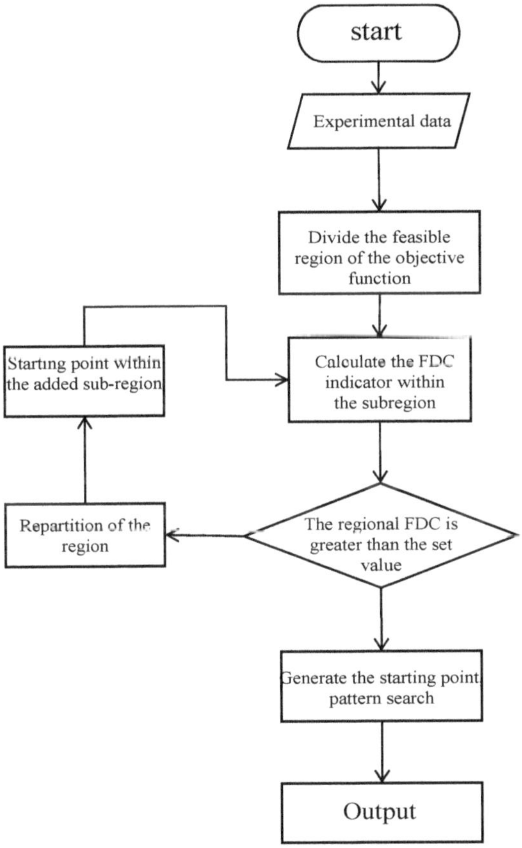

Figure 9. Algorithm flowchart.

Step 1: For the feasible domain of the objective function, the region is initially divided to analyze each subregion. Within each subdivided subregion, a set of sample points is uniformly generated with a fixed sampling interval *d*, and the extremum points within the sample set are designated as the optimal points P. Subsequently, the distance from each sample point to the optimal point P is calculated. The FDC index is then computed using Formula (1).

Step 2: For regions where the FDC index is relatively lower than the predefined threshold, further regional division is undertaken. The FDC value for each subregion is recalculated iteratively until the FDC value of a region exceeds the set threshold, at which point no additional division is performed.

Step 3: For each division in region P, N initial exploration points are randomly generated based on the value of FDC according to Formula (3). These points are then used as starting points to execute the pattern search algorithm in pursuit of the optimal solution. The goal is to locate the local optimal solution in each region. The selection of N is based on distributing the total number of initial points across each region according to a certain method, as shown in Formula (4).

$$N = \sum_{i=1}^{m} n_i \tag{4}$$

Here, N represents the total number of initial points, and m represents the number of sub-regions. The initial point n_0 can be calculated by substituting into Formula (4) from Formula (3).

4. Case Study

This section presents a case study to illustrate the application and effectiveness of the proposed method. Section 4.1 begins with theoretical test functions to assess the method's performance under controlled conditions. Section 4.2 then moves on to a practical engineering case, demonstrating how the method is applied in a real-world setting.

4.1. Theoretical Test Functions

Multiple functions with varying characteristics are employed to generate the surface morphologies of parts. Subsequently, the extremum search method proposed in this study is applied to identify extremum points within these generated morphologies. This process serves to validate the effectiveness of the proposed method by demonstrating its capability to accurately locate extremum points across diverse surface profiles. The proposed method was validated using Matlab R2023. The experiments were conducted on a computer with 32 GB of RAM and an NVIDIA RTX 2000 GPU. All algorithms and visualizations presented in this study were implemented using the Matlab programming language.

Function 1: Peaks function

$$f(x,y) = 3(1-x)^2 e^{-(x)^2-(y+1)^2} - 10\left(\frac{x}{5} - x^3 - y^5\right)e^{-x^2-y^2} - \frac{1}{3}e^{-(x+1)^2-y^2} \tag{5}$$

Function 2: Branin-Hoo function

$$f(x) = \left(x_2 - \frac{5.1x_1^2}{4\pi^2} + \frac{5x_1}{\pi} - 6\right)^2 + 10\left(1 - \frac{1}{8\pi}\right)\cos(x_1) + 10 \tag{6}$$

Function 3: Six Hump Camel Back function

$$f(x) = 4x_1^2 - 2.1x_1^4 + \frac{1}{3}x_1^6 + x_1x_2 - 4x_2^2 + 4x_2^4 \tag{7}$$

Function 4: Ackely function

$$f(x) = -20e^{\left(-0.2\sqrt{\frac{1}{n}\sum_{i=1}^{n} x_i^2}\right)} - e^{\left(\frac{1}{n}\sum_{i=1}^{n}\cos(2\pi x_i)\right)} + 20 + e \tag{8}$$

Function 5: Rastrigin function

$$f(x) = 10n + \sum_{i=1}^{n} \left[x_i^2 - 10\cos(2\pi x_i) \right] \quad (9)$$

Utilizing the improved pattern search algorithm presented in this paper, the surface extremum points for five test functions were calculated. Although Test Functions 1–5 selected in this study are explicit functions, they can still be treated as black-box functions for experimental purposes.

The surface morphology of the Peaks function is presented as Figure 10. The search results for the Peaks function are presented in Table 1 and Figure 11. By applying the improved pattern search algorithm, information regarding regional local extreme points, global extreme points, and regional ruggedness was obtained, as presented in the table above. This demonstrates the effectiveness of the ruggedness index and validates the method used for verifying local extreme points. Figure 10 is a surface morphology diagram. The color represents the height or depth value, with yellow indicating the peak value and blue indicating the trough value. Figure 11 is a distribution diagram of extreme points. The color differentiates different sub-regions or types of extreme values, where yellow represents the peak region and blue represents the trough region. In Figure 11, the local minima of each sub - region are marked with red asterisks.

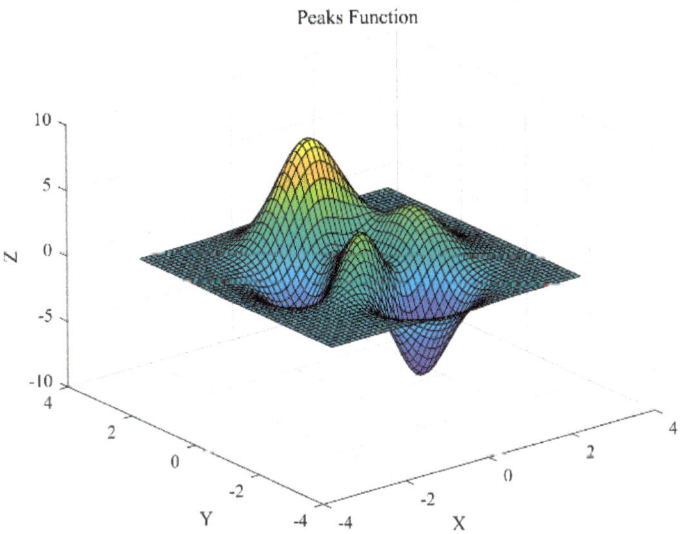

Figure 10. The surface morphology of Peaks function.

Table 1. The extreme points of Peaks function.

Local Minimum of the Region	x	y	f	FDC
P1	−1.58744	0.20134	−2.86052	−0.51618
P2	3.12570	2.20451	−3.04984	0.71009
P4	0.22828	−1.62553	−6.55113	0.62020
P4	3.00000	−0.20000	0.00041	−0.71606

The surface morphology of the Peaks function is presented as Figure 12. The search results for the Peaks function are presented in Table 2 and Figure 13. For the Branin-Hoo function, which has a relatively low local ruggedness, it is divided into four regions for search. The information of all extreme points and the ruggedness index within the feasible domain of this function are calculated. As shown in Figures 12 and 13, the Branin-Hoo function has a relatively low ruggedness, and the region division is relatively simple. By applying this method, all local extreme points can be directly obtained. Figure 12 is a surface morphology diagram. The color represents the height or depth value, with yellow indicating the peak value and blue indicating the trough value. Figure 13 is a distribution diagram of extreme points. The color differentiates different sub - regions or types of extreme values, where yellow represents the peak region and blue represents the trough region. In Figure 13, the red asterisks represent the extreme points found by the pattern search algorithm.

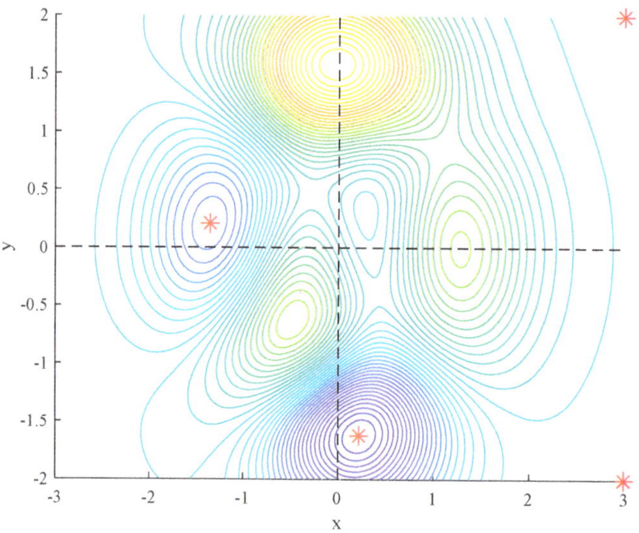

Figure 11. The extreme points of Peaks function.

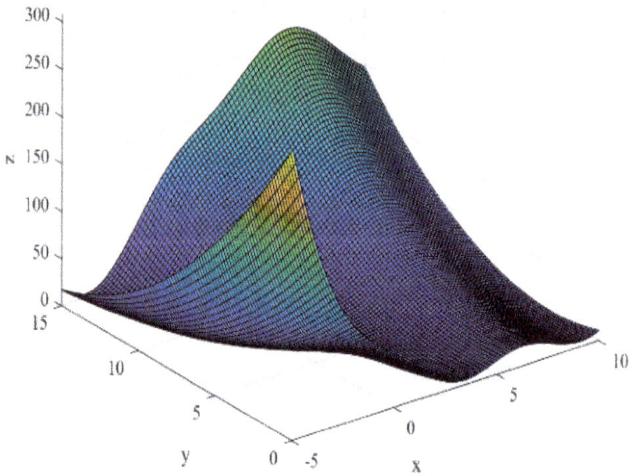

Figure 12. The surface morphology of Branin-Hoo function.

Table 2. The extreme points of Branin-Hoo function.

Local Minimum of the Region	x	y	f	*FDC*
P1	−4.34093	12.00251	130.34272	−0.93752
P3	9.68231	2.155532	200.15123	−0.92851
P3	2.03201	2.000000	190.00781	−0.88186

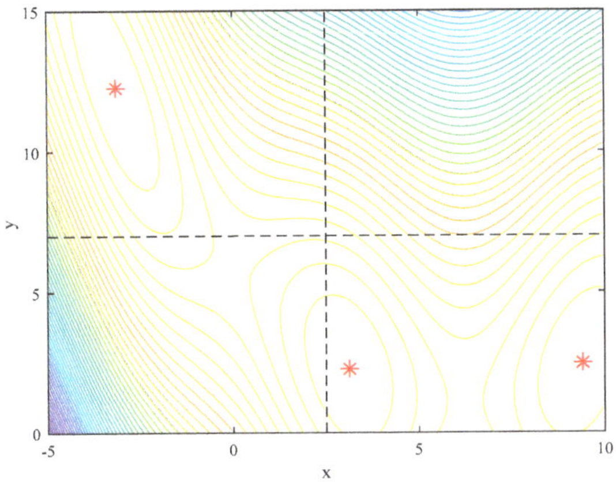

Figure 13. The extreme points of Branin-Hoo function.

The surface morphology of the Six Hump Camel Back function is presented as Figure 14. The search results for this function are presented in Table 2 and Figure 15. For the Six Hump Camel Back function, the feasible domain of the function is $[-3,3]^2$, and the function region is divided into nine equal parts. Assuming that the sampling intervals of x_1 and x_2 are both d, the feasible domain FDC index is calculated using Formula (1), as shown in Table 3. Figure 14 is a surface morphology diagram. The color represents the height or depth value, with yellow indicating the peak value and blue indicating the trough value. Figure 15 is a distribution diagram of extreme points. The color differentiates different sub - regions or types of extreme values, where yellow represents the peak region and blue represents the trough region. In Figure 15, red asterisks are also used to mark the extreme points located by the algorithm, verifying the effectiveness of the method.

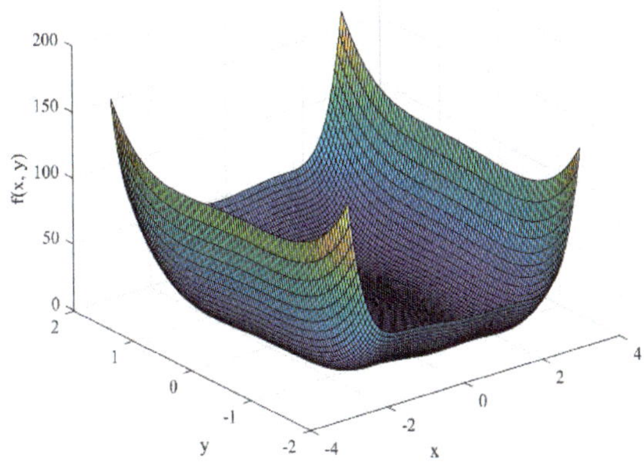

Figure 14. The surface morphology of Six Hump Camel Back function.

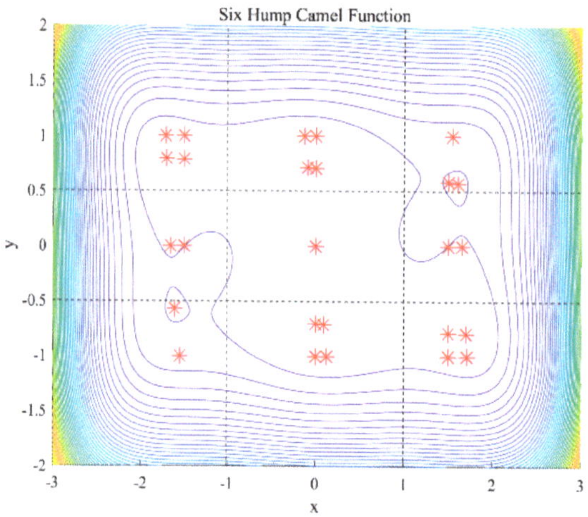

Figure 15. The extreme points of Six Hump Camel Back function.

Table 3. The FDC parameters of Six Hump Camel Back function.

Subinterval FDC	d = 0.1	d = 0.01	Number of Initial Points in Each Region
P1	0.9114	0.9473	10
P2	0.8961	0.9128	10
P3	0.8865	0.9117	10
P4	0.8511	0.9036	10
P5	0.8790	0.8993	10
P6	0.9167	0.9389	10
P7	0.8756	0.9109	10
P8	0.8663	0.8891	10
P9	0.8901	0.9207	10
Sample Size	100	1000	-

After dividing the feasible domain of the test function into nine equal parts and evaluating the FDC calculations with different sampling intervals h within each sub-feasible region, it was observed that the FDC values varied depending on the size of the sampling interval. Notably, smaller sampling intervals led to higher FDC accuracy. However, the analysis regarding the relationship between sample size and interval suggests that even larger sampling intervals are sufficient for a preliminary assessment of the subinterval's morphological complexity. By reducing the sampling interval to 0.1 and selecting a certain number of samples within each sub-region, the FDC values showed relative stability. Further reducing the sampling interval to 0.01, the FDC error range was limited to within 0.05. This indicates that the FDC values are not highly sensitive to changes in the sampling interval, and under normal circumstances, selecting a moderate number of sample points is sufficient for a reasonable FDC estimate. This stability is due to the fact that as long as the sample size is adequate to roughly map the surface complexity, the FDC value remains stable, proving that under the premise of ensuring experimental quality, the chosen sampling interval for $d = 0.1$ is already sufficient.

The surface morphology of the Ackely function is presented as Figure 16. The search results for this function are presented in Table 4 and Figure 17. For this function, due to its high ruggedness, as determined by direct computation, the test function is divided into nine equal segments within the feasible domain. Sampling is performed with a sampling interval of d = 0.1. The FDC of each sub-region is computed, and the local extreme points within each divided region are searched. The FDC calculation results and partial results of the extreme point search for this function are shown in Table 4. Figure 16 is a surface morphology diagram. The color represents the height or depth value, with yellow indicating the peak value and blue indicating the trough value. Figure 17 is a distribution diagram of extreme points. The color differentiates different sub-regions or types of extreme values, where yellow represents the peak region and blue represents the trough region. In Figure 17, red asterisks are also used to mark the extreme points located by the algorithm, verifying the effectiveness of the method.

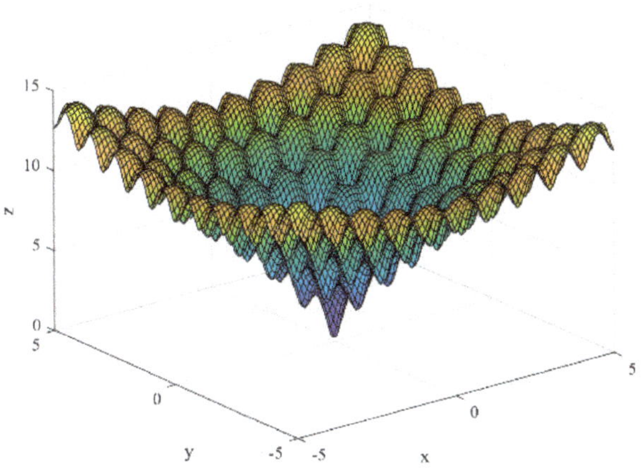

Figure 16. The surface morphology of Ackely function.

Table 4. The extreme points of Ackely function.

Regional Minimum	x	y	f	FDC
P1	−1.9604	−4.5200	10.8326	0.64247
P2	−3.1415	2.2750	0.39788	0.41217
P3	3.14159	1.1173	0.37880	0.67094
P4	0.00001	−2.1000	12.8520	0.94989
P5	−1.4247	−2.4750	0.39788	0.85449
P6	−3.2500	2.7500	13.1665	0.94751
P7	−1.8318	4.4752	0.39788	0.92398
P8	4.2486	2.47500	0.39788	0.91097
P9	4.0663	7.40092	14.19820	0.96738

The algorithm performs a separate search within the nine feasible regions after dividing test function 4. Table 4 only reports the coordinates of some local minimum points from the nine regions. From the FDC index values, it can be seen that with a sampling interval of d = 0.1, the FDC values are smaller in the P2 and P1 regions, indicating the presence of multiple local extremes in these regions. By further subdividing these regions and performing pattern search, more local minima can be found. On the other hand, the FDC absolute values in the P3–P9 regions are relatively large, indicating fewer local extrema in these areas. In total, the algorithm identified 60 local minima, validating the effectiveness of the method.

The surface morphology of the Ackely function is presented as Figure 18. The search results for this function are presented in Table 5 and Figure 19. Due to its high surface roughness and the presence of multiple local extremum points, it first defines the initial number of search points. Then, combined with the method proposed in this paper, the total feasible domain of the function is divided into nine sub-feasible regions. Parallel search is conducted within each sub-feasible region, and the FDC value for each region is calculated. Some of the results are shown in Table 5. Figure 18 is a surface morphology diagram. The color represents the height or depth value, with yellow indicating the peak value and blue indicating the trough value. Figure 19 is a distribution diagram of extreme points. The color differentiates different sub-regions or types of extreme values, where yellow represents the peak region and blue represents the trough region. In Figure 19, red asterisks are also used to mark the extreme points located by the algorithm, verifying the effectiveness of the method.

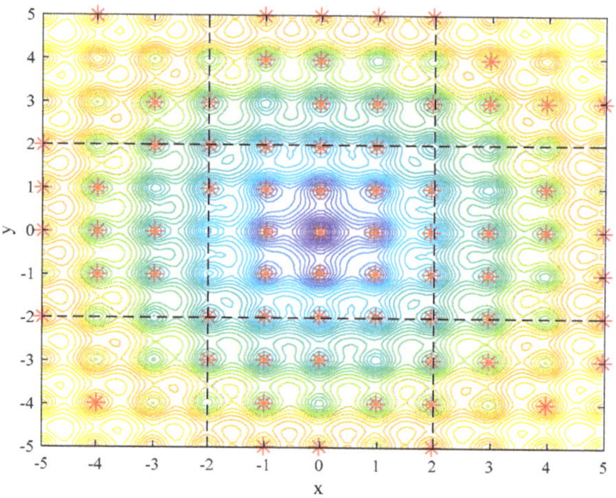

Figure 17. The extreme points of Ackely Function.

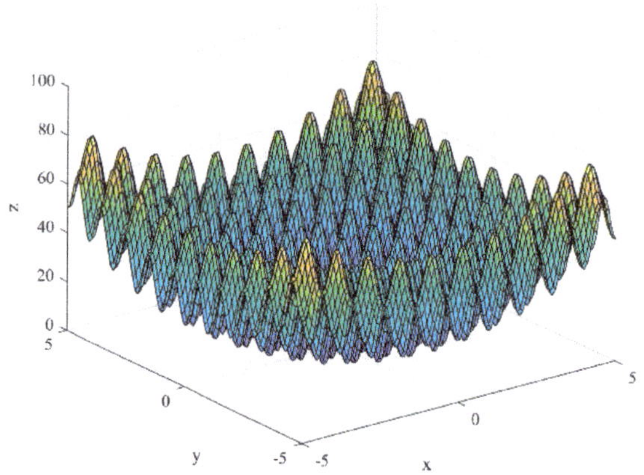

Figure 18. The surface morphology of Rastrigin function.

Due to the high ruggedness of this test function and the need to divide the domain into multiple regions, the method searches for a total of 90 local extremum points for the Rastrigin function. Only a subset of extremum points and some regional FDC index information are shown in the table. For regions P1 to P4, the FDC values are relatively small, so multiple subdivisions of these regions were made, and a parallel search algorithm was implemented to obtain local extremum points within these regions.

For the above test functions, the improved mode search algorithm proposed in this paper was applied. The feasible domains were divided for different functions, and FDC indices were calculated. A parallel search was conducted to find all local extremum points on the surfaces, which validated the effectiveness and robustness of the method.

Table 5. The extreme points of Ackely function.

Regional Minimum	x	y	f	FDC
P1	−1.58991	−1.77991	15.95966	0.54558
P2	1.52991	1.989912	25.83671	0.53961
P3	4.99561	−0.99496	14.98991	0.70103
P4	−3.71490	2.98485	9.94956	0.35330
P5	−0.51200	−4.5420	10.49309	0.67660
P6	3.53200	0.59360	6.37620	0.46190
P7	−4.99050	−4.9710	4.98991	0.51060
P8	1.58210	−4.5840	3.49600	0.69140
P9	3.53170	−3.67930	6.89030	0.44980

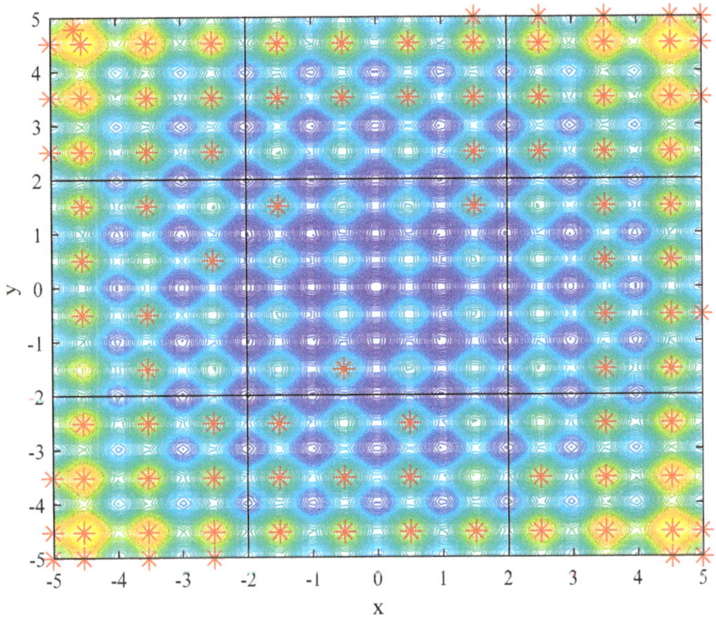

Figure 19. The extreme points of Rastrigin function.

4.2. Practical Engineering Case

We designed a physical experiment to validate the method proposed in this paper. As shown in Figure 20, the test sample is a simulated flat part with sealing requirements. It measures 200 mm × 100 mm, with a flatness tolerance of 0.01 mm. This experiment aims to demonstrate the application of the proposed method in surface quality evaluation.

As shown in Figure 21, the experimental procedure consists of four main steps. First, the NCA8107 Coordinate Measuring Machine (CMM), shown in Figure 22, is used to evenly measure 30 × 20 points on the surface of the test sample. Second, the Kriging model is applied to fit the measurement data and obtain the surface morphology model of the test sample, as shown in Figure 23. Third, the proposed method is used to search for extremum points on the surface morphology model. Finally, both the original measurement data and the extremum points are used for surface quality evaluation of the part.

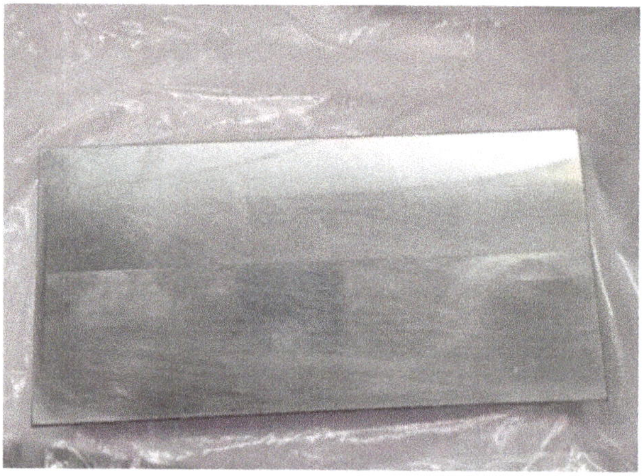

Figure 20. Physical model of the test sample.

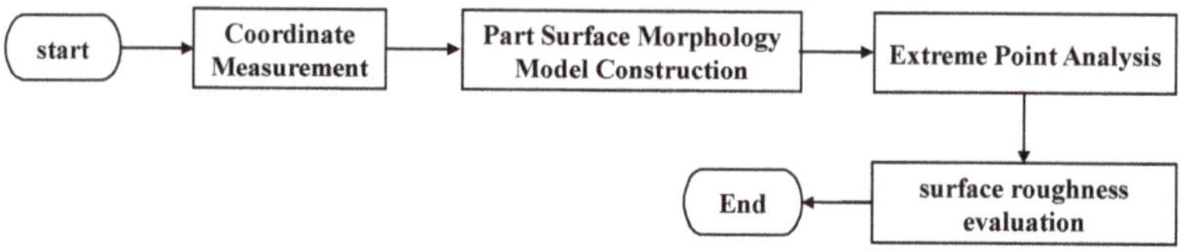

Figure 21. Experimental procedure.

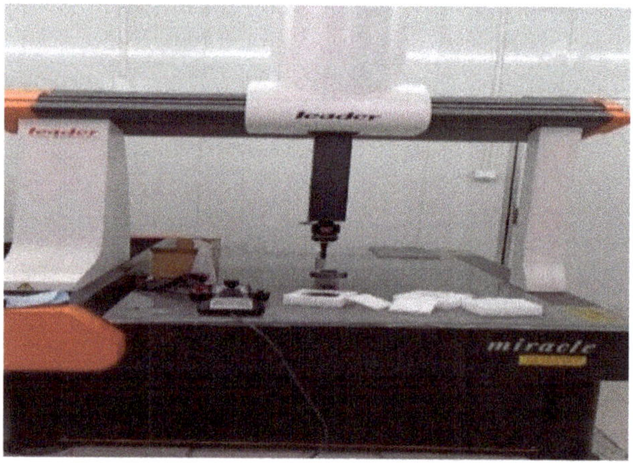

Figure 22. NCA8107 coordinate measuring machine (CMM).

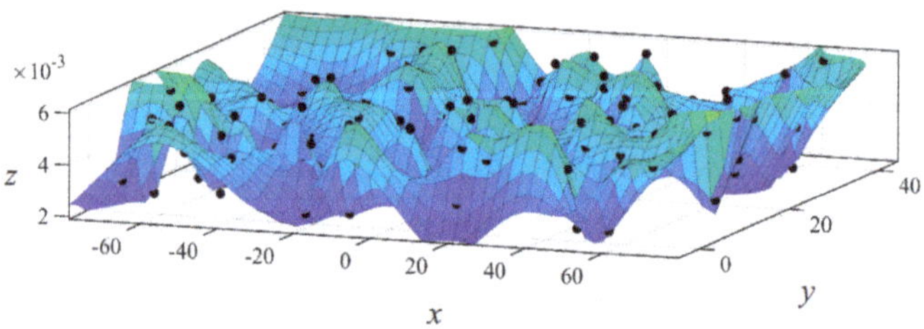

Figure 23. Surface morphology model of the test sample.

To analyze the impact of extremum points on surface quality evaluation, two methods were used for calculation. The first method directly utilizes the measurement data for surface quality evaluation. The second method combines the measurement data with the extremum points before performing the surface quality evaluation. The calculation results of both methods are shown in Table 6. From the results, it can be seen that the flatness evaluation value considering extremum points is 2.84% higher than the one without considering extremum points. This is because when only measurement data are considered, the influence of extremum points is ignored. Extremum points, typically representing peaks or valleys in the surface morphology model, have a direct impact on flatness.

Table 6. Flatness error evaluation results.

Methods	Flatness Error	Flatness Error as a Percentage of Design Tolerance
Without considering extremum points	7.90×10^{-3}	78.94%
Considering extremum points	8.18×10^{-3}	81.81%

5. Conclusions

This study proposes a fitness landscape-based method for extreme point analysis to extract surface morphology characteristic parameters of a part. The method conceptualizes the search for global extremum points as identifying optimal solutions. An adaptive value surface analysis approach, combined with an improved mode search algorithm, is introduced. First, the Fitness-Distance Correlation (FDC) and a roughness index are integrated to analyze surface roughness and determine the number and distribution of initial points based on roughness. The surface morphology's domain is partitioned using the FDC, and initial points are allocated according to the surface roughness, reducing optimization complexity. Finally, the roughness index further subdivides the region, and the mode search algorithm is implemented in parallel to enhance search efficiency. This method enables the extraction of surface parameters and global extremum points without the need for gradient information. Compared to existing methods, the proposed approach improves efficiency by reducing the required number of initial points and adapting the search process based on surface roughness, which addresses challenges in complex surface morphologies. Additionally, it eliminates the need for gradient information, making it suitable for highly intricate surfaces where gradient-based methods may not be effective.

The method is applicable to Industry 4.0 environments, such as coordinate measurement systems (CMSs) for assessing geometric shape or surface roughness. By incorporating roughness and adaptive value surface analysis, it provides precise surface characterization crucial for high-precision manufacturing. At the same time, the method requires the same input as existing surface roughness evaluation methods, making it highly compatible and easy to integrate into current workflows. In conclusion, the proposed method enhances search efficiency and robustness, offering a reliable solution for modern manufacturing applications. Its ability to analyze surface quality aligns with Industry 4.0 objectives, making it an essential tool for smart manufacturing and quality control.

Author Contributions: Conceptualization, W.T.; Methodology, W.T. and J.S.; Validation, W.T.; Writing—original draft, J.S. and W.T.; Writing—review and editing, J.S. All authors have read and agreed to the published version of the manuscript.

Funding: This research was funded by the Aviation Science Foundation under Grant 2022Z050111001, the National Natural Science Foundation of China under Grant 52105559, the Key Research and Development Program of Shaanxi under Grant 2024GX-YBXM-278, the Xi'an Science and Technology program under Grant 23GXFW0022, the Natural Science Foundation of Shaanxi Province under

Grant 2021JQ-680 and the Scientific Research Foundation for Doctor of Xi'an Polytechnic University, grant number BS201909.

Data Availability Statement: Data are contained within the article.

Conflicts of Interest: The authors declare no conflicts of interest.

References

1. Yan, X.; Ballu, A. Review and Comparison of Form Error Simulation Methods for Computer-Aided Tolerancing. *J. Comput. Inf. Sci. Eng.* **2019**, *19*, 010802. [CrossRef]
2. Schleich, B.; Wartzack, S. Evaluation of geometric tolerances and generation of variational part representatives for tolerance analysis. *Int. J. Adv. Manuf. Technol.* **2015**, *79*, 959–983. [CrossRef]
3. Jin, S.; Ding, S.; Li, Z.; Yang, F.; Ma, X. Point-based solution using Jacobian-Torsor theory into partial parallel chains for revolving components assembly. *J. Manuf. Syst.* **2018**, *46*, 46–58. [CrossRef]
4. Liu, Y.; Shen, H.; Zhao, G.; Jing, X.; Du, X. Improved Jacobian-Torsor model for optimizing mechanical product quality. *Proc. Inst. Mech. Eng. Part B J. Eng. Manuf.* **2024**. [CrossRef]
5. Schleich, B.; Wartzack, S. Novel approaches for the assembly simulation of rigid Skin Model Shapes in tolerance analysis. *Comput. Aided Des.* **2018**, *101*, 1–11. [CrossRef]
6. Ma, S.; Hu, T.; Xiong, Z. Precision assembly simulation of skin model shapes accounting for contact deformation and geometric deviations for statistical tolerance analysis method. *Int. J. Precis. Eng. Manuf.* **2021**, *22*, 975–989. [CrossRef]
7. Chang, Z.; Ren, Y.; Li, X.; Yan, K.; Wang, Z.; Zhao, Y. Assembly deformation prediction of aircraft composite panel considering manufacturing variation and actual assembly process. *Int. J. Adv. Manuf. Technol.* **2024**, *135*, 811–3825. [CrossRef]
8. Zhang, J.; Xia, H.; Liu, J.; Shen, H.; Luo, J.; Zhang, X. A multi-objective search-based approach for position and orientation deviations in assemblies with multiple non-ideal surfaces. *Adv. Eng. Softw.* **2025**, *199*, 103820. [CrossRef]
9. Qiu, C.; Liu, Z.Y.; Bu, W.H.; Tan, J.R. Hybrid dimension based modeling of part surface topography and identification of its characteristic parameters. *Appl. Surf. Sci.* **2012**, *258*, 7082–7093. [CrossRef]
10. Chan, Q.; Zhenyu, L.; Xiang, P.; Jianrong, T. A non-ideal geometry based prediction approach of fitting performance and leakage characteristic of precision couplings. *IEEE Access* **2018**, *6*, 58204–58212. [CrossRef]
11. Li, H.; Qiu, L.; Zhang, S.; Tan, J.; Wang, Z.; Liu, X. A surface modeling method for product virtual assembly based on the root mean square of the regional residuals. *Proc. Inst. Mech. Eng. Part B J. Eng. Manuf.* **2020**, *234*, 229–242. [CrossRef]
12. Pawlus, P.; Reizer, R.; Wieczorowski, M. Functional importance of surface texture parameters. *Materials* **2021**, *14*, 5326. [CrossRef]
13. Sun, Q.C.; Zhao, B.B.; Liu, X.; Mu, X.K.; Zhang, Y.L. Assembling deviation estimation based on the real mating status of assembly. *Comput. Aided Des.* **2019**, *115*, 244–255. [CrossRef]
14. Sun, X.M.; Bao, J.S.; Li, J.; Zhang, Y.M.; Liu, S.M.; Zhou, B. A digital twin-driven approach for the assembly-commissioning of high precision products. *Robotics Comput.-Integr. Manuf.* **2020**, *61*, 101839. [CrossRef]
15. Söderberg, R.; Wärmefjord, K.; Carlson, J.S.; Lindkvist, L. Toward a Digital Twin for real-time geometry assurance in individualized production. *CIRP Ann.* **2017**, *66*, 137–140. [CrossRef]
16. Tang, W.B.; Xu, G.S.; Zhang, S.J.; Jin, S.F.; Wang, R.X. Digital Twin-Driven Mating Performance Analysis for Precision Spool Valve. *Machines* **2021**, *9*, 157. [CrossRef]
17. Guo, F.; Liu, J.; Wang, Z.; Zou, F.; Zhao, X. Positioning error guarantee method with two-stage compensation strategy for aircraft flexible assembly tooling. *J. Manuf. Syst.* **2020**, *55*, 285–301. [CrossRef]
18. Liu, Y.; Ren, Y.; Lin, Q.; Yu, W.; Pan, W.; Su, A.; Zhao, Y. A digital twin-based assembly model for multi-source variation fusion on vision transformer. *J. Manuf. Syst.* **2024**, *76*, 478–501. [CrossRef]
19. Jia, K.; Wang, H.; Ren, D.; Liu, B.; Zhao, Q.; Hong, J. A general mathematic model framework for assembly process driven digital twin of assembly precision. *J. Manuf. Syst.* **2024**, *77*, 196–211. [CrossRef]
20. Liu, F.; Xu, G.; Liang, L.; Zhang, Q.; Liu, D. Least squares evaluations for form and profile errors of ellipse using coordinate data. *Chin. J. Mech. Eng.* **2016**, *29*, 1020–1028. [CrossRef]
21. Singh, D.; Arunachalam, N. A Novel Iterative Field Search Approach to Minimum Zone Circle for Roundness Error Estimation. *J. Comput. Inf. Sci. Eng.* **2023**, *23*, 041011. [CrossRef]
22. Wang, S.; Zheng, P.; Kong, Y.; Zhu, Y.; Wang, G.; Zhang, X. A novel method for roundness error based on searching the center of the circle. *Meas. Sci. Technol.* **2022**, *33*, 095008. [CrossRef]
23. Khlil, A.; Shi, Z.; Umar, A.; Ma, B. Improved algorithm for minimum zone of roundness error evaluation using alternating exchange approach. *Meas. Sci. Technol.* **2022**, *33*, 045003. [CrossRef]
24. Salhi, S.; Queen, N.M. A hybrid algorithm for identifying global and local minima when optimizing functions with many minima. *Eur. J. Oper. Res.* **2004**, *155*, 51–67. [CrossRef]

25. Wang, H.; Moon, I.; Yang, S.; Wang, D. A memetic particle swarm optimization algorithm for multimodal optimization problems. *Inf. Sci.* **2012**, *197*, 38–52. [CrossRef]
26. Gnandt, C.; Callies, R. CGRS—An advanced hybrid method for global optimization of continuous functions closely coupling extended random search and conjugate gradient method. *J. Comput. Appl. Math.* **2018**, *333*, 99–115. [CrossRef]
27. Minaee, S.; Boykov, Y.; Porikli, F.; Plaza, A.; Kehtarnavaz, N.; Terzopoulos, D. Image segmentation using deep learning: A survey. *IEEE Trans. Pattern Anal. Mach. Intell.* **2021**, *44*, 3523–3542. [CrossRef]
28. Koessler, E.; Almomani, A. Hybrid particle swarm optimization and pattern search algorithm. *Optim. Eng.* **2021**, *22*, 1539–1555. [CrossRef]

machines

MDPI

Article

Milling-Force Prediction Model for 304 Stainless Steel Considering Tool Wear

Changxu Wang [1], Yan Li [1,*], Feng Gao [1], Kejun Wu [1], Kan Yin [1], Peng He [1] and Yunjiao Xu [2]

[1] School of Mechanical and Precision Instrument Engineering, Xi'an University of Technology, Xi'an 710048, China; 15315216859@163.com (C.W.); gf2713@xaut.edu.cn (F.G.); bestafter@163.com (K.W.); 15934486412@163.com (K.Y.); hp19991624011@163.com (P.H.)

[2] School of Machinery and Automation, Weifang University, Weifang 262799, China; xyj18053647607@163.com

* Correspondence: liyangf@xaut.edu.cn

Abstract: The high-performance alloy, 304 stainless steel, is widely used in various industries. However, its material properties lead to severe tool wear during milling processes, significantly increasing milling force and adversely impacting machining quality and efficiency. Consequently, an accurate milling-force model is crucial for guiding the formulation and optimization of machining parameters. This paper presents a milling-force prediction model for 304 stainless steel that incorporates the effect of tool wear, based on the mechanistic modeling approach. Side-milling experiments on 304 stainless steel were conducted to analyze the relationship between milling force and tool wear, identify the model coefficients, and validate the prediction accuracy of the milling-force model. The results demonstrate that the model accurately predicts the milling forces of worn tools while side milling 304 stainless steel under various machining parameters and tool wear conditions.

Keywords: 304 stainless steel; tool wear; model coefficient identification; milling-force prediction

1. Introduction

Due to excellent mechanistic strength, heat resistance, ductility, and corrosion resistance [1,2], components made of 304 stainless steel are durable, reliable, and aesthetically appealing [3]. Consequently, it is widely utilized in the medical, chemical, energy, defense, and aerospace industries [4]. Because of its high mechanistic strength and tendency toward hardening, 304 stainless steel requires a significant cutting force during machining. Additionally, its low thermal conductivity causes heat to accumulate in the cutting zone, increasing the temperature at the cutting edge. The high affinity and ductility of 304 stainless steel further lead to chips adhering to the tool surface, forming a built-up edge (BUE), which contributes to rapid tool wear [5,6]. As the tool wears, the ploughing force on the flank face increases, significantly increasing cutting force [7]. When machining a workpiece with relatively low stiffness (especially thin-walled components), excessive cutting forces can deform the workpiece, resulting in machining errors [8]. An increase in cutting force can affect the elastic and plastic deformation of the workpiece surface as well as the cutting temperature, thereby influencing the surface machining quality [9]. An increase in cutting force results in higher cutting power, thereby contributing to an increase in the energy consumption of the machine tool [10]. In summary, the significant increase in cutting force caused by tool wear during actual machining can adversely affect both machining quality and efficiency. Bhushan et al. [11] calculated milling forces under various machining parameters using the developed cutting-force model and optimized parameters

https://doi.org/10.3390/machines13010072

with milling force as the objective. Liu et al. [12] processed the milling-force signals using the cutting-force model that considers tool wear, enabling the monitoring of tool wear. Du et al. [13] calculated the milling forces under different machining parameters using the milling-force model, and by considering the effect of milling forces on tool deflection, they analyzed the tool's motion trajectory to predict the surface roughness of the machined part. Therefore, it is evident that the milling-force model can be applied to optimize machining parameters, monitor tool wear, and predict surface quality. To improve machining quality and efficiency and mitigate the adverse effects caused by the significant increase in cutting force due to tool wear, establishing an accurate and easily applicable milling-force model that considers tool wear is essential.

At present, there are five main models of cutting force: the empirical model, analytical model, mechanistic model, finite element model, and data-driven model. The empirical model relies on data from cutting-force tests or actual production and uses statistical methods to establish the relationship between cutting force and machining parameters [14]. The analytical model employs orthogonal or oblique cutting theory to analyze the cutting process from both geometric and mechanistic perspectives and determine the cutting force, which was founded by Merchant's research [15,16]. The mechanistic model proposed by Altintas assumes the cutting force as the product of the area of cutting zone and the cutting-force coefficient, based on the analysis of cutting conditions, in which a cutting experiment is conducted to identify the cutting-force coefficients [17]. Commercial FEA (finite element analysis) software is used to visually simulate cutting forces under various cutting conditions [18]. The data-driven model uses neural networks or deep learning to analyze large datasets and establish the mapping relationship between cutting force and working conditions through an evolution learning mechanism [19].

Compared to the other four models, the mechanistic model of cutting force emphasizes the engagement relationship between the tool and workpiece during the cutting process, using numerical methods to efficiently handle complex cutting conditions. As a result, the mechanistic model is widely used to establish milling-force models that consider tool wear. Zhao et al. [20] replaced the actual wear with the wear width of the flank face and introduced a variable attenuation coefficient in the stress distribution function to develop a milling-force prediction model for worn ball-end tools considering the effect of rebound. Wojciechowski et al. [21] considered the effects of friction, rubbing, and ploughing mechanisms between the workpiece and tool flank face during machining and investigated the cutting forces in the ball-end milling of inclined surfaces. Wang et al. [22] developed a milling-force model for an asymmetric micro milling cutter, including the effect of tool wear on instantaneous undeformed chip thickness, shear/ploughing force and friction in the wear zone of the flank face. Gao et al. [23] derived an instantaneous undeformed chip-thickness model and proposed a milling-force prediction model for worn micro-end mills by accounting for the effects of flute radius and flank wear on tool runout. Liu et al. [24] developed a micro-milling-force model, combining the influences of tool runout, cutting-edge radius and tool wear by integrally calculating the instantaneous undeformed chip thickness, shear-force and ploughing-force coefficients. Wojciechowski et al. [25] considered the geometric errors, tool deflections, minimum uncut chip thickness, and chip-thickness accumulation in the machining system, and using a combined numerical-analytical approach, they evaluated the minimum uncut chip thickness, ploughing, and shearing forces, further developing a milling-force model for the micro-milling of AISI 1045 steel. Hou et al. [26] decomposed the cutting force of a worn tool into shear force and friction forces, explaining the distribution of friction stress on the flank face to build a milling-force model for a worn tool, with the model coefficients identified experimentally. Huang et al. [27], considering the complexity arising from tool geometry, applied a genetic algorithm to determine the coefficients of the cutting-force model, thereby establishing a cutting-force model for worn CBN tools to turn hardened 52100 bearing steel. Chinchanikar et al. [28] extended the Waldorf's orthogonal cutting-force modeling method into a three-dimensional model. By using an equivalent cutting-edge geometry to determine the shear angle, chip flow angle, and other model parameters, a regression analysis method was employed to establish a cutting-force model for turning quenched AISI 4340 steel with a worn tool. Patel et al. [29] conducted the full factorial turning experiments of hardened AISI D2 steel and extended an empirical model of cutting force based on Waldorf's theory considering tool wear. Toubhans et al. [30] analyzed the engagement area between the tool and the workpiece, developing a time-varying cutting-force model for tool wear in the turning of Inconel 718, considering the effect of cutting speed and feed rate on tool wear, and adding an additional time-dependent term to the cutting-force coefficient.

Based on the above research, cutting-force modeling considering tool wear has primarily focused on micro-milling and turning processes. However, in practical manufacturing, side milling serves as the primary machining method for most metals and some difficult-to-machine materials. Since 304 stainless steel is widely used and prone to tool wear during machining, this paper establishes a milling-force model considering tool wear, based on side-milling experiments conducted with 304 stainless steel. It also visually demonstrates the tool wear process and the variation of milling forces with tool wear. To improve the model's precision, this paper incorporates the effect of tool wear on the actual tool radius. A method to calculate method for the worn tool's actual radius has been derived and applied in the calculation of tool entry/exit angle and the instantaneous undeformed chip thickness. While most studies adopt Altintas' [17] quick identification method for milling-force coefficients, which necessitates slot milling conditions, it is important to note that slot milling can significantly affect tool life, especially when processing difficult-to-cut metals. Therefore, this study proposes a milling-force coefficient identification method, based on average milling forces, that is suitable for side-milling conditions, in conjunction with the established milling-force model. The remaining sections of this paper are organized as follows: In Section 2, the worn end mill is discretized into cutting-edge elements for cutting-force analysis. Based on the engagement relationship examined between tool and workpiece, and an instantaneous milling-force mechanistic model for the worn tool is derived. In Section 3, a method for identifying milling-force model coefficients using average milling forces is proposed. Side-milling experiments on 304 stainless steel were conducted to identify the model coefficients and validate the accuracy of the model. In Section 4, the study's findings, the model's limitations, and its potential industrial applications are discussed. Finally, Section 5 summarizes the conclusions.

2. Mechanistic Modeling of Milling Forces for Worn Tools

2.1. Geometrical Discretization Method for the Cutting Edge

Figure 1 illustrates the geometric relationship during the side-milling process using a four-flute end mill. A right-handed Cartesian coordinate system (O-XYZ) is established, positioning the origin O at the bottom center of the end mill. The Z axis aligns with the tool's axis, the Y axis extends along the tool's feed direction, and the X axis is perpendicular to the YZ plane. This coordinate system aligns with the machine tool's coordinate system.

The end mill is discretized into multiple cutting-edge elements along the Z axis. As an example, consider the cutting-edge element P shown in Figure 1. The position of any arbitrary cutting-edge element during the milling process is determined by the tool–workpiece engagement angle θ, which is calculated as follows:

$$\theta = \left(\varphi + \frac{2\pi}{N}(i-1) + \frac{z\tan\beta}{R_{ac}}\right)\text{mod}2\pi \quad (1)$$

where N is the total number of cutting edges, i is the index of the cutting edges, z is the height of the cutting-edge element from the bottom of the tool edge, β is the helix angle of the tool, R_{ac} is the actual radius of the tool, mod represents the modulus operation, and φ is the rotation angle of the end mill. The rotation angle φ can be calculated as follows:

$$\varphi = \omega t \tag{2}$$

where ω is the spindle speed and t is the milling time.

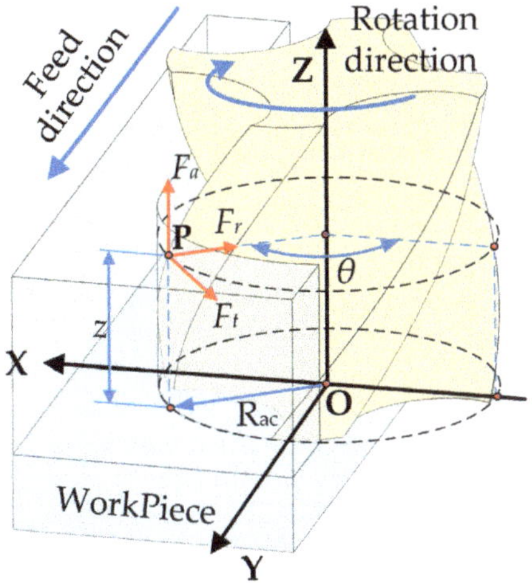

Figure 1. Geometric relationship during side-milling process of an end mill.

2.2. Force Analysis of Cutting-Edge Elements for Worn Tools

During the milling process, the cutting-edge elements experience tangential force dF_t, radial force dF_r, and axial force dF_a (where the tangential t, radial r, and axial a directions are defined as follows: the tangential direction t is along the direction of the cutting-edge velocity, the radial direction r points toward the tool axis along the tool radius, and the axial direction a is perpendicular to both the tangential and radial directions, oriented vertically along the tool axis).

When the cutting edge is sharp or exhibits no significant wear, the ploughing effect on the flank face is very weak, and there is almost no extrusion or friction between the flank face and the workpiece. This assumes that the tool is only subjected to the shearing force generated by the rake face [20,26], including the tangential shear force dF_{tc}, radial shear force dF_{rc}, and axial shear force dF_{ac}, as shown in Figure 2.

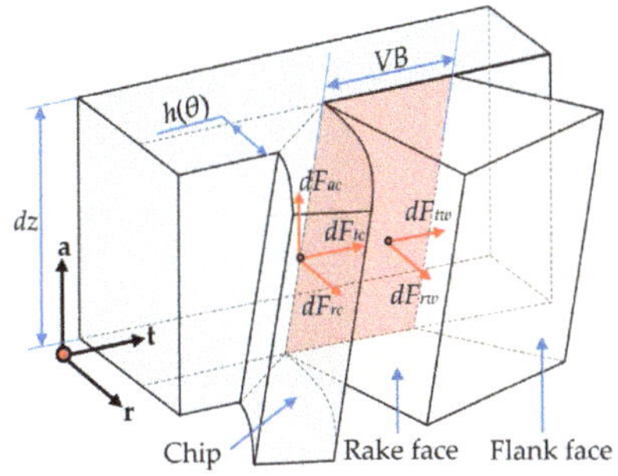

Figure 2. Forces on the cutting-edge element of a worn end mill.

When a significant wear area appears on the flank face (highlighted in red in Figure 2), extrusion and friction occur between the worn tool's flank face and the machined surface, resulting in a ploughing effect. Since the wear on the flank face does not generate an axial ploughing force [31], the ploughing force is decomposed into tangential contact force dF_{tw} and radial contact force dF_{rw}.

Based on the geometric relationship, the tangential force dF_t, radial force dF_r, and axial force dF_a are expressed as follows:

$$\begin{cases} dF_t &= dF_{tc} + dF_{tw} \\ dF_r &= dF_{rc} + dF_{rw} \\ dF_a &= dF_{ac} \end{cases} \tag{3}$$

Based on the mechanistic cutting-force model established by Altintas [17], the further derivation of Equation (3) leads to an expression for the cutting force on the cutting-edge element at any tool–workpiece engagement angle θ as follows:

$$\begin{cases} dF_t(\theta) &= K_{tc}h(\theta)dz + K_{tw}(VB)dz \\ dF_r(\theta) &= K_{rc}h(\theta)dz + K_{rw}(VB)dz \\ dF_a(\theta) &= K_{ac}(VB)h(\theta)dz \end{cases} \tag{4}$$

where $h(\theta)$ is the thickness of the instantaneous undeformed chip, VB is the tool flank-wear width, and K_{tc}, K_{rc} and $K_{ac}(VB)$ represent the tangential, radial and axial shear-force coefficients, respectively. Meanwhile, $K_{tw}(VB)$, and $K_{rw}(VB)$ represent the tangential and radial contact-force coefficients. $K_{ac}(VB)$, $K_{tw}(VB)$, and $K_{rw}(VB)$ are expressed as functions of VB. dz represents the height of the cutting-edge element, and the corresponding calculation equation is given as follows:

$$dz = \frac{R_{ac}d\theta}{\tan\beta} \tag{5}$$

2.3. *Analysis of Tool–Workpiece Engagement Relationship*

The instantaneous milling force of the end mill is the sum of the cutting forces from all cutting-edge elements. Since not all the cutting-edge elements engage in the milling process simultaneously, it is essential to calculate the tool entry and exit angle of the end mill to determine the range of the milling area. As the flank wear progresses, the actual radius of the tool decreases, which leads to changes in the tool entry angle, tool exit angle and instantaneous undeformed chip thickness, which in turn affect the milling force.

2.3.1. Calculation of Actual Radius for Worn End Mills

In order to analyze the variation of the actual radius of the end mill during the tool wear process, an orthogonal cutting-edge coordinate system $X_eO_eY_e$ is established at the bottom edge of the tool, as shown in Figure 3. The origin O_e is positioned at the cutting edge, with the Y_e axis directed from the bottom cutting edge to the top cutting edge. The X_e axis is perpendicular to the Y_e axis and tangent to the cutting-edge arc. The worn area of the tool is represented by segment OeAB, where segment O_eA corresponds to the rake face and segment O_eB corresponds to the flank face. According to the geometric relationship, as the flank face wears, the actual radius of the tool decreases. The calculation equation is as follows:

$$R_{ac} = \sqrt{(\frac{VB \times \sin\gamma}{\cos(\alpha+\gamma)})^2 + R^2 - \frac{2 \times VB \times R \times \cos\alpha \times \sin\gamma}{\cos(\alpha+\gamma)}} \tag{6}$$

where R is the radius of the sharp end mill, and α and γ represent the tool's rake angle and flank angle of the tool, respectively.

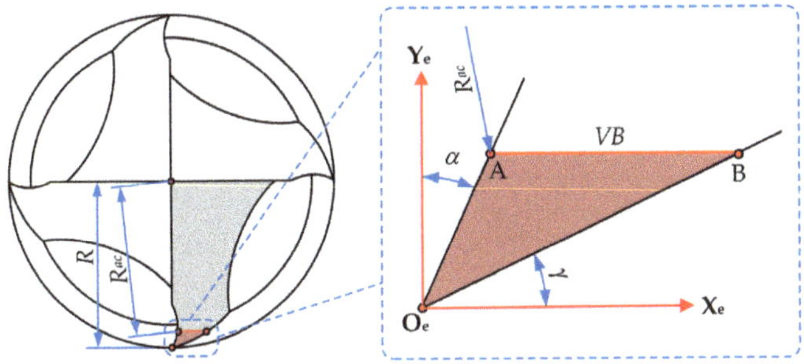

Figure 3. Actual radius of the worn end mill.

2.3.2. Calculation of Tool Entry and Exit Angle

In the side milling of 304 stainless steel, a small cutting width is typically used for climb milling. Figure 4 shows the geometric relationship of the tool–workpiece engagement during climb milling with a cutting width $a_e < R_{ac}$. The angle at which the cutting edge first engages the workpiece is defined as the tool entry angle θ_{en}. At this point, the instantaneous undeformed chip thickness starts from zero and increases. It reaches its maximum value at the transition angle θ_{tr} and then gradually decreases until it becomes zero at the exit angle θ_{ex}, when the tool exits the workpiece. The geometric relationship between the tool entry angle, transition angle and tool exit angle can be expressed as follows:

$$\begin{cases} \theta_{en} &= \pi - \arccos\left(1 - \frac{R_{ac}}{a_e}\right) \\ \theta_{tr} &= \pi - \arctan\left(\frac{\sqrt{R_{ac}{}^2 - (R_{ac} - a_e)^2} - f_z}{R_{ac} - a_e}\right) \\ \theta_{ex} &= \pi + \arcsin\left(\frac{f_z}{2 \times R_{ac}}\right) \end{cases} \tag{7}$$

where f_z is the feed per tooth.

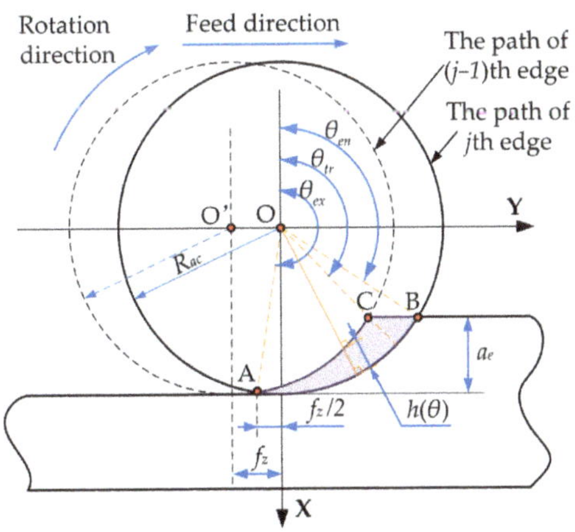

Figure 4. Geometric relationship of the end mill during climb milling.

2.3.3. Calculation of Instantaneous Undeformed Chip Thickness

In the milling process, the relative motion between the end mill and the workpiece can be decomposed into linear motion and rotation. Therefore, the motion of the cutting edge is described by a complex cycloidal motion. Due to the small feed per tooth and the high rotational speed of the end mill, the actual trajectory of the cutting edge can be assumed to be a circular arc [32]. As shown in Figure 4, AC represents the surface formed by the previous cutting edge removing the workpiece material, while AB represents the surface formed by the current cutting edge. The radial distance between two adjacent surfaces is referred to as the instantaneous undeformed chip thickness. Based on the geometric relationship shown in Figure 4, the instantaneous undeformed chip thickness can be derived as follows:

$$h(\theta) = \begin{cases} R_{ac} - \frac{R_{ac} - a_e}{\cos(\pi - \theta)} & \theta_{en} \le \theta \le \theta_{tr} \\ R_{ac} + f_z \sin\theta - \sqrt{{R_{ac}}^2 - {f_z}^2 \cos^2\theta} & \theta_{tr} \le \theta \le \theta_{ex} \end{cases} \tag{8}$$

2.4. Calculation of Instantaneous Milling Force

Based on the geometric relationship during side milling of the end mill shown in Figure 1, the tangential force dF_t, radial force dF_r and axial force dF_a acting on the cutting-edge element at the tool–workpiece engagement angle θ are transformed into the O-XYZ coordinate system. The calculation equation is as follows:

$$\begin{bmatrix} dF_x(\theta) \\ dF_y(\theta) \\ dF_z(\theta) \end{bmatrix} = J(\theta)T(\theta)\begin{bmatrix} dF_t(\theta) \\ dF_r(\theta) \\ dF_a(\theta) \end{bmatrix} \tag{9}$$

where $J(\theta)$ is a step function used to determine whether the cutting-edge element participates in the milling process, defined as

$$J(\theta) = \begin{cases} 1 & if\ \theta_{en} \le \theta \le \theta_{ex} \\ 0 & otherwise \end{cases} \tag{10}$$

$T(\theta)$ is the coordinate transformation matrix, defined as

$$T(\theta) = \begin{bmatrix} -\sin\theta & \cos\theta & 0 \\ -\cos\theta & -\sin\theta & 0 \\ 0 & 0 & 1 \end{bmatrix} \tag{11}$$

In the milling process, based on the tool–workpiece engagement relationship at different moments, the forces acting on each cutting-edge element within the milling range of each cutting edge are integrated. By summing the forces acting on each cutting edge, the instantaneous milling force acting on the end mill at the rotational angle φ is obtained as follows:

$$\begin{aligned} \begin{bmatrix} F_x \\ F_y \\ F_z \end{bmatrix}_{(\varphi)} = & \begin{bmatrix} \sum\limits_{i=1}^{N} \int_{\theta_{en}}^{\theta_{ex}} -\sin\theta J(\theta)h(\theta)dz & \sum\limits_{i=1}^{N} \int_{\theta_{en}}^{\theta_{ex}} \cos\theta J(\theta)h(\theta)dz & 0 \\ \sum\limits_{i=1}^{N} \int_{\theta_{en}}^{\theta_{ex}} -\cos\theta J(\theta)h(\theta)dz & \sum\limits_{i=1}^{N} \int_{\theta_{en}}^{\theta_{ex}} -\sin\theta J(\theta)h(\theta)dz & 0 \\ 0 & 0 & \sum\limits_{i=1}^{N} \int_{\theta_{esn}}^{\theta_{ex}} J(\theta)h(\theta)dz \end{bmatrix} \begin{bmatrix} K_{tc} \\ K_{rc} \\ K_{ac}(VB) \end{bmatrix} \\ & + \begin{bmatrix} \sum\limits_{i=1}^{N} \int_{\theta_{en}}^{\theta_{ex}} -\sin\theta J(\theta)dz & \sum\limits_{i=1}^{N} \int_{\theta_{ex}}^{\theta_{ex}} \cos\theta J(\theta)dz & 0 \\ \sum\limits_{i=1}^{N} \int_{\theta_{en}}^{\theta_{ex}} -\cos\theta J(\theta)dz & \sum\limits_{i=1}^{N} \int_{\theta_{ex}}^{\theta_{ex}} -\sin\theta J(\theta)hdz & 0 \\ 0 & 0 & \sum\limits_{i=1}^{N} \int_{\theta_{en}}^{\theta_{ex}} J(\theta)h(\theta)dz \end{bmatrix} \begin{bmatrix} K_{tw}(VB) \\ K_{rw}(VB) \\ 0 \end{bmatrix} \end{aligned} \tag{12}$$

3. Milling Experiment and Analysis

Side-milling experiments on 304 stainless steel were conducted on a three-axis machining center, using climb milling as the cutting method. The tool employed was a Heye and Sumitomo high-speed steel end mill, with the tool geometric parameters detailed in Table 1. The experimental setup and equipment layout are shown in Figure 5, with the models of the equipment listed in Table 2.

Table 1. Geometric parameters of the end mill.

Diameter (mm)	Number of Edges	Helix Angle (°)	Rake Angle (°)	Flank Angle (°)
8	4	30	12	12

Figure 5. Experimental setup and equipment layout.

Table 2. Experimental equipment and their models.

Equipment	Model
Machining center (Beijing Jingdiao Co., Ltd., Beijing, China)	JDHGT400A10SH
Three-component dynamometer (Kistler Co., Ltd, Winterthur, Switzerland)	Kistler 9257B
Charge amplifier (Kistler Co., Ltd., Winterthur, Switzerland)	Kistler 5070
Data acquisition instrument (Econ Co., Ltd., Hangzhou, China)	Yiheng MI7008
Industrial CCD camera (BeiYinHu Optics (Shenzhen) Co., Ltd., Shenzhen, China)	RZSP-2KCH

Measurement of milling forces: Milling-force signals were measured using a Kistler 9257B three-component dynamometer, processed by a Kistler 5070 charge amplifier, and saved to a laptop via a Yiheng MI7008 data acquisition instrument with a sampling frequency of 12 kHz.

Measurement of tool flank-wear widths (*VB*): The *VB* value was obtained through image processing the tool flank wear, captured by the RZSP-2KCH industrial CCD camera, in accordance with the ISO 8688-2: 1989 [33] standard. To minimize human error during the measurement process, each measurement was taken at a fixed position, with five measurements performed per blade, and the results were averaged.

3.1. Identification of Milling-Force Model Coefficients

3.1.1. Coefficient-Identification Method

Based on the calculation method for the average milling force of a single cutting edge proposed in reference [17], and combined with Equation (12), an equation for calculating the average cutting force of the entire tool over any given cycle can be further derived as

$$\begin{bmatrix} \overline{F_x} \\ \overline{F_y} \\ \overline{F_z} \end{bmatrix} = \frac{1}{\varphi_{fi} - \varphi_{st}} \int_{\varphi_{st}}^{\varphi_{fi}} \begin{bmatrix} F_x \\ F_y \\ F_z \end{bmatrix}_{(\varphi)} d\varphi \tag{13}$$

where φ_{st} is the spindle rotation angle at the start of milling and φ_{fi} is the spindle rotation angle at the end of milling.

It has been assumed that only the shearing effect of the rake face exists when the tool is sharp. Combined with Equations (12) and (13), it can be inferred that the tangential shear-force coefficient K_{tc}, the radial shear-force coefficient K_{rc} and the axial shear-force coefficient $K_{ac}(VB)$ when the tool is unworn can be identified by the average milling force of the sharp end mill. The equation is as follows:

$$\begin{bmatrix} K_{tc} \\ K_{rc} \\ K_{ac}(VB) \end{bmatrix} = \begin{bmatrix} \sum\limits_{i=1}^{N} \int_{\varphi_{st}}^{\varphi_{fi}} \int_{\theta_{en}}^{\theta_{ex}} -\sin\theta J(\theta) h(\theta) dz d\varphi & \sum\limits_{i=1}^{N} \int_{\varphi_{st}}^{\varphi_{fi}} \int_{\theta_{en}}^{\theta_{ex}} \cos\theta J(\theta) h(\theta) dz d\varphi & 0 \\ \sum\limits_{i=1}^{N} \int_{\varphi_{st}}^{\varphi_{fi}} \int_{\theta_{en}}^{\theta_{ex}} -\cos\theta J(\theta) h(\theta) dz d\varphi & \sum\limits_{i=1}^{N} \int_{\varphi_{st}}^{\varphi_{fi}} \int_{\theta_{en}}^{\theta_{ex}} -\sin\theta J(\theta) h(\theta) dz d\varphi & 0 \\ 0 & 0 & \sum\limits_{i=1}^{N} \int_{\varphi_{st}}^{\varphi_{fi}} \int_{\theta_{en}}^{\theta_{ex}} J(\theta) h(\theta) dz d\varphi \end{bmatrix}^{-1} \begin{bmatrix} \overline{F_x} \\ \overline{F_y} \\ \overline{F_z} \end{bmatrix} \tag{14}$$

Since only shear force acts in the axial direction [31], the axial shear-force coefficient $K_{ac}(VB)$ for the worn tool is also identified using Equation (14).

Tool wear leads to friction and extrusion between the tool flank face and the workpiece, generating ploughing force. The ploughing-force coefficients include the tangential contact-force coefficient $K_{tw}(VB)$ and the radial contact-force coefficient $K_{rw}(VB)$. The ploughing-force coefficients at different tool flank-wear widths are identified by the difference between the average cutting forces in the X and Y directions at the current tool flank-wear width and the sharp tool condition. By combining Equations (12) and (13), the ploughing-force coefficients can be identified by the following equation:

$$\begin{bmatrix} K_{tw}(VB) \\ K_{rw}(VB) \end{bmatrix} = \begin{bmatrix} \sum\limits_{i=1}^{N} \int_{\varphi_{st}}^{\varphi_{fi}} \int_{\theta_{en}}^{\theta_{ex}} -\sin\theta J(\theta) dz d\varphi & \sum\limits_{i=1}^{N} \int_{\varphi_{st}}^{\varphi_{fi}} \int_{\theta_{ex}}^{\theta_{en}} \cos\theta J(\theta) dz d\varphi \\ \sum\limits_{i=1}^{N} \int_{\varphi_{st}}^{\varphi_{fi}} \int_{\theta_{en}}^{\theta_{ex}} -\cos\theta J(\theta) dz d\varphi & \sum\limits_{i=1}^{N} \int_{\varphi_{st}}^{\varphi_{fi}} \int_{\theta_{ex}}^{\theta_{en}} -\sin\theta J(\theta) dz d\varphi \end{bmatrix}^{-1} \times \left[\begin{bmatrix} \overline{F_x}(VB) \\ \overline{F_y}(VB) \end{bmatrix} - \begin{bmatrix} \overline{F_{xs}} \\ \overline{F_{ys}} \end{bmatrix} \right] \tag{15}$$

where $\overline{F_x}(VB)$, $\overline{F_y}(VB)$ represent the average milling forces in the X and Y directions at the current tool flank-wear width, $\overline{F_{xs}}$, $\overline{F_{ys}}$ represent the average milling forces in the X and Y directions when the tool is sharp.

3.1.2. Experimental Identification Results

A side-milling experiment on 304 stainless steel was conducted to identify the coefficients K_{tc}, K_{ac}, $K_{ac}(VB)$, $K_{tw}(VB)$ and $K_{rw}(VB)$. The experimental parameters are shown in Table 3.

Table 3. Milling-force model coefficient-identification experiment.

Spindle Speed S (r/min)	Feed per Tooth f_z (mm/tooth)	Axial Depth a_p (mm)	Radial Depth a_e (mm)
750	0.04	3	0.5

The end mill performs climb milling along the Y-axis. The length of the workpiece in the Y direction is 100 mm. A milling cycle is defined as the tool's feeding along the Y-axis by 100 mm. A tool-rotation cycle is defined as a full 360° rotation of the tool. The flank wear is measured and the milling forces are recorded every time the flank wear increases by approximately 0.05 mm. The tool flank-wear widths and the corresponding milling forces are shown in Figure 6.

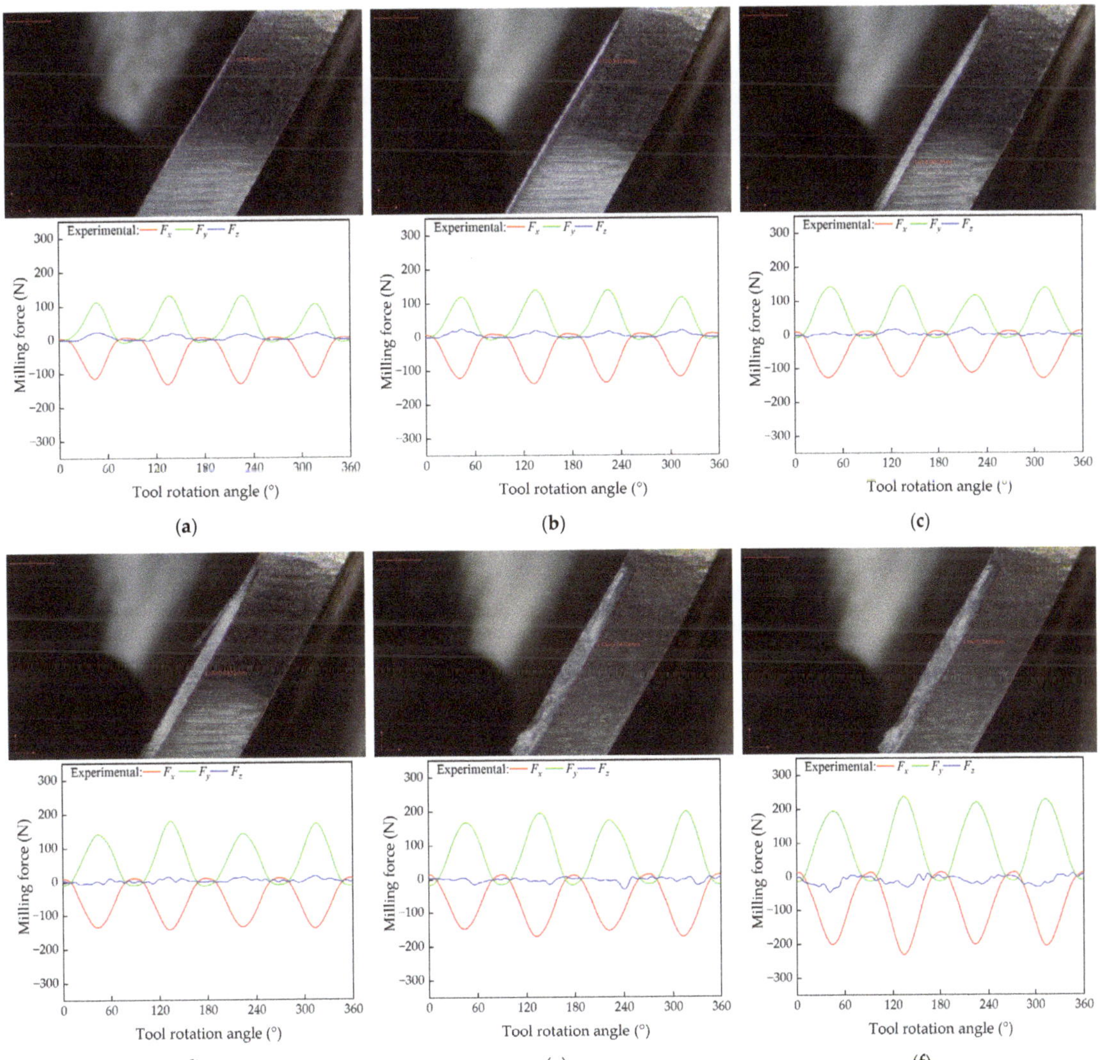

Figure 6. *Cont.*

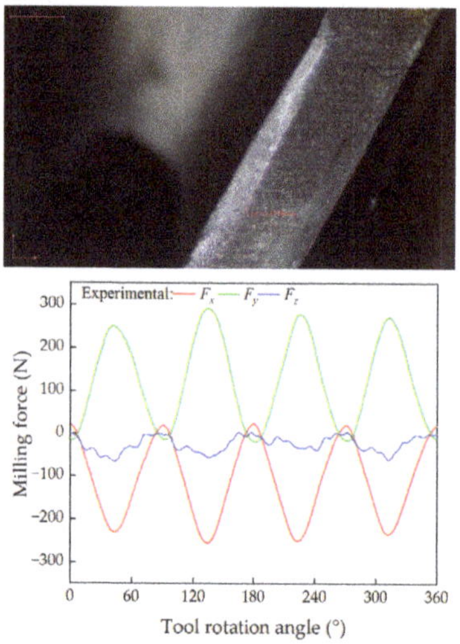

(**g**)

Figure 6. Tool flank-wear widths with corresponding milling forces: (**a**) VB = 0 mm; (**b**) VB = 0.0521 mm; (**c**) VB = 0.1038 mm; (**d**) VB = 0.1508 mm; (**e**) VB = 0.2013 mm; (**f**) VB = 0.253 mm; (**g**) VB = 0.3022 mm.

Equations (14) and (15) were used to identify the milling-force model coefficients. To minimize errors in the identification process and reduce randomness, milling forces were selected from five randomly chosen tool-rotation cycles (with each cycle being random and non-continuous) within a milling cycle. These milling-force data will be used to calculate the average milling force, which is used for the identification of the model coefficients. The identification results of $K_{ac}(VB)$, $K_{tw}(VB)$, and $K_{rw}(VB)$ under varying tool flank-wear widths are shown in Figure 7. These coefficients were then fitted as functions of the VB value. A summary of the identification results of all the milling-force model coefficients is provided in Table 4.

Table 4. Coefficients of the milling-force model.

	Coefficients of Shear Force
K_{tc} (N/mm^2)	5532.29
K_{rc} (N/mm^2)	2850.51
$K_{ac}(VB)$ (N/mm^2)	$-135760.81 \times VB^3 + 10268.67 \times VB^2 - 621.92 \times VB + 549.1$
	Coefficients of Ploughing Force
$K_{tw}(VB)$ (N/mm)	$3320.59 \times VB^3 - 339.45 \times VB^2 + 127.86 \times VB - 0.22$
$K_{rw}(VB)$ (N/mm)	$835.05 \times VB^3 + 238.77 \times VB^2 - 3.36 \times VB + 0.59$

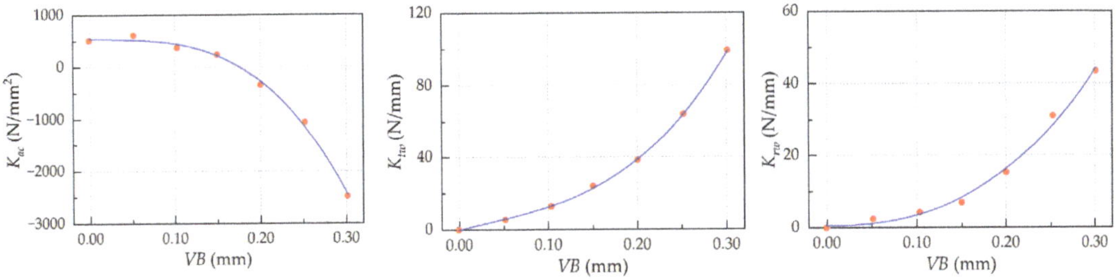

Figure 7. The variation of $K_{ac}(VB)$, $K_{tw}(VB)$ and, $K_{rw}(VB)$ with VB.

3.2. Validation of the Milling-Force Model

Two end mills were employed to conduct the following six experiments to validate the accuracy of the milling-force prediction model for worn end mills. The experimental parameters are shown in Table 5.

Table 5. Milling-force model validation experiments.

Test.	Spindle Speed S (r/min)	Feed per Tooth f_z (mm/tooth)	Axial Depth a_p (mm)	Radial Depth a_e (mm)	Tool Flank-Wear Width VB (mm)
1					0
2	830	0.05	3.5	0.6	0.1098
3					0.2109
4					0
5	900	0.03	4	0.7	0.1185
6					0.2003

Figure 8 shows the flank-wear images for each experiment, along with the comparison curves of the corresponding experimental milling forces and the predicted milling forces.

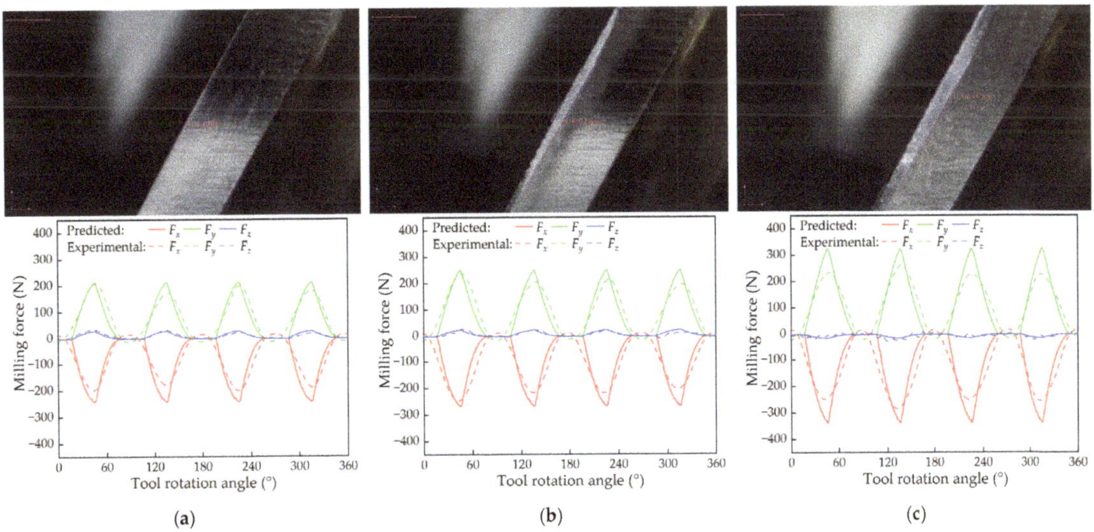

Figure 8. *Cont.*

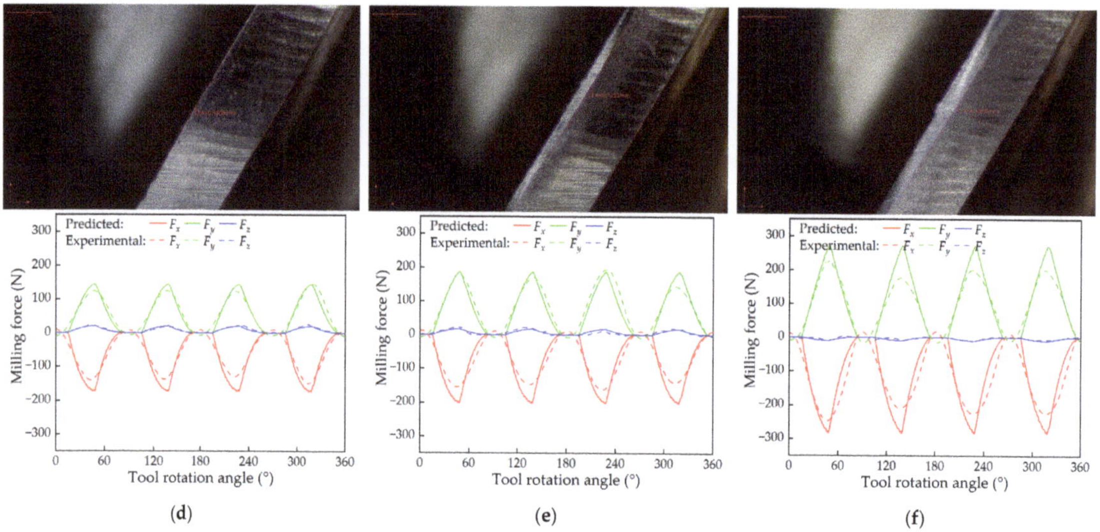

Figure 8. Tool flank-wear widths with corresponding milling forces: (**a**) Test.1; (**b**) Test.2; (**c**) Test.3; (**d**) Test.4; (**e**) Test.5; (**f**) Test.6.

Based on the results in the figure, it can be observed that the experimental milling-force curve aligns well with the predicted milling-force curve. A comparison of the average milling forces in the XYZ directions between the experimental and predicted models was conducted, and the corresponding errors were calculated. The results are shown in Figure 9.

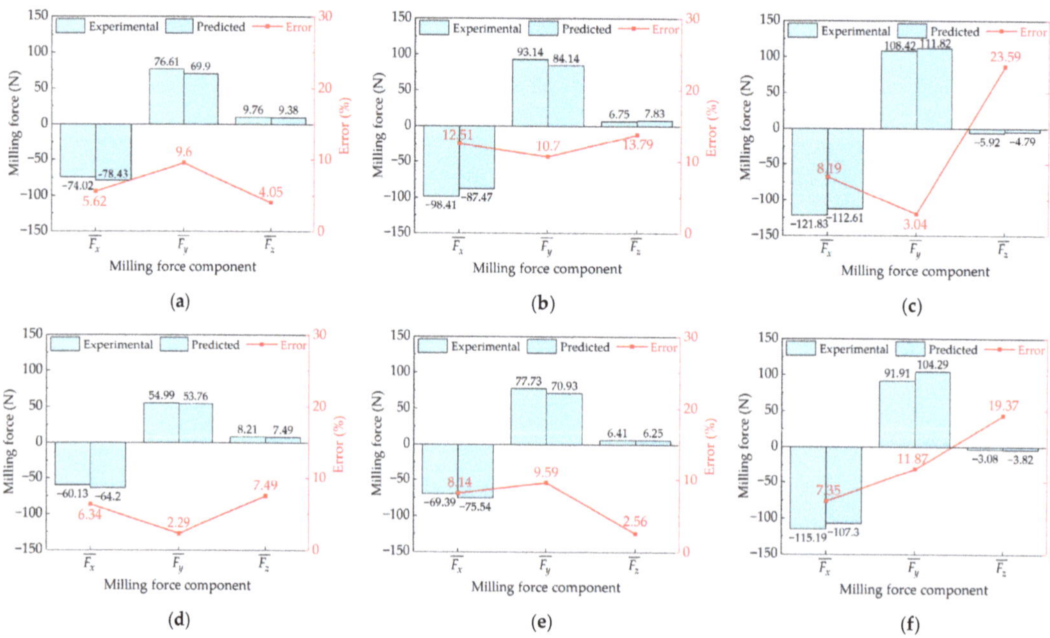

Figure 9. Comparison of experimental and predicted average milling forces: (**a**) Test.1; (**b**) Test.2; (**c**) Test.3; (**d**) Test.4; (**e**) Test.5; (**f**) Test.6.

From the results in the figure, it can be seen that the proposed model demonstrates strong predictive performance, with the comprehensive prediction errors of the average milling forces in the XYZ directions being 8.03%, 7.85%, and 11.81%, respectively. Therefore, the cutting-force prediction model for 304 stainless steel proposed in this paper, which accounts for tool wear, demonstrates reliable predictive accuracy.

4. Discussion

This study established a cutting-force model for milling 304 stainless steel, incorporating tool wear effects, based on side-milling experiments. Additionally, the study explored how milling forces vary with tool wear. This section discusses the research findings, existing limitations and potential industrial applications of the model.

1. To visually demonstrate the variation in milling forces with tool wear during the milling process, we used Equation (13) to calculate the average milling forces in the XYZ directions from the coefficient-identification experiment. The results are presented in Figure 10. From these results, it is clear that tool wear leads to an increase in the ploughing-force coefficient, which subsequently leads to a progressive rise in milling force. When the tool reaches its end of life ($VB = 0.3$), the increase in cutting force can exceed 100% compared to when the tool is sharp. This observation is consistent with the findings of most previous studies. Both the milling-force and ploughing-force coefficient increase at an accelerating rate as the tool wears. This behavior aligns with the findings in the study by Liu et al. [24]. Furthermore, the tangential contact-force coefficient is much larger than the radial contact-force coefficient—approximately twice as large. This pattern agrees with the findings of Chen et al. [34]. Their study focused on milling TC4; due to the differences in material properties, their study found that the tangential contact-force coefficient was approximately three-times larger than the radial contact-force coefficient. Therefore, the model proposed in this study is deemed reliable, although variations in milling force due to tool wear may differ across various tools and materials.
2. Discrepancies existed between the experimental and predicted values, especially when the tool flank wear was significant. These discrepancies are hypothesized to arise from several factors: On one hand, as tool wear progresses, especially when the tool flank-wear width (*VB*) becomes substantial, significant built-up-edge (BUE) formation and chip adhesion occur, as shown in Figure 11. These built-up edges and adhered chips replaced the original tool-edge geometry during the cutting process, significantly affecting the cutting force [30]. On the other hand, since neither the tool nor the workpiece is perfectly rigid, tool runout and workpiece deformation occur during the cutting process. The combined effects of these factors contribute to the discrepancies between the predicted and experimental cutting forces, especially when tool flank wear is significant.
3. Concerning the model's limitations, these primarily stem from two aspects:
    (a) Limitations in work condition universality due to the different material properties of various tools and workpiece materials, alongside their distinct contact characteristics. When the tool or material is changed, it is necessary to re-identify the coefficients in the milling-force model.
    (b) Limitations of simplifications and assumptions: In developing the model, certain factors (e.g., environmental temperature, cooling conditions, and the rigidity of the machine tool and clamping system) are often neglected or simplified to facilitate calculation and analysis. Consequently, the model may exhibit some errors. Therefore, when considering different machining

conditions, it is important to carefully evaluate which factors are necessary and which can be neglected to maintain the model's accuracy.

4. Regarding the potential industrial applications of the model, the following two scenarios merit discussion:
   (a) During the machining process: Tool wear leads to a significant increase in cutting force, subsequently affecting spindle torque, machine tool energy consumption, and vibration signals. Therefore, by monitoring these signal changes and combining them with the cutting-force model, tool wear can be estimated without stopping the machine, preventing the adverse effects of excessive tool wear on the machining process.
   (b) Before the machining process: Establishing a correlation between tool wear progression and machining parameters and applying the milling-force model to formulate and optimize the machining parameters will help to improve machining quality and efficiency.

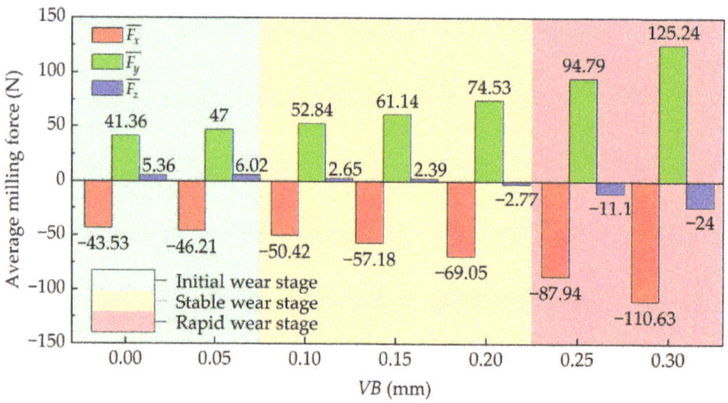

Figure 10. Variation of average milling force with tool flank-wear widths (*VB*).

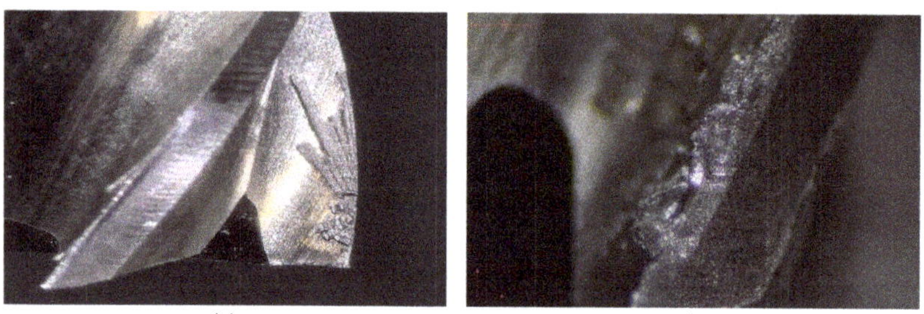

Figure 11. Chip adhesion and built-up edge (BUE) in the milling process: (**a**) Chip adhesion; (**b**) Built-up edge (BUE).

5. Conclusions

This paper develops a milling-force prediction model for 304 stainless steel that incorporates tool wear. Through conducting side-milling experiments on 304 stainless steel, the variation of milling force with tool wear was analyzed, and the accuracy of the proposed model was validated. The main conclusions are as follows:

1. Following the mechanistic modeling approach for cutting forces, the worn end mill was discretized into cutting-edge elements. The forces acting on these elements were studied. Considering the effect of tool wear on the actual radius of the end mill, the

tool–workpiece engagement relationship during the milling process was analyzed. A method for identifying the milling-force model coefficients under side-milling conditions has been developed. By integrating and summing, the instantaneous milling-force mechanistic model for the worn tool was formulated.

2. The side-milling experiments on 304 stainless steel revealed that with progressive wear on the flank face of the end mill, milling force increased at an accelerating rate. During the rapid-wear stage, the average milling force exerted by the worn end mill was more than twice that of the sharp tool. Therefore, developing a milling-force model that considers tool wear is crucial for the formulation and optimization of milling process parameters.
3. The established milling-force model can accurately predict the milling force of worn tools based on tool geometry and milling parameters. It can visually display cyclic fluctuations of milling force during tool engagement and disengagement under actual machining conditions.

The study proposes a milling-force prediction model for 304 stainless steel that incorporates tool wear considerations. The model demonstrates reliable predictive performance across various machining parameters and tool wear conditions and offers applicative value in relevant industrial settings.

Author Contributions: Conceptualization, Y.L. and C.W.; methodology, C.W.; software, K.W.; validation, K.W., K.Y. and P.H.; formal analysis, Y.L.; investigation, Y.X.; resources, F.G.; data curation, Y.X.; writing—original draft preparation, C.W.; writing—review and editing, C.W.; visualization, C.W.; supervision, Y.L.; project administration, Y.L.; funding acquisition, F.G. All authors have read and agreed to the published version of the manuscript.

Funding: This research was funded by the Shaanxi Province Modern Equipment Green Manufacturing Collaborative Innovation Center Project (No. 23JY054) and 'Leader Unveiling' projects for the industrialization of key core technologies in Shaanxi Province (No. 2205-610361-04-02-940891).

Data Availability Statement: Data are contained within the article.

Acknowledgments: Thanks to Xi'an University of Technology Shaanxi Provincial Key Laboratory of Manufacturing equipment to provide equipment support.

Conflicts of Interest: The authors declare no conflicts of interest.

References

1. Zhang, Z.; Zhang, H.; Liu, X.; Wang, T.; Huang, Q.; Liao, X. Effect of roll surface topography on microstructure and mechanical properties of 304 stainless steel ultra-thin strip. *J. Manuf. Process.* **2023**, *108*, 764–778. [CrossRef]
2. Korkut, I.; Kasap, M.; Ciftci, I.; Seker, U. Determination of optimum cutting parameters during machining of AISI 304 austenitic stainless steel. *Mater. Des.* **2004**, *25*, 303–305. [CrossRef]
3. Boillot, P.; Peultier, J. Use of stainless steels in the industry: Recent and future developments. *Procedia Eng.* **2014**, *83*, 309–321. [CrossRef]
4. Lo, K.H.; Shek, C.H.; Lai, J.K.L. Recent developments in stainless steels. *Mater. Sci. Eng. R Rep.* **2009**, *65*, 39–104. [CrossRef]
5. Kaladhar, M.; Subbaiah, K.V.; Rao, C.H.S. Machining of austenitic stainless steels—A review. *Int. J. Mach. Mach. Mater.* **2012**, *12*, 178–192. [CrossRef]
6. França, P.H.; Barbosa, L.M.; Fernandes, G.H.; Machado, Á.R.; Martins, P.S.; da Silva, M.B. Internally cooled tools: An eco-friendly approach to wear reduction in AISI 304 stainless steel machining. *Wear* **2024**, *554–555*, 205490. [CrossRef]
7. Wojciechowski, S.; Krajewska-Śpiewak, J.; Maruda, R.W.; Krolczyk, G.M.; Nieslony, P.; Wieczorowski, M.; Gawlik, J. Study on Ploughing Phenomena in Tool Flank Face–Workpiece Interface Including Tool Wear Effect during Ball-End Milling. *Tribol. Int.* **2023**, *181*, 108313. [CrossRef]
8. Ge, G.; Xiao, Y.; Lv, J.; Du, Z. A Non-Iterative Compensation Method for Machining Errors of Thin-Walled Parts Considering Coupling Effect of Tool-Workpiece Deformation. *Manuf. Lett.* **2024**, *41*, 287–295. [CrossRef]
9. Liu, C.; Huang, Z.; Huang, S.; He, Y.; Yang, Z.; Tuo, J. Surface roughness prediction in ball screw whirlwind milling considering elastic-plastic deformation caused by cutting force: Modelling and verification. *Measurement* **2023**, *220*, 113365. [CrossRef]

10. Shi, K.N.; Liu, N.; Liu, C.L.; Ren, J.X.; Yang, S.S.; Tan, W.C. Indirect approach for predicting cutting force coefficients and power consumption in milling process. *Adv. Manuf.* **2022**, *10*, 101–113. [CrossRef]
11. Bhushan, R.K. GA approach for optimization of parameters in machining Al alloy SiC particle composite for minimum cutting force. *J. Alloys Metall. Syst.* **2023**, *1*, 100002. [CrossRef]
12. Liu, T.; Song, J.; Zhang, K.; Liu, Q.; Chen, F. Tooth-wise monitoring of the asymmetrical tool wear in micro-milling based on the chip thickness reconstruction and cutting force signal. *Mech. Syst. Signal Process.* **2024**, *208*, 111004. [CrossRef]
13. Du, Y.; Song, Q.; Liu, Z. Prediction of Micro Milling Force and Surface Roughness Considering Size-Dependent Vibration of Micro-End Mill. *Int. J. Adv. Manuf. Technol.* **2022**, *119*, 5807–5820. [CrossRef]
14. Davoodi, B.; Tazehkandi, A.H. Experimental investigation and optimization of cutting parameters in dry and wet machining of aluminum alloy 5083 in order to remove cutting fluid. *J. Clean. Prod.* **2014**, *68*, 234–242. [CrossRef]
15. Merchant, M.E. Mechanics of the metal cutting process. I. Orthogonal cutting and a type 2 chip. *J. Appl. Phys.* **1945**, *16*, 267–275. [CrossRef]
16. Merchant, M.E. Mechanics of the metal cutting process. II. Plasticity conditions in orthogonal cutting. *J. Appl. Phys.* **1945**, *16*, 318–324. [CrossRef]
17. Altintas, Y. *Manufacturing Automation: Metal Cutting Mechanics, Machine Tool Vibrations, and CNC Design*; Cambridge University Press: Cambridge, UK, 2012.
18. Yang, D.; Zhang, Y.; Wang, R.; Wei, F.; Zeng, L.; Liu, M. Finite Element Modeling and Optimization Analysis of Cutting Force in Powder Metallurgy Green Compacts. *Processes* **2023**, *11*, 3186. [CrossRef]
19. Forootan, M.; Akbari, J.; Ghorbani, M. A real-time intelligent method to identify mechanistic cutting force coefficients in 3-axis ball-end milling process using stochastic gradient decent: The mechanistic network. *Int. J. Adv. Manuf. Technol.* **2023**, *129*, 2949–2968. [CrossRef]
20. Zhao, M.; Cai, Y.; Wang, N.; Song, Y.; Wang, H. The research of cutting force prediction for worn ball-end tool considering spring back. *Int. J. Adv. Manuf. Technol.* **2023**, *125*, 5619–5629. [CrossRef]
21. Wojciechowski, S.; Maruda, R.W.; Nieslony, P.; Krolczyk, G.M. Investigation on the edge forces in ball end milling of inclined surfaces. *Int. J. Mech. Sci.* **2016**, *119*, 360–369. [CrossRef]
22. Wang, W.; Tang, Y.; Guo, X.; Zhang, K.; Liu, T.; Wang, C. Modeling of cutting force in micro-milling considering asymmetric tool wear. *Int. J. Adv. Manuf. Technol.* **2024**, *133*, 1597–1608. [CrossRef]
23. Gao, S.; Duan, X.; Zhu, K.; Zhang, Y. Influence of tool flank wear considering tool edge radius on instantaneous uncut chip thickness and cutting force in micro-end milling. *Int. J. Adv. Manuf. Technol.* **2024**, *133*, 1639–1650. [CrossRef]
24. Liu, T.; Liu, Y.; Zhang, K. An improved cutting force model in micro-milling considering the comprehensive effect of tool runout, size effect, and tool wear. *Int. J. Adv. Manuf. Technol.* **2022**, *120*, 659–668. [CrossRef]
25. Wojciechowski, S.; Matuszak, M.; Powałka, B.; Madajewski, M.; Maruda, R.W.; Królczyk, G.M. Prediction of cutting forces during micro end milling considering chip thickness accumulation. *Int. J. Mach. Tools Manuf.* **2019**, *147*, 103466. [CrossRef]
26. Hou, Y.; Zhang, D.; Wu, B.; Luo, M. Milling force modeling of worn tool and tool flank wear recognition in end milling. *IEEE/ASME Trans. Mechatron.* **2014**, *20*, 1024–1035. [CrossRef]
27. Huang, Y.; Liang, S.Y. Modeling of cutting forces under hard turning conditions considering tool wear effect. *J. Manuf. Sci. Eng.* **2005**, *127*, 262–270. [CrossRef]
28. Chinchanikar, S.; Choudhury, S.K. Cutting force modeling considering tool wear effect during turning of hardened AISI 4340 alloy steel using multi-layer TiCN/Al_2O_3/TiN-coated carbide tools. *Int. J. Adv. Manuf. Technol.* **2016**, *83*, 1749–1762. [CrossRef]
29. Patel, V.D.; Gandhi, A.H. Modeling of cutting forces considering progressive flank wear in finish turning of hardened AISI D2 steel with CBN tool. *Int. J. Adv. Manuf. Technol.* **2019**, *104*, 503–516. [CrossRef]
30. Toubhans, B.; Fromentin, G.; Viprey, F.; Karaouni, H.; Dorlin, T. Machinability of Inconel 718 during turning: Cutting force model considering tool wear, influence on surface integrity. *J. Mater. Process. Technol.* **2020**, *285*, 116809. [CrossRef]
31. Teitenberg, T.M.; Bayoumi, A.E.; Yucesan, G. Tool wear modeling through an analytic mechanistic model of milling processes. *Wear* **1992**, *154*, 287–304. [CrossRef]
32. Martellotti, M.E. An analysis of the milling process. *Trans. Am. Soc. Mech. Eng.* **1941**, *63*, 677–695. [CrossRef]
33. *ISO 8688-2:1989*; Tool Life Testing in Milling—Part 2: End Milling. ISO: Geneve, Switzerland, 1989.
34. Chen, X.; Zhang, Z.; Wang, Q.; Zhang, D.; Luo, M. A new method for prediction of cutting force considering the influence of machine tool system and tool wear. *Int. J. Adv. Manuf. Technol.* **2022**, *120*, 1843–1852. [CrossRef]

MDPI AG
Grosspeteranlage 5
4052 Basel
Switzerland
Tel.: +41 61 683 77 34

Machines Editorial Office
E-mail: machines@mdpi.com
www.mdpi.com/journal/machines

www.ingramcontent.com/pod-product-compliance
Lightning Source LLC
Chambersburg PA
CBHW061443210926
51850CB00008B/1593
9783725861491